An Introduction to Number Theory

HAROLD M. STARK

An Introduction to Number Theory

The MIT Press
Cambridge, Massachusetts, and London, England

Sixth Printing, 1989

First MIT Press paperback edition, 1987

Original edition published by Markham Publishing Company, 1970
Copyright © 1970 by Harold M. Stark

Printed and bound in the United States
of America by Edwards Brothers.

Library of Congress Cataloging in Publication Data

Stark, Harold M   1939–
    An introduction to number theory.

    Originally published by Markham Pub. Co., Chicago.
    Bibliography: p.
    Includes index.
    1. Numbers, Theory of. I. Title.
QA241.S72 1978   512′.7   78-2744
ISBN 0–262–69060–8

PREFACE

The majority of students who take number theory courses are mathematics majors who will not become number theorists. In fact, many of them will go out into the world and teach mathematics in high school and junior college. It is hoped that this book will satisfy the needs of these students. To this end, the text is intentionally wordier than the average number theory text. There are more discussions of proofs and more worked out examples than will usually be found elsewhere. In particular, we sometimes make use of very large numbers. After all, the percentage of positive integers that are less than a trillion is rather small. It seems rather unfair, then, to constantly use these small numbers in examples.

In addition, there are hundreds of homework exercises of all levels for the student to sink his teeth into. The exercises fall into two categories. In the first category we find the problems at the end of the sections. These are mostly of a computational nature designed to help the student become familiar with the material of the section, although there are questions about the proofs in the book and occasional "theory" problems. A few of the more difficult problems in these sections have been marked with an asterisk. In the second category we have miscellaneous exercises at the end of each chapter. These are, on the average, of a decidedly more difficult character than the section problems, although there are some, such as the perpetual calendar problems in Chapter 3, that are merely too long to include in the sections. The answers to these problems are not included in the answer section. Because of the difficulty of these problems, they should be used sparingly outside of honors courses.

The textual material is fairly standard in Chapters 1, 2, 3, 5, and 6, and somewhat nonstandard in Chapters 4, 7, and 8. The last section of Chapter 6, marked by an asterisk, is somewhat difficult and may be skipped; it is not used later. Chapter 4 is not essential to the subject of number theory, but in my experience students find magic squares very interesting and the material will give them good practice using congruences. Nevertheless, if time is very critical in a course, the last half (or even all) of the chapter could be skipped.

Chapter 7 presents continued fractions from a geometric viewpoint. While not the standard treatment, I believe it to be clearer and it certainly leads more naturally to higher dimensional generalizations such as the Jacobi algorithm than does the usual approach. The natural means for treating this topic is two-dimensional vectors. The necessary material on vectors is in Appendix A, but most students will have had it in calculus courses. The last two sections of Chapter 7 treat the expansion of quadratic irrationalities and their connection with the Fermat–Pell equation. These two sections, both marked by an asterisk, use 2×2 matrices and might well be skipped in the average course. The only material used from these sections later is the fact that the infinitely many solutions to the Fermat–Pell equation imply that there are infinitely many units in a real quadratic field. To cover this somewhat, it is shown in Section 5.4 that if there is one nontrivial solution to the Fermat–Pell equation, then there are infinitely many.

Chapter 8 treats the topic of quadratic fields. This material is designed to make the student think about some of the "obvious" concepts that he has taken for granted earlier. It will also illustrate some of the simplest applications of algebraic number theory to elementary number theory problems. The teacher may wish to illustrate the material throughout by use of some particular field. The following fields, all Euclidean, offer an ample choice of any level of complexity that the teacher may desire: $Q(\sqrt{-1})$, $Q(\sqrt{-2})$, $Q(\sqrt{-3})$, $Q(\sqrt{2})$, $Q(\sqrt{5})$.

The law of quadratic reciprocity has been intentionally omitted. It is my belief that even for the serious number theory student, the law of quadratic reciprocity is better taught in a second course of number theory than a first course. After the student has worked with quadratic Diophantine equations and quadratic fields, then he can appreciate the usefulness of the quadratic reciprocity law. As a result, only in the miscellaneous exercises of Chapter 8 is the law even mentioned and there the student can prove many special cases using the theory of quadratic fields.

A bibliography is included at the end of the book. References to specific results given in the book may be found here. This will be mostly of interest to specialists. I for one become very frustrated when I find some advanced result that I had not known stated in an elementary book without reference. I hope this will not happen to readers here.

Finally, I wish to express my thanks to Prof. D. J. Lewis, Prof. James Schafer, my wife Betty, Miss Ramelle Myers, and the unnamed reviewers who read earlier versions of the book. Their comments have greatly improved the final result.

Harold M. Stark

Massachusetts Institute of Technology
Cambridge, Massachusetts

CONTENTS

PREFACE   *v*

Chapter 1   AN INTRODUCTION TO NUMBER THEORY   1
   1.1   An Introduction to Number Theory   1
   1.2   Some Elementary Properties of Divisibility   11

Chapter 2   THE EUCLIDEAN ALGORITHM AND UNIQUE
              FACTORIZATION   16
   2.1   The Euclidean Algorithm   16
   2.2   The Fundamental Theorem of Arithmetic   26
   2.3   Applications of the Fundamental Theorem   33
   2.4   Multiplicative Functions   36
   2.5   Linear Diophantine Equations   44

Chapter 3   CONGRUENCES   51
   3.1   Introduction   51
   3.2   Fundamental Properties of Congruences   54
   3.3   Linear Congruence Equations   66
   3.4   Reduced Residue Systems and Euler's ϕ Function   77
   3.5   More on Euler's ϕ Function   82
   3.6   Polynomial Congruences   86
   3.7   Primitive Roots   97

Chapter 4   MAGIC SQUARES   118
   4.1   The Uniform Step Method   118
   4.2   Filled and Magic Squares   123
   4.3   Diabolic and Symmetric Squares   132
   4.4   Historical Comments   137

Chapter 5 DIOPHANTINE EQUATIONS   145
   5.1 Introduction   145
   5.2 The Use of Congruences in Solving Diophantine
          Equations   148
   5.3 Pythagorean Triples   151
   5.4 Fermat's Method of Descent   155

Chapter 6 NUMBERS, RATIONAL AND IRRATIONAL   164
   6.1 Rational Numbers   164
   6.2 Irrational Numbers   170
   6.3 Liouville's Theorem and Transcendental Numbers   172

Chapter 7 CONTINUED FRACTIONS FROM A
         GEOMETRIC VIEWPOINT   181
   7.1 Introduction   181
   7.2 The Continued Fraction Algorithm   185
   7.3 Computation of a_n 196
   7.4 The Best Approximations   209
   7.5 A Commentary of Proof by Picture   223
   7.6 Periodic Continued Fractions   226
   7.7 The Fermat–Pell Equation and the Continued
          Fraction Expansion of $\sqrt{d}$ 239

Chapter 8 QUADRATIC FIELDS   257
   8.1 Introduction   257
   8.2 Quadratic Fields and Quadratic Integers   258
   8.3 Divisibility and Factorization into Primes   270
   8.4 Unique Factorization and Euclidean Domains   281
   8.5 Applications of Quadratic Fields to Diophantine
          Equations   299
   8.6 Historical Comments   305

Appendix A TWO-DIMENSIONAL VECTORS   317

Appendix B TWO BY TWO MATRICES   322

Appendix C FIELDS   331

      BIBLIOGRAPHY   334
      ANSWERS TO SELECTED EXERCISES   341
      LIST OF SYMBOLS   344
      INDEX   345

Chapter 1

AN INTRODUCTION TO NUMBER THEORY

1.1. An Introduction to Number Theory

The theory of numbers is concerned with properties of numbers, particularly properties of the integers, $0, \pm 1, \pm 2, \pm 3, \ldots$. It may be asked: What properties can numbers have? After all, they may be added, subtracted, multiplied, and divided; what else is there? It is the purpose of this section to illustrate some of the answers to this question. Many of the results indicated here will be proved in later chapters, but proofs of some are too advanced and cannot be included here.

One main subdivision of elementary number theory deals with multiplicative properties of integers. Fundamental to questions in this area is the notion of divisibility.

Definition. If a and b are integers, $a \neq 0$, and if there is an integer c such that $b = ac$, then we say that a **divides** b, and we write $a|b$. If a does not divide b, then we write $a \nmid b$.

Thus, although $\frac{7}{5} = 1.4$, the quotient is not an integer and thus $5 \nmid 7$. Other examples are

$$2|18, \quad 1|42, \quad 3|(-6), \quad -7|49, \quad 9 \nmid 80, \quad -6 \nmid 31.$$

Certain positive integers, such as 1, 2, 3, 13, and 10 006 721,[1] have the property that the only positive integers that divide them are themselves and 1. These numbers were called the prime numbers by the ancients, but more and more it has become advantageous to exclude 1 from this list, and thus the modern definition of a prime number is

[1] The usual commas that separate thousands and millions are too confusing. It is customary not to use them.

1

Definition. An integer greater than one whose only positive divisors are itself and one is called a **prime number**. An integer greater than one which is not a prime number is said to be **composite**.

All sorts of questions immediately spring to mind. How many primes are there? The answer is infinitely many. This means that there is no last prime. Or, alternatively, it can be thought of as meaning that there are more than 1 million primes, more than 1 billion primes, more than 1 trillion primes, in fact, more primes than any number that you care to name. This fact was known by Euclid over 2000 years ago, and his proof will be given shortly. What is the nth prime? For any given n, this question can always be answered in a finite amount of time. For example, the 664 999th prime is 10 006 721.[2] But in the sense of giving a formula which yields the nth prime for all n, this has never been done. Is there a formula which at least gives only primes? No one has ever found one. Centuries ago, it was believed that if n is an integer, then

$$n^2 + n + 41$$

is always a prime number. It is for $n = 0, 1, 2, 3, \ldots, 39$, but it fails to be for $n = 40$ and it fails obviously for $n = 41$ (there is a factor of 41 in both cases). We will see in Chapter 3 that no polynomial can give only primes. Fermat (1601–1665) conjectured that the numbers

$$F_n = 2^{2^n} + 1$$

are primes for all integers $n \geq 0$. He checked this for $n = 0, 1, 2, 3, 4$ and found that the corresponding F_n's, 3, 5, 17, 257, and 65 537, are indeed primes. Since

$$F_5 = 4\,294\,967\,297,$$

Fermat did not attempt to verify his conjecture any further. Fermat undoubtedly had good reasons for believing his conjecture, nevertheless, he was wrong. Euler (1707–1783) found in 1732 that $641 | F_5$ and hence F_5 is composite. Since then it has been discovered that several additional F_n's are composite. In fact, no F_n with $n > 4$ has yet been proved to be a prime.

How many primes are there less than a given integer n? Legendre (1752–

---

[2] I fear that this was done the hard way. All the primes from 2 to 10 006 721 were listed by D. N. Lehmer in 1914—before the days of computers (see the bibliography at the end of the book). There were 664 999 of them. Why did he stop here rather than at the next prime? Lehmer defined 1 to be a prime also, and thus, in his terminology, he stopped at the 665 000th prime. His goal was to list all primes from 1 to 10 000 000; at 5000 primes per page, he reached the last prime under 10 000 000 on page 133 and then simply completed the page.

1833) and, after him, Gauss (1777–1855) conjectured that the answer is approximately

$$\frac{n}{1 + \dfrac{1}{2} + \dfrac{1}{3} + \cdots + \dfrac{1}{n}}.$$

The fact that the ratio of this number to the number of primes less than n gets nearer and nearer 1 as n gets larger is known as the **prime number theorem**. It was first proved in 1896 by Hadamard (1865–1963) and de la Vallée Poussin (1866–1962), and even today its proof is far from simple.

How can we tell whether or not a given integer is a prime? There is no general method, although it is sometimes possible to find a small factor of the given number by trial and error and hence show that it is composite. For example, is

$$32\ 589\ 158\ 477\ 190\ 044\ 731$$

a prime number? How would you tell? There are some general methods for answering this question theoretically, but they are completely unsuited for practical computations. For example, we will show in Chapter 3 that a number n is a prime if and only if

$$n | [(n - 1)! + 1].$$

For example, $5|(4! + 1)$ and hence is a prime, while $6 \nmid (5! + 1)$ and hence is not a prime. This is an interesting property of primes, but it is totally useless for verifying that the 20-digit number above is or is not a prime.

Another main category in number theory is given by additive questions. The most familiar question of this type to the reader is the problem of writing a perfect square as the sum of two perfect squares. Because of the Pythagorean theorem, this problem is equivalent to the problem of finding right triangles with integral sides. The reader has probably seen the 3,4,5 and 5,12,13 right triangles. With the exception of these triangles and triangles similar to them [such as twice the 3,4,5 triangle (6,8,10) or five times the 5,12,13 triangle (25,60,65)], the reader may not have seen others. But there are others. Both the 3,4,5 triangle and the 5,12,13 triangles have the hypotenuse being one unit longer than one of the sides. If the sides are a and b and the hypotenuse is $b + 1$, then by the Pythagorean theorem

$$a^2 + b^2 = (b + 1)^2$$

or

(1)
$$a^2 = (b + 1)^2 - b^2$$
$$= b^2 + 2b + 1 - b^2$$
$$= 2b + 1.$$

Since $2b + 1$ is an odd number, that is, since $2 \nmid (2b + 1)$ (since $b + \frac{1}{2}$ is not an integer), we see that a^2 is an odd number and so we put

(2)
$$a = 2n + 1.$$

Then we see from (1) that

(3)
$$b = \frac{a^2 - 1}{2}$$
$$= \frac{(2n + 1)^2 - 1}{2}$$
$$= \frac{4n^2 + 4n + 1 - 1}{2}$$
$$= 2n^2 + 2n.$$

In equations (2) and (3), we have a and b in terms of n. Thus, we expect that

$$(2n + 1), \quad (2n^2 + 2n), \quad (2n^2 + 2n + 1)$$

is a Pythagorean triplet; that is, $2n + 1$ and $2n^2 + 2n$ are the sides of a right triangle and $2n^2 + 2n + 1$ is the hypotenuse. This may be easily checked by merely verifying that

$$(2n + 1)^2 + (2n^2 + 2n)^2 = (2n^2 + 2n + 1)^2$$

for all n. When $n = 1$ and $n = 2$, we get the 3,4,5 triangles and the 5,12,13 triangles. When $n = 3$, $n = 4$, and $n = 5$, we get the triangles

$$7,24,25; \quad 9,40,41; \quad 11,60,61.$$

We may continue plugging in different values of n as long as we desire. As the 8,15,17 triangle shows, this method does not give all right triangles with integral sides, but it does go far beyond the old standards 3,4,5 and 5,12,13. The problem of finding all right triangles with integral sides boils down to finding all solutions to the equation

$$x^2 + y^2 = z^2$$

in positive integers. Such an equation is called a **Diophantine equation** in honor of the Greek mathematician Diophantus (4th century A.D.?), who first investigated the problem of finding integral solutions to equations, particularly the cases with more unknowns than equations. In Chapter 5 we will study some of the simpler methods of solving such equations, but we will also run into them in other chapters.

Fermat generalized the Pythagorean equation by looking at the equation

$$(4) \qquad x^n + y^n = z^n,$$

where n is an integer greater than or equal to 3. In the margin of his copy of the works of Diophantus, Fermat stated that he had a truly wondrous proof of the fact that, unlike the case of $n = 2$, when $n \geq 3$, equation (4) has no solutions where x, y, and z are all nonzero integers. Unfortunately, Fermat continued, the margin was not big enough to hold the proof. This result has come to be known as **Fermat's last theorem**, or **Fermat's great theorem** (as opposed to Fermat's lesser theorem, which we find in Chapter 3). It is unfortunate that Fermat left no hints as to his method of proof because no one has been able to prove his theorem since! In fact, it is one of the two or three most famous unsolved mathematical problems today.[3] The question naturally arises: Did Fermat really have a proof of his theorem? There are those who argue that Fermat did have a proof of his theorem and note that the wisdom of the ancients far exceeded that of the mere mortal man of today. Then there are the more cynical who believe that Fermat must have made one of the mistakes that many after him have made. This question is as much fun to argue as any philosophic or theologic question, and, like them, there are no facts to contradict one's arguments.

As another example of an additive question, we have Goldbach's conjecture made in 1742 that every even integer greater than 2 is the sum of two primes. For example,

$$4 = 2 + 2, \quad 6 = 3 + 3, \quad 8 = 5 + 3, \quad 20 = 13 + 7, \quad 100 = 83 + 17.$$

This conjecture has been verified by Pipping for all even numbers less than 100 000, but no one has been able to prove it.

In this book, we will prove many seemingly obvious theorems. Perhaps a word is in order on why we bother. There are two reasons why something is obvious: First, it may sound very reasonable and, second, it may have been

---

[3] Before the reader attempts to solve this problem, he should finish reading the book. If he still insists on solving the problem after that, I request that he not send his solution to me, as I am not qualified to judge the correctness of so difficult a work. The reader should be warned that thousands of people have submitted solutions to this problem and none has been anywhere near correct. An announcement by an amateur that he has solved the problem is greeted with the same skepticism as an announcement by a sailor that he has seen a sea serpent.

verified so often by personal experience that it no longer seems questionable (it is hard to doubt something that has worked a million times in a row). How obvious should something be before we accept it as true? A few examples may help answer this question.

Let us investigate the conjecture that any odd number which is not divisible by either 3 or 5 is a prime. The odd numbers less than 20 which are not divisible by 3 or 5 are 7, 11, 13, 17, and 19, all primes. The odd numbers between 20 and 40 not divisible by 3 or 5 are 23, 29, 31, and 37, also all primes. Perhaps we should believe the conjecture and try to prove it. But wait! In the next set of 20, we get the numbers 41, 43, 47, 49, 53, and 59, and $49 = 7 \cdot 7$ is not a prime. Thus the conjecture is false and having seen a counterexample, it is easy to construct others. The remaining counterexamples less than 100 are $77 = 7 \cdot 11$ and $91 = 7 \cdot 13$. Perhaps, then, we should not accept something as true until it has been verified past 100.

Twenty-five centuries ago, the Chinese gave what they believed was an infallible rule for determining primality. Their rule stated that n is a prime if and only if

$$n|(2^n - 2).$$

For example, $2^7 - 2 = 126 = 7 \cdot 18$ and 7 is a prime, while $2^{10} - 2 = 1022$, which is not divisible by the composite number 10. It is doubtful that the Chinese had any reason to believe their rule other than the fact that it seemed to work. Owing to the complexity of the number $2^n - 2$ when n is large, it is hard to believe that the Chinese verified their rule for very many n. And yet the Chinese rule was believed to be true for more than 23 centuries, and it has been verified for all n up to 300. Further, Fermat showed that the Chinese were correct when n is a prime. But in spite of all this, the Chinese were wrong. It can be shown that their rule fails for $n = 341 = 11 \cdot 31$. I would not advise checking this statement, since $2^{341} - 2$ has 103 digits. Besides, results of Fermat, Euler, and Gauss, presented in Chapter 3, will make it trivial that

$$341|(2^{341} - 2).$$

Let us examine another conjecture. By the number of prime factors of an integer n, we mean the number of factors (whether distinct from each other or not) when n is written as a product of primes. For example, $12 = 2 \cdot 2 \cdot 3$ has three prime factors by this definition, and $16 = 2 \cdot 2 \cdot 2 \cdot 2$ has four. We shall say that 1 has zero prime factors and that a prime has one. Let O_n be the number of positive integers less than or equal to n which have an odd number of prime factors and let E_n be the number of positive integers less than or equal to n which have an even number of prime factors. For example, $O_{12} = 7$ (the numbers being 2, 3, 5, 7, 8, 11, and 12) and $E_{12} = 5$ (the numbers being 1, 4, 6, 9, and 10). A product of two primes is clearly greater than either of its prime factors. A product of four primes has smaller divisors which are products of three primes, and so on. Thus, in some sense, the numbers which are products of an odd number of primes come earlier in the sequence of positive integers than the numbers which are products of an even number of primes. This leads us to suspect that there are at least as many numbers less than n with an odd number of prime factors as there are numbers with an even number of prime factors. In other words, we have the conjecture that

$$O_n \geq E_n.$$

This is known as the *Polya conjecture*, after G. Polya (1887–), who in 1919 conjectured that if $n \geq 2$, then $O_n \geq E_n$ ($O_1 = 0$ and $E_1 = 1$, but after this, we come to the primes before we come to the product of two primes). The Polya conjecture sounds reasonable without even experimentally verifying it. But since the Polya conjecture had many important consequences in advanced number theory, it was checked experimentally and it was found to be true for the first million positive integers. Is it any wonder, then, that the majority of mathematicians were confident that the Polya conjecture would eventually be proved? But they were wrong. In 1958, Haselgrove showed that there are infinitely many n for which

$$O_n < E_n.$$

The smallest known counterexample to Polya's conjecture was found by R. S. Lehman in 1962, and it is

$$n = 906\ 180\ 359,$$

at which

$$O_n = E_n - 1.$$

Perhaps the word "obvious" is beginning to lose its meaning, but to make sure, we give one last example. We see that $x = 1, y = 0$ satisfies the Diophantine equation

(5) $$x^2 - 1141y^2 = 1.$$

We might ask, does equation (5) have any solution in positive integers? We see from (5) that

$$x = \sqrt{1141y^2 + 1}.$$

Thus the question is: Is $1141y^2 + 1$ ever a perfect square? This may be checked experimentally. It turns out that the answer is no for all positive y less than 1 million. In view of the previous example, perhaps we should experiment further. The answer is still no for all y less than 1 trillion (1 million

million, or 10^{12}). We go overboard and check all y up to 1 trillion trillion (10^{24}). Again the answer is, no. No one in his right mind would really believe that there could be a positive y such that $\sqrt{1141y^2 + 1}$ is an integer if there is no such y less than 1 trillion trillion. But there is. In fact, there are infinitely many of them, the smallest among them having 26 digits. If you still do not believe this, we will prove in Chapter 7 that there are infinitely many such y and give a method whereby you may start from scratch and find the smallest positive value of y in less than an hour (with a desk calculator).

In later chapters, we will discuss two widely believed conjectures of Euler, both of which have been shown to be false within the last ten years. Thus it is that the mathematician refuses to accept a statement as true, no matter how plausible it is, until it is proved. In this vein, it is interesting to note that we have used an obvious result in two of the examples above. How do we know that every factorization of n into primes has the same number of prime factors? How do we know that if two primes greater than 5 are multiplied, the result will not be divisible by 3 or 5? Maybe it has never occurred to you to ask these questions, but the chances are that your beliefs are based on experience with rather small numbers. In view of the previous examples, we will prove these rather obvious statements in Chapter 2 as part of a general theorem on factorization. In Chapter 8, we will reexamine these "obvious" concepts from a more advanced standpoint.

There are other questions that may be asked about numbers that are neither multiplicative nor additive in character. For example, consider the number $\pi = 3.14159\ldots$. Pause at this point and think of a fraction which you associate with π. I suspect that you have thought of the number twenty-two sevenths. Assuming this to be true, let us ask why you associate this particular fraction with π. The first answer is that you were taught it in school. But why did your teachers pick $\frac{22}{7}$? Presumably, the answer is that $\frac{22}{7}$ is close to π and it is easier to work with $\frac{22}{7}$ than 3.14159 .... But why sevenths? If it is ease of operation that we desire, $\frac{31}{10}$ is close to π and by far easier to use. Were your teachers being sadistic in making you always divide by 7 rather than by 10, or is $\frac{22}{7}$ somehow a better representative of π than $\frac{31}{10}$? The answer is that $\frac{22}{7}$ is a far better approximation to π than $\frac{31}{10}$. In fact, in a sense to be explained in Chapter 7, $\frac{22}{7}$ is one of the best approximations to π by fractions.

How close can we expect a fraction with denominator q to come to a given real number α? We can partially answer this question here. We consider all the fractions with denominator q:

$$\ldots, \frac{-5}{q}, \frac{-4}{q}, \frac{-3}{q}, \frac{-2}{q}, \frac{-1}{q}, \frac{0}{q}, \frac{1}{q}, \frac{2}{q}, \frac{3}{q}, \frac{4}{q}, \frac{5}{q}, \ldots.$$

Either α is one of these fractions or α lies between two consecutive fractions. In either case, there are two consecutive numerators n and $n + 1$ such that

$$\frac{n}{q} \leq \alpha \leq \frac{n + 1}{q}.$$

Now the following theorem becomes reasonable.

Theorem 1.1. Given a real number α and a positive integer q, there is an integer p such that

$$\left| \alpha - \frac{p}{q} \right| \leq \frac{1}{2q}.$$

Proof. As noted above, there is an integer n such that

$$\frac{n}{q} \leq \alpha \leq \frac{n + 1}{q}.$$

Therefore, either

$$\frac{n}{q} \leq \alpha \leq \frac{n}{q} + \frac{1}{2q} \left(= \frac{n + 1}{q} - \frac{1}{2q} \right)$$

or

$$\frac{n + 1}{q} - \frac{1}{2q} \leq \alpha \leq \frac{n + 1}{q}.$$

In the first case, α is within $1/2q$ of n/q and we take $p = n$; in the second case, α is within $1/2q$ of $(n + 1)/q$ and we take $p = n + 1$.   ▲[4]

Theorem 1.1 is the best that we can do for an arbitrary denominator. For example, with denominator $q = 2$, we can come no closer to $\frac{3}{4}$ than the $1/2q$ of the theorem. With the denominator $q = 10$, we come slightly closer to π ($|\pi - \frac{31}{10}| = .04159\ldots$) than the $\frac{1}{20} = .05$ guaranteed by the theorem. On the other hand, with the denominator $q = 7$, we come considerably closer to π than the $\frac{1}{14} = .0714\ldots$ of the theorem, since

$$|\pi - \tfrac{22}{7}| ,= .0012 \ldots .$$

In other words, $\frac{22}{7}$ is roughly $.0714\ldots/.0012\ldots \approx 60$ times closer to π than what is guaranteed by Theorem 1.1. Thus it appears that for certain exceptional denominators, we can find far better fractional approximations to real

---

[4] We use the symbol ▲ to signify that we have reached the end of a proof.

numbers than what is guaranteed by Theorem 1.1. Let us illustrate the difference between the ordinary denominator and the exceptional denominator. The closest approximation to π with denominator 117 is given by the fraction 368/117 (it barely satisfies Theorem 1.1). The closest approximation to π with denominator 113 is given by the fraction 355/113. Suppose we wish to compute the circumference of the earth from its diameter. If we use 368/117 in place of π, we will make an error of about $29\frac{1}{2}$ miles, while if we use 355/113 in place of π, we will make an error of only $11\frac{1}{2}$ feet! In Chapter 7 we will learn how to find the exceptional fractional approximations to a real number.

Let us completely change the subject. In 1693, De la Loubère gave a rule for inserting the numbers $1, 2, \ldots, n^2$ into an $n \times n$ square (n odd) so that the sums of the numbers of any two rows or columns are the same (this square array is commonly called a magic square). We illustrate the Loubère method in Figure 1.1 with a 5×5 square. Place 1 anywhere in the square and then move diagonally upward to the right, inserting the numbers 2, 3, and 4 as you go. We pretend that the 5×5 squares on the boundary of our square are copies of the original. Thus when 4 is put in the lower-right-hand corner of the square above, we should also put 4 in the lower-right-hand corner of the original square. We always move diagonally upward when possible.

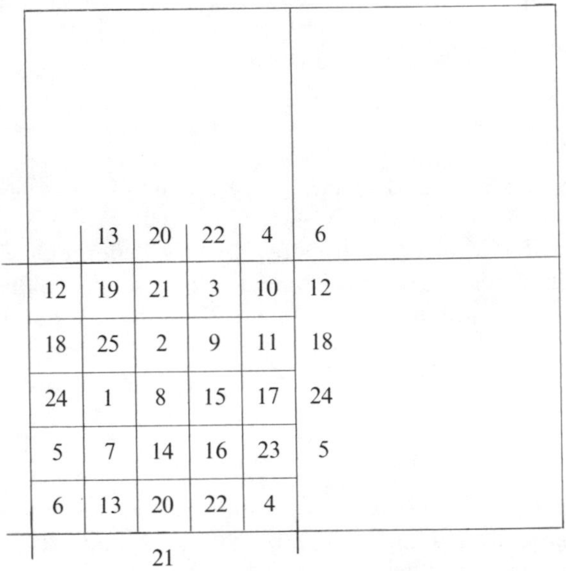

12	19	21	3	10
18	25	2	9	11
24	1	8	15	17
5	7	14	16	23
6	13	20	22	4

Figure 1.1. The constructive process and the finished square. The numbers in any row or column add up to 65.

When we are blocked by a previous entry, as in going from 5 to 6 or from 10 to 11, we drop down one square instead and then continue diagonally upward from there. In Chapter 4 we will apply the theory of Chapter 3 to show that the Loubère method does what it claims to do for all odd n. In the meantime, the reader may enjoy trying it out for other odd n.

1.2. Some Elementary Properties of Divisibility

In this section we derive some of the most used properties of divisibility that do not depend on the factorization of a number into primes. Throughout the rest of the book, unless otherwise mentioned, the small Roman letters $a, b, c, \ldots$ (with the possible exception of x, y, z) will stand for integers. The letter p, except in Chapter 7, will be reserved for primes. Small Greek letters $\alpha, \beta, \gamma, \ldots$ will stand for real numbers, except in Chapter 8, where they may be complex as well.

Number theory could be deduced from a small set of axioms but we shall not take this approach here. There are in particular two basic facts about integers that we shall use throughout the book. The first states that any nonempty set of positive integers contains a smallest member. Known as the **well-ordering principle**, this property of integers will be used implicitly time and again (for example, it is used in the paragraphs immediately before and after the statement of Theorem 1.5). The second fact is known as the **division algorithm** (logically, it is a consequence of the well-ordering principle). It states that if a and b are positive integers, then there are unique integers q and r such that

$$a = bq + r, \qquad 0 \leq r < b.$$

It is called an algorithm because the ordinary method of long division of a by b produces the quotient q and remainder r.

Theorem 1.2. If a, b, d, r, s are integers, $d \neq 0$, and $d|a$, $d|b$, then $d|(ra + sb)$. It follows that $d|(a + b)$, $d|(a - b)$, $d|ra$.

Proof. By definition, there are integers e and f such that

$$a = de, \qquad b = df.$$

Thus

$$ra + sb = rde + sdf$$
$$= d(re + sf),$$

where $re + sf$ is also an integer. Therefore, $d|(ra + sb)$. The special cases have $r = s = 1, r = 1$ and $s = -1, b = a$ and $s = 0$, respectively.   ▲

Theorem 1.3. If a, b, c are integers, $a \neq 0$, $b \neq 0$ and $a|b$, $b|c$, then $a|c$.

Proof. By definition, there are integers d and e such that

$$b = ad, \qquad c = be.$$

Therefore,

$$c = ade = a(de)$$

and hence $a|c$.   ▲

Theorem 1.4. If a, b, and k are integers, $a \neq 0$, $k \neq 0$, then $a|b$ if and only if $ak|bk$.

Proof. If $a|b$, then there is an integer c such that

$$b = ac.$$

Therefore,

$$bk = (ak)c$$

and hence $ak|bk$. Conversely, if $ak|bk$, then there is an integer c such that

$$bk = (ak)c.$$

Thus, since $k \neq 0$,

$$b = ac$$

and hence $a|b$.   ▲

As an example of the application of Theorem 1.3, let n be an integer greater than 1 and let m be the smallest divisor of n which is greater than 1 (this is n itself if n is a prime). Then m is a prime. For if m were composite, we would have an integer k, smaller than m but greater than 1, which divides m. Thus by Theorem 1.3, $k|n$, since $k|m$ and $m|n$. Thus k is a divisor of n which is greater than 1 and smaller than the smallest such divisor, m. This is a contradiction and hence m is a prime. By the way, it follows from this that every positive integer greater than 1 has a prime divisor.

Theorem 1.5. If n is an integer greater than 1, then either n is a prime or n is a finite product of primes.

Proof. If the theorem is false, then there are composite numbers which are not representable as a product of a finite number of primes. Let N be the smallest such number. Thus if $1 < n < N$, the theorem is true for n. Let p be a prime divisor of N. Since N is composite,

$$1 < \frac{N}{p} < N.$$

But this means that the theorem is true for N/p and hence there are primes $p_1, p_2, \ldots, p_k$ such that

$$\frac{N}{p} = p_1 p_2 \cdots p_k.$$

Therefore,

$$N = p p_1 p_2 \cdots p_k$$

is a product of a finite number of primes also. This is a contradiction and thus the theorem is true for all n. ▲

Let us illustrate Theorem 1.2 by giving Euclid's proof of a classic result.

Theorem 1.6. There are infinitely many primes.

Proof. Suppose to the contrary that there are only k primes, $p_1, p_2, \ldots, p_k$ and that all other integers greater than 1 are composite. Let

$$n = p_1 p_2 \cdots p_k + 1$$

and let p be a prime divisor of n (it is possible that $p = n$). Then p is one of the numbers $p_1, p_2, \ldots, p_k$ and hence $p|(p_1 p_2 \cdots p_k)$. Since $p|n$, Theorem 1.2 tells us that

$$p|(n - p_1 p_2 \cdots p_k).$$

But

$$n - p_1 p_2 \cdots p_k = 1$$

and $p \nmid 1$ since $p > 1$. This is a contradiction and hence there are infinitely many primes. ▲

In the examples

$$3 = 2 + 1, \quad 7 = 2 \cdot 3 + 1, \quad 31 = 2 \cdot 3 \cdot 5 + 1,$$

1 plus the product of the first k primes is a prime ($k = 1, 2, 3$). The obvious conjecture that this always occurs is false. The first counterexample is

$$2 \cdot 3 \cdot 5 \cdot 7 \cdot 11 \cdot 13 + 1 = 30\,031 = 59 \cdot 509,$$

which is composite.

EXERCISES
1. Show that if $a \neq 0$, then $a|0$ and $a|a$.
2. Show that if $d \neq 0$, $d|a$, then $d|(-a)$ and $-d|a$.
3. What properties of integers do you use to show that if $n > 1$, then $n \nmid 1$?
4. Show that if $a|b$ and $b|a$, then either $a = b$ or $a = -b$.
5. List all the divisors of 12.
6. List all the numbers which divide both 24 and 36 (compare your answer with your answer to the previous problem).

MISCELLANEOUS EXERCISES
1. Show that if n is composite, then there exists a prime $p \leq \sqrt{n}$ such that $p|n$. (*Hint:* Consider what happens when two numbers greater than $\sqrt{n}$ are multiplied.)
2. Use the idea of problem 1 to test the numbers 91, 103, and 343 as to whether they are prime or composite.
3. Write down the numbers from 1 to 40. Starting with $2 \cdot 2$, cross out every second number: 4, 6, 8, 10, ... Starting with $2 \cdot 3$, cross out every third number: 6, 9, 12, 15, ... Starting with $2 \cdot 5$, cross out every fifth number: 10, 15, 20, ... Use the result of problem 1 to show that the numbers that are not crossed out (except for 1) are exactly the set of primes less than 40.
4. Generalize the result of problem 3 to show how you would find all primes less than or equal to a given integer n. Show that in using this method, it is not necessary to know the primes less than $\sqrt{n}$ beforehand, since after the multiples of the jth prime have been crossed out, the next number remaining after the jth prime is the $(j + 1)$st prime. This method is known as the *Sieve of Eratosthenes*, and its generalizations have been used to construct the modern tables of primes. (For example, the 168 primes less than 1000 will produce all the primes less than 1 000 000.)
5. We will constantly be talking about the smallest (or first) integer of a set of positive integers. Show that there is no such thing as the smallest (or first) positive rational number.

6. Starting with 1 in the lower-left-hand corner, construct the 3×3 and 4×4 squares given by the Loubère method. Verify that all rows and columns have the sum 15 in the 3×3 square and that all the columns of the 4×4 square have the sum 34 but that the rows add up to 32 and 36.

7. Show that if $p \nmid n$ for all primes $p \le \sqrt[3]{n}$, then n is either a prime or a product of two primes.

8. Let p and q be two consecutive odd members of the sequence of primes $2, 3, 5, 7, 11, \ldots$. Show that every factorization of $p + q$ into primes involves at least three (not necessarily distinct) primes. As an example, $7 + 11 = 2 \cdot 3 \cdot 3$.

9. Note that

$$\underline{1}5^2 = \underline{2}25, \underline{3}5^2 = \underline{1}225, \underline{8}5^2 = \underline{7}225, \underline{10}5^2 = \underline{11}025$$

(the underlined portions are for emphasis only). Find a rule for squaring an integer ending in 5 and prove that it works.

10. Note that

$$\frac{1}{7} = \frac{7}{49} \approx \frac{7}{50} = \frac{.7}{5}.$$

$$
\begin{array}{r}
.1\ 4\ 2\ 8\ 5\ 7 \\
5)\overline{.7\ 1\ 4\ 2\ 8\ 5\ 7} \\
5 \\
\overline{2\ 1} \\
2\ 0 \\
\overline{1\ 4} \\
1\ 0 \\
\overline{4\ 2} \\
4\ 0 \\
\overline{2\ 8} \\
2\ 5 \\
\overline{3\ 5} \\
3\ 5 \\
\overline{7}
\end{array}
$$

(repeats)

The modification of the usual division process for .7/5 shown above gives $\frac{7}{49}$ exactly. Use this illustration as a guide to find a simplified method for evaluating the decimal expansion of a/b when $0 < a < b$ and b ends in the digit 9. Illustrate your method with $\frac{1}{19}$ and prove that your method always works.

Chapter 2

THE EUCLIDEAN ALGORITHM AND UNIQUE FACTORIZATION

2.1. The Euclidean Algorithm

Consider the set of all common divisors of the two integers a and b. If $a = b = 0$, then the set of common divisors of a and b is the set of all nonzero integers. If not both a and b are 0, then there are only a finite number of common divisors of a and b, one of which is always 1, and thus there will be a greatest member of this set and it will be positive.

Definition. Let a and b be integers, not both zero. Let d be the largest number in the set of common divisors of a and b. Then we call d the **greatest common divisor** of a and b and we write

$$d = (a,b).$$

For example,

$$(6,4) = 2, \quad (3,5) = 1, \quad (-9,3) = 3, \quad (-6,-4) = 2, \quad (4,0) = 4, \quad (5,5) = 5.$$

Since any divisor of an integer n is also a divisor of $-n$, we see that if a and b are not both zero,

$$(a,b) = (|a|,|b|).$$

Hence we will restrict ourselves at first to finding the greatest common divisor of positive integers.

Let us illustrate the general process by an example. Let

$$d = (54,21).$$

By Theorem 1.2, d also divides

$$12 = 54 - 2 \cdot 21$$

and thus is a common divisor of 12 and 21. By Theorem 1.2, d divides

$$9 = 21 - 12,$$

and therefore d is a common divisor of 9 and 12. By Theorem 1.2 again, d divides

$$3 = 12 - 9.$$

Since $3|9$, we stop at this point. Now that we know that $d|3$, we know that $d \leq 3$. We turn these equations around and using Theorem 1.2 again each time see that 3 divides

$$12 = 3 + 9,$$

and then 3 divides

$$21 = 9 + 12,$$

and then 3 divides

$$54 = 12 + 2 \cdot 21.$$

Thus $3|21$, $3|54$, and $(21, 54) = d \leq 3$. Therefore,

$$(21,54) = 3.$$

We find also that we may use the above equations to write 3 as a linear combination of 21 and 54. Using each equation successively we get

$$12 = 54 - 2 \cdot 21,$$
$$9 = 21 - 12 = 21 - (54 - 2 \cdot 21) = 3 \cdot 21 - 54,$$
$$3 = 12 - 9 = (54 - 2 \cdot 21) - (3 \cdot 21 - 54) = 2 \cdot 54 - 5 \cdot 21.$$

Everything done above is perfectly general. Let d_{-2} and d_{-1} be positive integers. The ordinary division algorithm for

$$\frac{d_{-2}}{d_{-1}}$$

gives a quotient a_0 and remainder d_0 such that

$$d_{-2} = a_0 d_{-1} + d_0, \qquad 0 \leq d_0 < d_{-1}.$$

If d_0 is 0 we stop; otherwise the division algorithm for

$$\frac{d_{-1}}{d_0}$$

gives a quotient a_1 and remainder d_1 such that

$$d_{-1} = a_1 d_0 + d_1, \qquad 0 \leq d_1 < d_0.$$

If $d_1 \neq 0$, we continue onward, getting, successively,

$$d_0 = a_2 d_1 + d_2, \qquad 0 \leq d_2 < d_1,$$
$$d_1 = a_3 d_2 + d_3, \qquad 0 \leq d_3 < d_2,$$
$$\vdots$$
$$d_{k-2} = a_k d_{k-1} + d_k, \qquad 0 \leq d_k < d_{k-1},$$

where it is assumed that $d_j \neq 0$ if $j < k$. Since

$$d_{-1} > d_0 > d_1 > d_2 > d_3 > \cdots > d_{k-1} > d_k \geq 0,$$

it is clear that, sooner or later, some d_j will equal zero and, in fact, since each d_j is at least one smaller than the d_j before it, we will come to a $d_j = 0$ with $j < d_{-1}$. (Actually, it will happen much sooner than $j = d_{-1} - 1$; we are concerned here only with the fact that it does happen.) If $d_{k+1} = 0$, then

$$d_{k-1} = a_{k+1} d_k.$$

Thus, we may put these equations together as

(1)
$$d_{-2} = a_0 d_{-1} + d_0, \qquad 0 < d_0 < d_{-1}$$
$$d_{-1} = a_1 d_0 + d_1, \qquad 0 < d_1 < d_0,$$
$$\vdots$$
$$d_{k-2} = a_k d_{k-1} + d_k, \qquad 0 < d_k < d_{k-1},$$
$$d_{k-1} = a_{k+1} d_k.$$

Theorem 2.1. If d_{-2} and d_{-1} are positive integers and d_k is found from the process of equations (1), then

$$(d_{-2}, d_{-1}) = d_k.$$

Further, we may find integers r and s in a systematic way from equations (1) such that

$$r d_{-2} + s d_{-1} = d_k.$$

Proof. Let $d = (d_{-2}, d_{-1})$. When we put (1) in the form

$$d_0 = d_{-2} - a_0 d_{-1},$$
$$d_1 = d_{-1} - a_1 d_0,$$
$$d_2 = d_0 - a_2 d_1,$$
$$\vdots$$
$$d_k = d_{k-2} - a_k d_{k-1},$$

we see from Theorem 1.2 that $d|d_0$, and then $d|d_1$, $d|d_2, \ldots, d|d_k$. Therefore,

$$(2) \qquad\qquad\qquad d \le d_k.$$

On the other hand, by starting at the last of equations (1) and working up, we see from Theorem 1.2 that in succession,

$$d_k|d_{k-1}, d_k|d_{k-2}, \ldots, d_k|d_2, d_k|d_1, d_k|d_0, d_k|d_{-1}, d_k|d_{-2}.$$

Thus d_k is a common divisor of d_{-1} and d_{-2} and, therefore, by the definition of the greatest common divisor,

$$d_k \le d.$$

This, combined with equation (2), says that

$$d_k = d,$$

as desired.

It is most convenient to give an inductive proof of the last part of the theorem. The main idea is that if we can express d_{j-2} and d_{j-1} as combinations of d_{-2} and d_{-1}, then we may use the equation

$$d_j = d_{j-2} - a_j d_{j-1}$$

to express d_j as a combination of d_{-2} and d_{-1} also. The actual induction is somewhat awkward since d_{j-2} and d_{j-1} are involved in getting the result for d_j. We may put things in the usual form for induction by complicating our induction hypothesis. Let S_n be the statement: There are integers r_{n-2}, s_{n-2}, r_{n-1}, and s_{n-1} such that

$$d_{n-2} = r_{n-2} d_{-2} + s_{n-2} d_{-1},$$
$$d_{n-1} = r_{n-1} d_{-2} + s_{n-1} d_{-1}.$$

Our goal is to prove that S_{k+1} is true since the second part of S_{k+1} says that there are integers r_k and s_k such that

$$d_k = r_k d_{-2} + s_k d_{-1}.$$

First, we note that

$$d_{-2} = 1 \cdot d_{-2} + 0 \cdot d_{-1},$$

$$d_{-1} = 0 \cdot d_{-2} + 1 \cdot d_{-1},$$

and thus S_0 is true. We now prove that if $0 \le n \le k$ and S_n is true, then S_{n+1} is true. Suppose that S_n is true so that there are integers $r_{n-2}, s_{n-2}, r_{n-1}$, and s_{n-1} such that

(3)
$$d_{n-2} = r_{n-2}d_{-2} + s_{n-2}d_{-1},$$

$$d_{n-1} = r_{n-1}d_{-2} + s_{n-1}d_{-1}.$$

We see from (1) that

$$d_n = d_{n-2} - a_n d_{n-1},$$

and if we substitute (3) into this, we get

$$\begin{aligned} d_n &= (r_{n-2}d_{-2} + s_{n-2}d_{-1}) - a_n(r_{n-1}d_{-2} + s_{n-1}d_{-1}) \\ &= (r_{n-2} - a_n r_{n-1})d_{-2} + (s_{n-2} - a_n s_{n-1})d_{-1} \\ &= r_n d_{-2} + s_n d_{-1}, \end{aligned}$$

where we have put

$$r_n = r_{n-2} - a_n r_{n-1},$$

$$s_n = s_{n-2} - a_n s_{n-1}.$$

Thus we have integers r_{n-1}, s_{n-1}, r_n, and s_n such that

$$d_{n-1} = r_{n-1}d_{-2} + s_{n-1}d_{-1},$$

$$d_n = r_n d_{-2} + s_n d_{-1},$$

which is the statement S_{n+1}. Thus S_{n+1} follows from S_n. Since S_0 is true, S_1 follows from S_0, and then S_2 follows from S_1, S_3 from $S_2, \ldots$, until finally S_{k+1} follows from S_k. ▲

Definition. The use of equations (1) for finding the greatest common divisor is called the **Euclidean algorithm**.

Euclid, of course, did not use algebraic manipulations but rather stated the whole process geometrically. We will say more about this in Chapter 7. In actual practice, when we wish to write d_k in terms of d_{-2} and d_{-1}, it is advantageous to proceed from the bottom of equations (1) rather than from the top.

As another example of the Euclidean algorithm, let us calculate $(53, 77)$ and express it as a linear combination of 53 and 77. We see that

$$77 = 1 \cdot 53 + 24,$$
$$53 = 2 \cdot 24 + 5,$$
$$24 = 4 \cdot 5 + 4,$$
$$5 = 1 \cdot 4 + 1,$$
$$4 = 4 \cdot 1,$$

and thus

$$(53, 77) = 1.$$

Working backward we see that

$$1 = 5 - 1 \cdot 4$$
$$= 5 - 1 \cdot (24 - 4 \cdot 5) = 5 \cdot 5 - 1 \cdot 24$$
$$= 5 \cdot (53 - 2 \cdot 24) - 1 \cdot 24 = 5 \cdot 53 - 11 \cdot 24$$
$$= 5 \cdot 53 - 11 \cdot (77 - 1 \cdot 53) = 16 \cdot 53 - 11 \cdot 77.$$

In the example above, 53 and 77 have 1 as their greatest common divisor and hence they have no common factors other than 1 and -1. Such a fact is sufficiently important to give it a name.

Definition. Let a and b be integers, not both zero. If

$$(a, b) = 1,$$

then we say that a and b are **relatively prime**.

Another way of putting this is to say that a and b are relatively prime if and only if 1 and -1 are their only common divisors.

The result on linear combinations will be very useful both here and later, and thus we will extend it to all integers and not just positive integers.

Theorem 2.2. Let a and b be integers, not both zero. Then there exist integers r and s such that

$$ar + bs = (a, b).$$

Proof. We take the cases of neither a and b are zero and one of a and b is zero separately. Suppose that $b = 0$. Then

$$(a,0) = (|a|, 0) = |a|$$

and

$$a(\pm 1) + 0(0) = |a|,$$

where the $+1$ is used if $a > 0$ and -1 is used if $a < 0$. In like manner, if $a = 0$, then

$$0(0) + b(\pm 1) = |b| = (0, b),$$

where the $+1$ is used if $b > 0$ and -1 is used if $b < 0$.

We may now restrict our attention to the case that neither a nor b is zero and, in this case, both $|a|$ and $|b|$ are positive. By Theorem 2.1, there are integers r and s such that

$$r|a| + s|b| = (|a|, |b|) = (a,b).$$

Since

$$a = \pm|a|, \qquad b = \pm|b|,$$

we see that

$$(\pm r)a + (\pm s)b = (a,b),$$

for an appropriate choice of signs.   ▲

As an example, we saw earlier that

$$2 \cdot 54 + (-5) \cdot 21 = 3 = (54, 21)$$

and thus

$$(-2) \cdot (-54) + (-5) \cdot 21 = 3 = (-54, 21),$$

$$2 \cdot 54 + 5 \cdot (-21) = 3 = (54, -21),$$

$$(-2) \cdot (-54) + 5 \cdot (-21) = 3 = (-54, -21).$$

One result of Theorem 2.2 is the following.

Theorem 2.3. Let $d = (a,b)$. Then n is a common divisor of a and b if and only if $n|d$.

Proof. If $n|d$, then since $d|a$ and $d|b$, Theorem 1.3 says that $n|a$ and $n|b$, and hence any divisor of d is a divisor of a and b. Conversely, if $n|a$ and $n|b$, then by Theorem 1.2, n divides

$$ar + bs = d,$$

where r and s are the integers given by Theorem 2.2. Thus a common divisor of a and b is also a divisor of d. ▲

This result is a much more useful property of the greatest common divisor than its definition as the largest of the common divisors. We will now assemble several of the other most used properties of greatest common divisors.

Theorem 2.4. Let $(a,b) = d$ and let k be an arbitrary integer. Then

(a) $(a, b + ka) = (a,b)$.

(b) $(ak,bk) = |k|(a,b)$ $(k \neq 0)$.

(c) $\left(\dfrac{a}{d}, \dfrac{b}{d}\right) = 1$.

Proof. If $n|a$, $n|b$, then by Theorem 1.2, $n|(b + ka)$. Thus any divisor of a and b is a divisor of a and $b + ka$. Conversely, if $n|a$, $n|(b + ka)$, then $n|[(b + ka) - ka]$, and thus n is a common divisor of a and b. Hence the set of divisors of a and b is also the set of divisors of a and $b + ka$, and therefore the greatest member of this set is the greatest common divisor of a and $b + ka$ as well as the greatest common divisor of a and b. This proves (a). Let

$$(ak,bk) = n,$$

and, for the moment, let k be positive. Since $d|a$, $d|b$, we see that $dk|ak$, $dk|bk$ and thus, by Theorem 2.3,

$$dk|n.$$

Thus there is a positive integer m such that

(4) $(ak, bk) = dkm.$

As a result,

$$dmk|ak, \qquad dmk|bk.$$

It follows from Theorem 1.4 that

$$dm|a, \qquad dm|b,$$

then from Theorem 2.3 that

$$dm|d \cdot 1,$$

and finally from Theorem 1.4 again that

$$m \mid 1.$$

Thus $m = \pm 1$ and since $m > 0$, $m = 1$. Equation (4) is thus

$$(ak,bk) = dk = k(a,b) = |k|(a,b),$$

which proves (b) when $k > 0$. Now if $k < 0$, then $-k = |k| > 0$, and therefore

$$(ak,bk) = (-ak,-bk) = (a|k|,b|k|) = |k|(a,b),$$

and thus (b) is true. Last, since $d > 0$, it follows from (b) that

$$d = (a,b) = \left(d \cdot \frac{a}{d}, d \cdot \frac{b}{d}\right) = d\left(\frac{a}{d},\frac{b}{d}\right),$$

which yields, after dividing both sides by d,

$$1 = \left(\frac{a}{d},\frac{b}{d}\right). \qquad \blacktriangle$$

We may also define the greatest common divisor of more than two integers. We will use this concept for three integers in Chapter 5, and so we present here the necessary details and leave further results to the problems.

Definition. If a, b, and c are integers, not all zero, and d is the largest of the common divisors of a, b, and c, then we say that d is the **greatest common divisor** of a, b, and c and we write

$$d = (a,b,c).$$

Since $1 \mid a$, $1 \mid b$, $1 \mid c$, we see that (a,b,c) is positive. As an example,

$$(4,8,10) = 2.$$

Theorem 2.5. If $(a,b,c) = d$, then

$$\left(\frac{a}{d},\frac{b}{d},\frac{c}{d}\right) = 1.$$

Proof. Let

$$\left(\frac{a}{d},\frac{b}{d},\frac{c}{d}\right) = n$$

so that $n \geq 1$. Thus

$$n \left| \frac{a}{d}, \quad n \right| \frac{b}{d}, \quad n \left| \frac{c}{d} \right.$$

and, by Theorem 1.4,

$$dn|a, \quad dn|b, \quad dn|c.$$

Hence dn is a common divisor of a, b, and c and hence is less than or equal to the greatest common divisor of a, b, and c:

$$dn \leq d.$$

We divide this by d and find that

$$n \leq 1.$$

Hence $n = 1$.   ▲

We note that it is possible to have

$$(a,b,c) = 1$$

even though no two of the numbers a, b, and c are relatively prime. For example,

$$(6,10,15) = 1$$

even though

$$(6,10) = 2, \quad (6,15) = 3, \quad (10,15) = 5.$$

Definition. Let $a_1, a_2, \ldots, a_n$ be nonzero integers. We say that these numbers are **pairwise relatively prime** if the greatest common divisor of each pair of these integers is 1.

For example, the integers 4, 15, and 77 are pairwise relatively prime since

$$(4,15) = (4,77) = (15,77) = 1,$$

while the integers 4, 15, 77, and 91 are not pairwise relatively prime since

$$(77,91) = 7.$$

EXERCISES

1. Show that if a, b, c are pairwise relatively prime, then

$$(a,b,c) = 1.$$

2. Use the Euclidean algorithm to find the greatest common divisor of (a) 77 and 91, (b) 182 and 442, and (c) 2311 and 3701.
3. Express (17,37) as a linear combination of 17 and 37.
4. Express (399,703) as a linear combination of 399 and 703.
5. Find integers r and s such that $547r + 632s = 1$.
6. Find integers r and s such that $398r + 600s = 2$.
*7. Find integers r and s such that $922r + 2163s = 7$.
*8. Are there integers r and s such that $1841r + 3647s = 1$? Why?
9. Show that if there is no prime p such that $p|a$, $p|b$, then

$$(a, b) = 1.$$

10. In the proof of Theorem 2.1, why did we restrict the proof that S_n implies S_{n+1} to $0 \leq n \leq k$?
11. Are the integers 101, 209, 283, and 341 pairwise relatively prime?
12. Show that if p is a prime and a an integer, then either $(a,p) = 1$ or $(a,p) = p$.
13. Use Theorem 2.4(c) to show that a fraction m/n can always be reduced to lowest terms.
14. Let $\alpha_j = d_{j-2}/d_{j-1}$. Show that the Euclidean algorithm of equation (1) takes the form

$$\alpha_0 = a_0 + \frac{1}{\alpha_1}, \qquad a_0 < \alpha_0 < a_0 + 1,$$

$$\alpha_1 = a_1 + \frac{1}{\alpha_2}, \qquad a_1 < \alpha_1 < a_1 + 1,$$

$$\alpha_k = a_k + \frac{1}{\alpha_{k+1}}, \qquad a_k < \alpha_k < a_k + 1,$$

$$\alpha_{k+1} = a_{k+1}.$$

2.2. The Fundamental Theorem of Arithmetic

The fundamental theorem of arithmetic, otherwise known as the unique factorization theorem, states that if you and I independently write an integer greater than 1 as a product of primes, we will get the same result except for the order in which the primes are written in the two products. This theorem will be used constantly throughout the rest of the book and well deserves its name. There are times that the following milder-sounding theorems will suffice in the applications; they are not really milder since they will be used to prove the fundamental theorem later in this section.

Theorem 2.6. If $(n,a) = 1$ and $n|ab$, then $n|b$.

Proof. Since $(n,a) = 1$, by Theorem 2.2, there are integers r and s such that

$$nr + as = 1.$$

Thus

$$nrb + abs = b.$$

Since $n|n$ and $n|ab$, $n|[n(rb) + (ab)s]$, which is to say $n|b$.   ▲

Theorem 2.7. If $(a,m,n) = 1$ (note that this is true whenever two of the numbers a, m, n are relatively prime), then

$$(a,mn) = (a,m) \cdot (a,n).$$

In particular, if $(a,m) = (a,n) - 1$, then $(a,mn) = 1$.

Proof. Let

$$d = (a,mn), \qquad d_1 = (a,m), \qquad d_2 = (a,n).$$

We then wish to show that $d = d_1 d_2$. By Theorem 2.2, there are integers r, s, t, and u such that

$$ar + ms = d_1, \qquad at + nu = d_2.$$

Therefore,

$$(ar + ms)(at + nu) = d_1 d_2;$$

that is,

$$a(art + rnu + mst) + mn(su) = d_1 d_2.$$

It follows from the definition of d and Theorem 1.2 that $d|d_1 d_2$. Hence

(5) $$d \leq d_1 d_2.$$

In order to prove the opposite inequality, we need to prove that

$$(d_1,d_2) = 1.$$

This is done as follows. Let $(d_1,d_2) = e \geq 1$. Then $e|d_1$, $e|d_2$ and thus by the definition of d_1 and d_2 and by Theorem 2.3, $e|a$, $e|m$, $e|n$. Thus e is a common divisor of a, m, n, and if $e > 1$, this contradicts the fact that $(a, m, n) = 1$. Hence $e = 1$, as desired. But now, note that $d_1|a$, $d_1|m$ (by definition) and thus, by Theorem 2.3, $d_1|d$. In like manner, $d_2|a$, $d_2|n$, and thus $d_2|d$. But this may be written $d_2|d_1 \cdot (d/d_1)$. Since $(d_2,d_1) = 1$, it follows from Theorem 2.6 that $d_2|(d/d_1)$, and then it follows from Theorem 1.4 that $d_1d_2|d$. Therefore,

$$d_1d_2 \leq d,$$

and comparing this with (5) we see that $d = d_1d_2$.   ▲

The next theorem is usually proved by using Theorem 2.6, but it is somewhat simpler to use Theorem 2.7.

Theorem 2.8. If p is a prime and $p|(a_1a_2 \cdots a_k)$, then for some j, $1 \leq j \leq k$, $p|a_j$. As a special case, if $p|a^k$, then $p|a$.

Proof. We note that the only positive divisors of p are 1 and p. If a is an arbitrary integer, then since $(a,p)|p$, we see that $(a,p) = 1$ or $(a,p) = p$. In the second case $p|a$. Thus if a is an integer such that $p\nmid a$, then

$$(p,a) = 1.$$

Now suppose that p divides none of the numbers $a_1, a_2, \ldots, a_k$. Then

$$(p,a_1) = 1, (p,a_2) = 1, \ldots, (p,a_k) = 1.$$

By Theorem 2.7,

$$(p,a_1a_2) = 1.$$

By Theorem 2.7, again,

$$(p,a_1a_2a_3) = 1.$$

After the $(k - 1)$st application of Theorem 2.7, we find that

$$(p,a_1a_2 \cdots a_k) = 1.$$

But this contradicts the fact that $p > 1$ is a common divisor of p and $a_1a_2 \cdots a_k$. Hence p divides one of the numbers $a_1, a_2, \ldots, a_k$, as desired.   ▲

Theorem 2.9. (The Fundamental Theorem of Arithmetic, or the Unique Factorization Theorem for Positive Integers). Suppose that $n > 1$ and

$$n = p_1p_2p_3 \cdots p_r = q_1q_2 \cdots q_s,$$

where $p_1, p_2, \ldots, p_r, q_1, q_2, \ldots, q_s$ are primes. Then $r = s$ and the two factorizations of n are the same apart from the order of the factors.

Proof. Suppose that the theorem is false. Then the theorem is false for certain values of n, and we will let N be the smallest of these. Thus we shall assume that the theorem is true for all integers n between 1 and N but that the theorem is false for $n = N$. We will show that this leads to a contradiction. Suppose that

$$N = p_1 p_2 \cdots p_r = q_1 q_2 \cdots q_s,$$

where $p_1, p_2, \ldots, p_r, q_1, q_2, \ldots, q_s$ are primes.

The theorem is clearly true for primes and thus N must be composite and hence $r \geq 2$, $s \geq 2$. Since the order of the factors is not important, we may assume that they have been written so that

(6)
$$p_r \geq p_j, \qquad 1 \leq j \leq r - 1.$$
$$q_s \geq q_j, \qquad 1 \leq j \leq s - 1.$$

We will first show that $p_r = q_s$. If this is false, then either $p_r > q_s$ or $q_s > p_r$. We will show here that $p_r > q_s$ is false; the proof that $q_s > p_r$ is false is identical and in fact may be given from our proof by interchanging the letters p and q and interchanging r and s. If $p_r > q_s$, then, by (6),

$$p_r > q_j, \qquad 1 \leq j \leq s.$$

Therefore, $p_r \nmid q_j$ for any of the q_j's. But by Theorem 2.8, this is a contradiction, since

$$p_r | (q_1 q_2 \cdots q_s),$$

the product being N. Thus the inequality $p_r > q_s$ is false and, in like manner, $q_s > p_r$ is false. Hence

$$p_r = q_s,$$

and therefore

(7)
$$\frac{N}{p_r} = p_1 p_2 \cdots p_{r-1} = q_1 q_2 \cdots q_{s-1}.$$

Since $r \geq 2$, $s \geq 2$, there is at least one prime in each of the factorizations of N/p_r in (7) and thus

$$1 < \frac{N}{p_r} < N.$$

As a result, the theorem holds for $n = N/p_r$ and therefore

$$r - 1 = s - 1,$$

and the factorization $q_1 q_2 \cdots q_{s-1}$ of N/p_r is the same factorization as $p_1 p_2 \cdots p_{r-1}$ except possibly for the order of the factors. It follows that $r = s$ and the two factorizations of N as $p_1 p_2 \cdots p_r$ and $q_1 q_2 \cdots q_r$ are the same except possibly for the order of the factors. Thus the theorem is true for N and this contradicts the definition of N. Thus the theorem is true for all $n > 1$.   ▲

As the example

$$2 = 1 \cdot 2 = 1 \cdot 1 \cdot 2 \cdot 1 \cdot 1 \cdot 1$$

shows, the fundamental theorem would be false if 1 were a prime. This is one reason why 1 is not considered a prime. As we see from the examples,

$$18 = 2 \cdot 3 \cdot 3, \qquad 36 = 2 \cdot 2 \cdot 3 \cdot 3, \qquad 64 = 2 \cdot 2 \cdot 2 \cdot 2 \cdot 2 \cdot 2,$$

it frequently happens that certain primes occur more than once in the factorization of a composite number. In such cases, it is customary to use exponents,

$$18 = 2 \cdot 3^2, \qquad 36 = 2^2 \cdot 3^2, \qquad 64 = 2^6,$$

and in general we will write

$$n = p_1^{a_1} p_2^{a_2} \cdots p_k^{a_k}.$$

When n is written this way, we will always assume that the numbers $p_1, p_2, \ldots, p_k$ are distinct primes and, unless otherwise stated, that $a_1 > 0$, $a_2 > 0, \ldots, a_k > 0$. The unique factorization theorem in this form says that if

$$n = p_1^{a_1} p_2^{a_2} \cdots p_k^{a_k} = q_1^{b_1} q_2^{b_2} \cdots q_m^{b_m}$$

(where the q_j are also primes and the b_j are positive), then $k = m$ and, in some order, the primes $p_1, p_2, \ldots, p_k$ and $q_1, q_2, \ldots, q_m$ are the same with the corresponding exponents being equal also. The following result is an immediate corollary of either Theorem 2.8 or 2.9, but it is one which will be used time and again.

Theorem 2.10. Suppose that the factorization of n into primes is given as

$$n = p_1^{a_1} p_2^{a_2} \cdots p_k^{a_k}$$

and that p is a prime such that $p|n$. Then for some j in the range $1 \le j \le k$, $p = p_j$.

Proof. By Theorem 2.8, $p|p_j$ for some j. Since $p > 1$ and the only positive divisors of p_j are 1 and p_j, it must be that $p = p_j$.   ▲

We may easily find the greatest common divisor of two (or more) integers if we know their factorizations into primes. For example, from the factorizations

$$2600 = 2 \cdot 2 \cdot 2 \cdot 5 \cdot 5 \cdot 13$$

$$10\,140 = 2 \cdot 2 \cdot 3 \cdot 5 \cdot 13 \cdot 13,$$

it is easy to see that

$$(2600, 10\,140) = 2 \cdot 2 \cdot 5 \cdot 13 = 260,$$

particularly if we write the above factorizations as

$$2600 = (2 \cdot 2 \cdot 5 \cdot 13) \cdot 2 \cdot 5,$$

$$10\,140 = (2 \cdot 2 \cdot 5 \cdot 13) \cdot 3 \cdot 13.$$

This process is perfectly general; its only disadvantage for large numbers is that you must know how they factor into primes.

Theorem 2.11. Suppose that

$$n = p_1 p_2 \cdots p_k q_1 q_2 \cdots q_i,$$

$$m = p_1 p_2 \cdots p_k r_1 r_2 \cdots r_j,$$

where $p_1, p_2, \ldots, p_k, q_1, q_2, \ldots, q_i, r_1, r_2, \ldots, r_j$ are primes such that none of the q's are equal to any of the r's. (If $k = 0$, we interpret the product $p_1 p_2 \cdots p_k$ as 1 and similarly for $i = 0$ and $j = 0$.) Then

$$(n,m) = p_1 p_2 \cdots p_k.$$

Proof. Let

$$d = p_1 p_2 \cdots p_k.$$

Then $d|n$, $d|m$ and hence by Theorem 2.3, $d|(n,m)$. Thus there is a positive integer a such that

$$(n,m) = da.$$

Therefore,

$$da|dq_1 q_2 \cdots q_i, \qquad da|dr_1 r_2 \cdots r_j$$

and hence, by Theorem 1.4,

$$a|q_1 q_2 \cdots q_i, \qquad a|r_1 r_2 \cdots r_j.$$

Our goal is to prove that $a = 1$. If $a > 1$, then there is a prime p which divides a, and it must also divide $q_1 q_2 \cdots q_i$ and $r_1 r_2 \cdots r_j$. By Theorem 2.10, p is one of the q_m's and is also one of the r_m's. Thus the primes $q_1, \ldots, q_i$ have a prime in common with the primes $r_1, \ldots, r_j$, which is contrary to the hypothesis of the theorem. Hence $a = 1$ and therefore

$$(n,m) = d. \qquad \blacktriangle$$

EXERCISES

In problems 1–6, find the greatest common divisor of m and n by means of Theorem 2.11 and check your result by using the Euclidean algorithm. You may assume that the factorizations given of m and n are factorizations into primes.

1. $m = 143 = 11 \cdot 13$, $n = 187 = 11 \cdot 17$.
2. $m = 231 = 3 \cdot 7 \cdot 11$, $n = 561 = 3 \cdot 11 \cdot 17$.
3. $m = 588 = 2 \cdot 2 \cdot 3 \cdot 7 \cdot 7$, $n = 7546 = 2 \cdot 7 \cdot 7 \cdot 7 \cdot 11$.
4. $m = 119\,790 = 2 \cdot 3 \cdot 3 \cdot 5 \cdot 11 \cdot 11 \cdot 11$, $n = 42\,900 = 2 \cdot 2 \cdot 3 \cdot 5 \cdot 5 \cdot 11 \cdot 13$.
5. $m = 830\,407 = 823 \cdot 1009$, $n = 919\,199 = 911 \cdot 1009$.
6. $m = 9797 = 97 \cdot 101$, $n = 14\,507 = 89 \cdot 163$.
7. What can you conclude about the four numbers $1\,456\,813$, $1\,468\,823$, $1\,476\,221$, and $1\,488\,391$ given that $1\,456\,813 \cdot 1\,488\,391 = 1\,468\,823 \cdot 1\,476\,221$? Justify your conclusions.
8. Suppose that

$$n = p_1^{a_1} p_2^{a_2} \cdots p_k^{a_k}, \qquad m = p_1^{b_1} p_2^{b_2} \cdots p_k^{b_k}.$$

(Any two positive integers may be written this way with the same primes if we allow zero exponents.) If $\min\{a,b\}$ means the smaller of a and b (or their common value if they are equal), show that

$$(n,m) = p_1^{\min\{a_1,b_1\}} p_2^{\min\{a_2,b_2\}} \cdots p_k^{\min\{a_k,b_k\}}.$$

9. Suppose that

$$n = p_1^{a_1} p_2^{a_2} \cdots p_k^{a_k}, \qquad m = p_1^{b_1} p_2^{b_2} \cdots p_k^{b_k},$$

where zero exponents are allowed. Prove that $n \mid m$ if and only if

$$a_1 \le b_1, a_2 \le b_2, \ldots, a_k \le b_k.$$

*10. Show that Theorem 2.6 can be proved from Theorem 2.9 without use of the material of Section 2.1.
11. Show that Theorem 2.8 can be proved from Theorem 2.9.
12. Show that $\log_{10} 2$ is irrational. (*Hint*: Let $\log_{10} 2 = n/m$ and show that $2^m = 10^n$.)

13. Show that if p is a prime and $p|a^n$, then $p^n|a^n$.
14. How many zeros are there at the end of $100!$?
15. Give an example of four positive integers such that any three of them have a common divisor greater than 1, although only ± 1 divide all four of them.

2.3. Applications of the Fundamental Theorem

This is actually a misleading section heading since it is usually Theorems 2.6, 2.7, and 2.8 that are used in applications rather than Theorem 2.9. But as none of these theorems could be true without unique factorization, the section heading accurately describes the fact that the results in this section depend on the unique factorization property of the positive integers. Our first application will be an application of the fundamental theorem itself. Although it seems very mild, Theorem 2.12 will be of great importance in Chapter 5.

Theorem 2.12. Suppose that a and b are relatively prime positive integers and

$$ab = c^n.$$

Then there are positive integers d and e such that

$$a = d^n, \qquad b = e^n.$$

Proof. If $a = 1$, then we may let $d = 1$, $e = c$; if $b = 1$, then we may let $d = c, e = 1$. Thus we may restrict our attention to the case that $a > 1, b > 1$. Since $(a,b) = 1$, the prime factors of a and b are distinct. Thus we may set

$$a = p_1^{a_1} p_2^{a_2} \cdots p_r^{a_r}, \qquad b = p_{r+1}^{a_{r+1}} \cdots p_{r+s}^{a_{r+s}},$$

where $p_1, p_2, \ldots, p_{r+s}$ are distinct primes, $r \geq 1$, $s \geq 1$. Suppose that the prime decomposition of c is given by

$$c = q_1^{b_1} q_2^{b_2} \cdots q_k^{b_k}.$$

Then

$$p_1^{a_1} p_2^{a_2} \cdots p_{r+s}^{a_{r+s}} = q_1^{nb_1} q_2^{nb_2} \cdots q_k^{nb_k}.$$

By Theorem 2.9, $k = r + s$, the primes q_j are the same as the primes p_j (except for order), and the corresponding exponents are the same. Thus we may renumber the q's so that

$$q_j = p_j, \qquad 1 \leq j \leq r + s$$

and then

$$a_j = nb_j, \qquad 1 \leq j \leq r + s.$$

Hence

$$a = (p_1^{b_1} p_2^{b_2} \cdots p_r^{b_r})^n,$$
$$b = (p_{r+1}^{b_{r+1}} p_{r+2}^{b_{r+2}} \cdots p_{r+s}^{b_{r+s}})^n. \qquad \blacktriangle$$

The reader may have learned that $\sqrt{2}$ is irrational. This is a special case of the converse of a far more general result,

Theorem 2.13. Suppose that a and n are positive integers and $\sqrt[n]{a}$ is rational. Then $\sqrt[n]{a}$ is an integer.

Proof. Since $\sqrt[n]{a}$ is rational (and positive), there are positive integers r and s such that

$$\sqrt[n]{a} = \frac{r}{s}.$$

We may even assume that $(r,s) = 1$, since we may otherwise divide the numerator and denominator by (r,s). We will show that $s = 1$. If $s > 1$, then there is a prime p which divides s and then p divides

$$as^n = r^n.$$

By Theorem 2.8, p also divides r, and this contradicts the fact that $(r,s) = 1$. Hence $s = 1$ and therefore

$$\sqrt[n]{a} = r,$$

an integer. $\blacktriangle$

As an example of this theorem, since $1 < \sqrt{2} < 2$, $\sqrt{2}$ is not an integer and hence not rational. As another example, since

$$2^3 < 10 < 3^3,$$

it follows that

$$2 < \sqrt[3]{10} < 3,$$

and thus $\sqrt[3]{10}$ is not an integer. Therefore, $\sqrt[3]{10}$ is irrational.

The next application is actually only a preliminary result which is necessary in the proof of Theorem 2.15 in the next section (such a result is sometimes called a lemma).

Theorem 2.14. Suppose that m and n are relatively prime positive integers. If d is positive and $d|mn$, then there are unique positive integers d_1 and d_2 such that

$$d = d_1 d_2, \qquad d_1|m, \quad d_2|n.$$

Conversely, if $d_1|m$ and $d_2|n$, then $d_1 d_2|mn$.

Proof. Suppose that $d|mn$. We first show that there exists at least one such pair of integers d_1, d_2. Let

(8) $d_1 = (d,m), \qquad d_2 = (d,n).$

Since $d|mn$, we see that $(d,mn) = d$. Further, since $(m,n) = 1$, we may apply Theorem 2.7 to get

$$d = (d,mn) = (d,m) \cdot (d,n) = d_1 d_2.$$

By definition, $d_1|m$, $d_2|n$, and thus d_1 and d_2 have the desired properties. Now suppose that d_1' and d_2' are positive integers with the properties that

$$d = d_1' d_2', \qquad d_1'|m, \quad d_2'|n.$$

Then we see from the definitions of d_1 and d_2 as greatest common divisors in (8) that

(9) $d_1' \leq d_1, \qquad d_2' \leq d_2$

and therefore

(10) $d_1' d_2' \leq d_1 d_2 = d.$

The only way that equality may hold in (10) is that equality holds in both of (9) and hence

$$d_1' = d_1, \qquad d_2' = d_2.$$

This proves that the representation of d in the form of the theorem is unique. The converse follows from the definition of divisibility.   ▲

EXERCISES

1. Prove that $\sqrt[3]{3}$ is irrational.
2. Prove that $\sqrt[5]{5}$ is irrational.
3. Prove that if $n \geq 2$, then $\sqrt[n]{n}$ is irrational. (*Hint :* Show that if $n \geq 2$, then $2^n > n$.)
*4. Verify that $\sqrt{2} + \sqrt{3}$ is a root of the equation

$$x^4 - 10x^2 + 1 = 0.$$

Use the methods in the proof of Theorem 2.13 to show that the only possible rational roots of this equation are $x = 1$ and $x = -1$, neither of which are roots. Conclude that the roots of this equation are all irrational.

5. The following numbers were once offered as a counterexample to Theorem 2.14: $m = 2^2 \cdot 3 \cdot 5$, $n = 7 \cdot 11$, $d = 11$ (the claim being that, as may be seen from the factorizations, there is no value of d_1 that will do). Is this really a counterexample?

2.4. Multiplicative Functions

Before we actually give a definition of multiplicative functions, we will present two examples.

Definition. Let n be a positive integer. We let $d(n)$ be the number of positive integers which divide n (including 1 and n itself). We let $\sigma(n)$ be the sum of the positive divisors of n (including 1 and n).

In Figure 2.1, we have evaluated $d(n)$ and $\sigma(n)$ for n in the range 1 to 20.

n	1	2	3	4	5	6	7	8	9	10	11	12	13	14	15	16	17	18	19	20
$d(n)$	1	2	2	3	2	4	2	4	3	4	2	6	2	4	4	5	2	6	2	6
$\sigma(n)$	1	3	4	7	6	12	8	15	13	18	12	28	14	24	24	31	18	39	20	42

Figure 2.1

It is clear that $d(n) = 2$ if and only if n is a prime and likewise $\sigma(n) = n + 1$ if and only if n is a prime. We should not expect a simple formula for either $d(n)$ or $\sigma(n)$, since then we could immediately decide from it whether or not a given integer is a prime. We will see shortly, however, that if we already know the factorization of n into primes, then there are simple formulas for $d(n)$ and $\sigma(n)$.

We see from the figure that there are times that $d(nm)$ [or $\sigma(mn)$] can be determined from $d(n)$ and $d(m)$ [or $\sigma(m)$ and $\sigma(n)$] by multiplication. For example,

$$d(2 \cdot 5) = \quad 4 = d(2) \cdot d(5),$$

$$d(3 \cdot 4) = \quad 6 = d(3) \cdot d(4),$$

$$\sigma(2 \cdot 9) = 39 = \sigma(2) \cdot \sigma(9),$$

$$\sigma(4 \cdot 5) = 42 = \sigma(4) \cdot \sigma(5).$$

On the other hand, this cannot always be done, as the following examples illustrate;

$$d(3 \cdot 6) = 6 \neq 8 = d(3) \cdot d(6),$$

$$\sigma(4 \cdot 4) = 31 \neq 49 = \sigma(4) \cdot \sigma(4).$$

In the above examples, we have had success in saying that

$$d(mn) = d(m) \cdot d(n),$$

$$\sigma(mn) = \sigma(m) \cdot \sigma(n),$$

in every case that $(m,n) = 1$. We will prove this to be true shortly. These examples motivate the following definition.

Definition. If the function $f(n)$ is defined for all positive integers n, then we say that $f(n)$ is **multiplicative** if for all pairs of relatively prime positive integers m and n,

$$f(mn) = f(m) \cdot f(n).$$

If this is true for all pairs of positive integers, relatively prime or not, then we say that $f(n)$ is **completely multiplicative**.

As the examples above show, the concept of completely multiplicative functions eliminates some functions of interest that the broader concept of multiplicative function is able to consider. Examples of completely multiplicative functions are $f(n) = n$ and the constant function, $f(n) = 1$. The usefulness of a multiplicative function is that if we know what it is at prime powers, then we know what it is for all positive integers by multiplication; for example, if $d(n)$ is multiplicative, then

$$d(126) = d(2 \cdot 3^2 \cdot 7) = d(2) \cdot d(3^2) \cdot d(7) = 2 \cdot 3 \cdot 2 = 12.$$

It is usually, but not always, easy to evaluate multiplicative functions at prime powers, and this leads to general formulas for all integers. The methods of showing that $d(n)$ and $\sigma(n)$ are multiplicative are virtually identical. As a result, we will prove a more general result which will be useful later and from which we may instantly show that $d(n)$ and $\sigma(n)$ are multiplicative.

It may be useful to review the summation notation before continuing. The reader is no doubt familiar with the notation

$$\sum_{n=a}^{b} f(n),$$

which is defined for $a \leq b$ as

$$\sum_{n=a}^{b} f(n) = f(a) + f(a + 1) + f(a + 2) + \cdots + f(b - 1) + f(b).$$

It frequently happens that we do not wish to add $f(n)$ for all n in an interval, but that we wish to add $f(n)$ for all n restricted in a certain manner. In this case, the restriction is usually put under the summation sign. For example,

$$\sum_{\substack{n=1 \\ n \text{ even}}}^{6} f(n) = f(2) + f(4) + f(6),$$

$$\sum_{\substack{p=4 \\ p \text{ prime}}}^{9} f(p) = f(5) + f(7),$$

$$\sum_{\substack{n=1 \\ n \text{ a} \\ \text{perfect} \\ \text{square}}}^{17} f(n) = f(1) + f(4) + f(9) + f(16) = \sum_{m=1}^{4} f(m^2),$$

$$\sum_{\substack{n=1 \\ n|60}}^{19} f(n) = f(1) + f(2) + f(3) + f(4) + f(5) + f(6) + f(10) + f(12) + f(15).$$

This last type of sum occurs particularly often in number theory. In such cases, it is usual to drop the range of summation and, if necessary, add a new restriction under the summation sign. For example, the last sum above may be written

$$\sum_{\substack{n|60, \\ 1 \leq n \leq 19}} f(n).$$

It is always assumed in this notation that we are speaking of only the positive divisors of an integer. Thus the preceding sum may just as well be written

$$\sum_{\substack{n|60 \\ n \leq 19}} f(n),$$

and this is always done by mathematicians. Other examples of this notation are

$$\sum_{d|10} f(d) = f(1) + f(2) + f(5) + f(10),$$

$$\sum_{\substack{d|12,\\ d<12}} f(d) = f(1) + f(2) + f(3) + f(4) + f(6),$$

$$\sum_{d|10} f(d)g\left(\frac{10}{d}\right) = f(1)g(10) + f(2)g(5) + f(5)g(2) + f(10)g(1),$$

$$\sum_{\substack{d_1|10,\\ d_2|10,\\ d_1 d_2 = 10}} f(d_1)g(d_2) = f(1)g(10) + f(2)g(5) + f(5)g(2) + f(10)g(1),$$

$$\sum_{\substack{d_1|3,\\ d_2|10}} f(d_1 d_2) = f(1 \cdot 1) + f(1 \cdot 2) + f(1 \cdot 5) + f(1 \cdot 10)$$
$$+ f(3 \cdot 1) + f(3 \cdot 2) + f(3 \cdot 5) + f(3 \cdot 10),$$

$$\sum_{d|30} f(d) = f(1) + f(2) + f(5) + f(10) + f(3) + f(6) + f(15)$$
$$+ f(30).$$

Notice that the third and fourth sums are the same, as are the fifth and sixth.

We are going to prove that if $f(n)$ is multiplicative, then so is the function $g(n)$ defined by

$$g(n) = \sum_{d|n} f(d).$$

Let us first illustrate the proof with a numerical example. We will show that $g(30) = g(3)g(10)$.

$$\begin{aligned}
g(3)g(10) &= \sum_{d_1|3} f(d) \cdot \sum_{d_2|10} f(d_2)\\
&= [f(1) + f(3)] \cdot [f(1) + f(2) + f(5) + f(10)]\\
&= f(1)f(1) + f(1)f(2) + f(1)f(5) + f(1)f(10)\\
&\quad + f(3)f(1) + f(3)f(2) + f(3)f(5) + f(3)f(10)\\
&= f(1 \cdot 1) + f(1 \cdot 2) + f(1 \cdot 5) + f(1 \cdot 10)\\
&\quad + f(3 \cdot 1) + f(3 \cdot 2) + f(3 \cdot 5) + f(3 \cdot 10)\\
&= f(1) + f(2) + f(5) + f(10) + f(3) + f(6) + f(15) + f(30)\\
&= \sum_{d|30} f(30)\\
&= g(30).
\end{aligned}$$

In the summation notation, this string of equalities may be written

$$g(3)g(10) = \sum_{d_1|3} f(d_1) \cdot \sum_{d_2|10} f(d_2)$$

$$= \sum_{\substack{d_1|3 \\ d_2|10}} f(d_1)f(d_2)$$

$$= \sum_{\substack{d_1|3 \\ d_2|10}} f(d_1 d_2)$$

$$= \sum_{d|30} f(d)$$

$$= g(30).$$

This is exactly what will be done in general.

Theorem 2.15. If $f(n)$ is multiplicative, and $g(n)$ is defined as

$$g(n) = \sum_{d|n} f(d),$$

then $g(n)$ is multiplicative.

Proof. Suppose that $n > 0$, $m > 0$, $(m,n) = 1$. Our goal is to show that

$$g(m)g(n) = g(mn).$$

We begin by finding $g(m)g(n)$,

(11) $$g(m)g(n) = \sum_{d_1|m} f(d_1) \sum_{d_2|n} f(d_2)$$

$$= \sum_{\substack{d_1|m \\ d_2|n}} f(d_1)f(d_2).$$

If $d_1|m$, $d_2|n$, and n and m have no common factors greater than 1, it follows that d_1 and d_2 have no common factors greater than 1, or, in other words, $(d_1,d_2) = 1$. Thus by the definition of multiplicative functions, if d_1 and d_2 are positive and $d_1|m$, $d_2|n$, then

$$f(d_1)f(d_2) = f(d_1 d_2).$$

Thus the expression in (11) becomes

(12) $$g(m)g(n) = \sum_{\substack{d_1|m \\ d_2|n}} f(d_1 d_2).$$

By Theorem 2.14, the set of numbers $d_1 d_2$, where d_1 and d_2 are positive divisors of m and n, is exactly the set of positive divisors of mn, and no duplications occur. Therefore, we may put (12) in the form

$$g(m)g(n) = \sum_{d|mn} f(d)$$
$$= g(mn),$$

and hence $g(n)$ is multiplicative. ▲

As immediate consequences of Theorem 2.15, we have

Theorem 2.16. The functions $d(n)$ and $\sigma(n)$ are multiplicative.

Proof. The function $f(n) = 1$ is multiplicative and

$$d(n) = \sum_{d|n} f(d),$$

since the sum on the right adds 1 as many times as there are positive divisors of n. By Theorem 2.15, $d(n)$ is multiplicative.

The function $f(n) = n$ is multiplicative and

$$\sigma(n) = \sum_{d|n} f(d).$$

Hence $\sigma(n)$ is also multiplicative. ▲

The facts that $d(n)$ and $\sigma(n)$ are multiplicative enable us to evaluate them in terms of the factorization of n into primes.

Theorem 2.17. If the factorization of n into primes is given by

$$n = p_1^{a_1} p_2^{a_2} \cdots p_k^{a_k},$$

then

$$d(n) = (a_1 + 1)(a_2 + 1) \cdots (a_k + 1)$$

and

$$\sigma(n) = (1 + p_1 + p_1^2 + \cdots + p_1^{a_1})(1 + p_2 + p_2^2 + \cdots + p_2^{a_2}) \cdots$$
$$\cdots (1 + p_k + p_k^2 + \cdots + p_k^{a_k})$$
$$= \frac{p_1^{a_1+1} - 1}{p_1 - 1} \frac{p_2^{a_2+1} - 1}{p_2 - 1} \cdots \frac{p_k^{a_k+1} - 1}{p_k - 1}.$$

Proof. If p is a prime and $a \geq 1$, then the divisors of p^a are $1, p, p^2, \ldots, p^a$. Hence

$$d(p^a) = a + 1$$

and

$$\sigma(p^a) = 1 + p + p^2 + \cdots + p^a$$
$$= \frac{p^{a+1} - 1}{p - 1}.$$

The last equality utilized the formula for the sum of a finite number of terms of a geometric progression (first term equals 1, and common ratio equals p). The result of the theorem now follows from the multiplicative properties of $d(n)$ and $\sigma(n)$ derived in Theorem 2.16.   ▲

As an example, the factorization of 20 into primes is

$$20 = 2^2 \cdot 5^1$$

and hence

$$d(20) = (2 + 1)(1 + 1) = 6,$$

$$\sigma(20) = \left(\frac{2^3 - 1}{2 - 1}\right)\left(\frac{5^2 - 1}{5 - 1}\right) = \frac{7}{1} \cdot \frac{24}{4} = 42.$$

These results are of course the same as those listed in Figure 2.1.

The function $\sigma(n)$ has been an object of interest since before the time of Euclid. The ancients considered the function

$$\sigma(n) - n = \sum_{\substack{d\mid n \\ d < n}} d,$$

or, in words, the sum of the positive divisors of n other than n itself. This function is not multiplicative, as the example

$$\sigma(6) - 6 = 6 \neq 1 \cdot 1 = (\sigma(2) - 2) \cdot (\sigma(3) - 3)$$

shows. This is the reason that $\sigma(n)$ is usually investigated today rather than $\sigma(n) - n$. Certain integers n (such as $n = 6$) have the property that

$$\sigma(n) - n = n.$$

The ancients believed that such numbers had mystical properties and called them **perfect numbers**. A somewhat larger example than 6 of a perfect number is $2^{11212}(2^{11213} - 1)$,[1] which has 6751 digits. Euler knew the form of all even perfect numbers. He showed that an even perfect number must be of the following form (Euclid had shown such numbers to be perfect):

$$n = 2^{p-1}(2^p - 1),$$

[1] As this is being written, $2^{11213} - 1$ is the largest number that has been proved to be a prime.

where both p and $(2^p - 1)$ are primes. To this day, it is not known if there are infinitely many perfect numbers, nor is it known if there are any odd perfect numbers. It is known that there are no odd perfect numbers less than 10^{20}, but in view of the examples in Section 1.1, this should not be regarded as conclusive evidence that there are no odd perfect numbers.

EXERCISES

1. Verify that 6, 28, and 496 are perfect numbers.
2. The Greeks defined the numbers m and n to be **amicable** if

$$\sigma(m) - m = n, \qquad \sigma(n) - n = m.$$

The amicable numbers 220 and 284 were known to Pythagoras. Verify that they are amicable.

3. Find $\sigma(n) - n$ for $n = 1184$ and $n = 1210$.
4. Find $\sigma(n) - n$ for $n = 12\,496 = 2^4 \cdot 11 \cdot 71$, $n = 14\,288 = 2^4 \cdot 19 \cdot 47$, $n = 15\,472 = 2^4 \cdot 967$, $n = 14\,536 = 2^3 \cdot 23 \cdot 79$, and $n = 14\,264 = 2^3 \cdot 1783$. These numbers were found by Poulet in 1918.
5. Prove that if $2^p - 1$ is a prime, then

$$n = 2^{p-1}(2^p - 1)$$

is a perfect number.

6. Find an integer n less than or equal to 70 such that $d(n) = 12$.
7. Find an integer n such that

$$\sigma(n) = 546.$$

8. Let $\sigma_2(n)$ be the sum of the squares of the positive divisors of n. Show that $\sigma_2(n)$ is multiplicative and show that if

$$n = p_1^{a_1} p_2^{a_2} \cdots p_k^{a_k},$$

then

$$\sigma_2(n) = \frac{p_1^{2a_1+2} - 1}{p_1^2 - 1} \frac{p_2^{2a_2+2} - 1}{p_2^2 - 1} \cdots \frac{p_k^{2a_k+2} - 1}{p_k^2 - 1}$$

9.* Show that if n is a perfect square, then $\sigma(n) | \sigma_2(n)$ (see problem 8). Give an example of an integer n such that $\sigma(n) \nmid \sigma_2(n)$.

10.* We seem to have done more in problem 5 than Euler's result says that we can do since we did not require that p be a prime number. Prove that if $2^n - 1$ is a prime, then n is a prime. (*Hint*: If n is composite, show $2^n - 1$ may be factored.)

2.5. Linear Diophantine Equations

With this title, we could investigate 20 linear equations in 37 unknowns, but we will stick to one equation in two unknowns. Our aim is to either find all integers x and y which satisfy the equation

$$ax + by = c$$

(a, b, and c are integers) or show that there are none.

Theorem 2.18. Suppose that a and b are nonzero integers and $d = (a,b)$. If $d \nmid c$, then the equation

(13) $$ax + by = c$$

has no integral solutions. If $d|c$, then the equation has infinitely many solutions. If $x = x_0$, $y = y_0$ is one integral solution to (13), then all integral solutions to (13) are given by

(14)
$$x = x_0 + t\frac{b}{d},$$

$$y = y_0 - t\frac{a}{d},$$

where t is an integer.

Proof. Since $d|a$, $d|b$, it follows that $d|(ax + by)$ for all integers x and y. Thus if

$$ax + by = c,$$

then $d|c$. Hence (13) has no solutions if $d \nmid c$. Suppose now that $d|c$. Then there is an integer e such that

$$c = de.$$

By Theorem 2.2, there are integers r and s such that

$$ar + bs = d.$$

Hence

$$a(re) + b(se) = de = c$$

and (13) has an integral solution. If

(15) $$ax_0 + by_0 = c,$$

then

$$a\left(x_0 + t\frac{b}{d}\right) + b\left(y_0 - t\frac{a}{d}\right) = ax_0 + by_0 = c,$$

and hence (13) has infinitely many solutions, among them being the infinitely many given in (14).

It remains to show that every solution of (13) can be put in the form of (14). Suppose that x and y are integers which satisfy (13). If we subtract (15) from (13), we get

$$a(x - x_0) + b(y - y_0) = 0$$

or

(16)
$$\frac{a}{d}(x - x_0) = -\frac{b}{d}(y - y_0).$$

Thus

$$\frac{b}{d}\bigg|\frac{a}{d}(x - x_0), \qquad \text{but} \quad \left(\frac{b}{d}, \frac{a}{d}\right) = 1$$

by Theorem 2.4, and hence, by Theorem 2.6,

$$\frac{b}{d}\bigg|(x - x_0).$$

Thus there is an integer t such that

(17)
$$x - x_0 = t\frac{b}{d}.$$

If we substitute this in (16), we get

$$\frac{a}{d} \cdot t\frac{b}{d} = -\frac{b}{d}(y - y_0)$$

and hence

$$y - y_0 = -t\frac{a}{d}.$$

From this and (17), we get

$$x = x_0 + t\frac{b}{d},$$

$$y = y_0 - t\frac{a}{d},$$

and thus every solution to (13) may be written in the form of (14).   ▲

When a and b are small, it is frequently possible to find a solution to (13) by inspection. Otherwise the Euclidean algorithm gives us a systematic method of finding a particular solution to (13) (assuming that $d|c$). For example, suppose that we wish to solve the equation

$$(18) \qquad\qquad 12x + 25y = 331.$$

We first use the Euclidean algorithm to express (12,25) in terms of 12 and 25. Since

$$25 = 2 \cdot 12 + 1,$$
$$12 = 12 \cdot 1,$$

we see that $(12,25) = 1$ and

$$-2 \cdot 12 + 1 \cdot 25 = 1.$$

If we multiply this through by 331, we find

$$12(-662) + 25(331) = 331$$

and thus

$$x = -662, \qquad y = 331$$

is a particular solution to (18). The general solution to (18) is then given as

$$(19) \qquad\qquad x = -662 + 25t, \qquad y = 331 - 12t.$$

It is interesting to note that (18) has a unique solution in nonnegative integers. If $x \geq 0$, then, by (19),

$$-662 + 25t \geq 0$$
$$25t \geq 662$$
$$t \geq \tfrac{662}{25} = 26 + \tfrac{12}{25}.$$

If $y \geq 0$, then by (19),

$$331 - 12t \geq 0$$
$$331 \geq 12t$$
$$27 + \tfrac{7}{12} = \tfrac{331}{12} \geq t.$$

Thus if $x \geq 0$ and $y \geq 0$, then

$$26 + \tfrac{12}{25} \leq t \leq 27 + \tfrac{7}{12},$$

and the only integer in this range is $t = 27$. If we put $t = 27$ into (19), then we get

$$x = 13, \qquad y = 7$$

as the only solution to (18) in nonnegative integers.

The fact that some Diophantine equations have unique solutions in positive integers leads to the possibility of being able to completely solve certain "word problems" without having as many equations available as unknowns. For example, consider the following problem : Jimmy bought a certain number of regular-size comic books at 12 cents apiece and a certain number of "giant"-size comic books at 25 cents apiece. If Jimmy spent $3.31 altogether, how many comic books did he buy? If we let x be the number of regular size and y the number of giant-size comic books that Jimmy bought, then x and y are related by equation (18). Clearly, x and y are restricted to being nonnegative integers by the problem and thus, as was shown above, $x = 13$, $y = 7$. Therefore Jimmy bought 20 comic books altogether.

There is one last item that we will discuss here since it sometimes causes confusion. Since $x = 13$, $y = 7$ is a solution to (18), it follows from Theorem 2.18 that all solutions to (18) can be written in the form

(20)
$$x = 13 + 25T,$$
$$y = 7 - 12T.$$

This seems to contradict what we derived in (19), but it does not. Equations (19) and (20) are connected by the relation

$$T = t - 27.$$

For example, the solution $x = 38$, $y = -5$ to (18) is given by $t = 28$ in (19) and $T = 1$ in (20). The form of the final answer depends on which particular solution is used in expressing it. There are infinitely many ways of writing the solution to (18). The reader should remember this if his answer to some of the exercises differs from the answer given at the back of the book.

EXERCISES

In problems 1–6, either find all integral solutions to the given equation or show that it has none.

1. $3x + 2y = 1$.
2. $3x - 2y = 1$.
3. $17x + 14y = 4$.
4. $33x - 12y = 9$.
5. $91x + 221y = 15$.
6. $401x + 503y = 20$.

In problems 7–10, find all solutions in positive integers to the given equation or show that there are none.

7. $23x - 7y = 1$.

8. $9x + 11y = 79$.

9. $39x + 47y = 4151$.

10. $5x + 6y = 50$. (You may check your answer by looking at the stars on an American flag.)

11. Harold and Betty find that their house is $\frac{39}{12}$ of a Harold length plus $\frac{59}{10}$ of a Betty length long. They find this hard to remember and would much prefer integral Harold and Betty lengths. Can this be done if Harold is 6 feet tall and Betty is 5 feet tall?

12. A grocer sells a 1-gallon container of milk for 79 cents and a $\frac{1}{2}$-gallon container of milk for 41 cents. At the end of the day he sold \$63.58 worth of milk. How many 1-gallon and $\frac{1}{2}$-gallon containers did he sell?

*13. A teacher has been designing a word problem for a number theory examination. Thus far he has decided on the following outline for his problem. A furniture dealer used to sell 49 black-and-white TV sets a week at \$70 apiece with a profit of 30 percent on each set. Since the advent of color TV, black-and-white TV sets became cheaper at the wholesale level, but the dealer has kept his price at \$70 a set and now he makes 40 percent on each set. However, in order to capture a greater share of the color-TV market, the dealer has reduced his profit on color-TV sets to 19 percent per set, which enables him to sell color-TV sets for only \$300 apiece. The dealer's total sales in TV sets last week was d dollars. How did his profits last week compare with the days before color TV?

The teacher desires to have the answer come out that the dealer made \$13 more with his present arrangements. What value of d should he use in his problem to get the desired answer? With this value of d, will the teacher's students get a unique answer?

MISCELLANEOUS EXERCISES

1. Show that $(a,b,c) = ((a,b),c)$ provided that a and b are not both 0.

2. Show that if m and n are positive, then

$$\frac{mn}{(m,n)}$$

is the least common multiple of m and n (that is, the smallest positive integer divisible by both m and n).

3. Show that if

$$a_n x^n + a_{n-1} x^{n-1} + \cdots + a_1 x + a_0 = 0,$$

where $a_0, a_1, \ldots, a_n$ are integers and $x = r/s$ is rational, $(r,s) = 1$, then $s|a_n$ and $r|a_0$. (In applications, it must be remembered that r/s may be negative.) In particular, show that if $a_n = 1$ and x is rational, then x is an integer.

4. If $ax^2 + bx + c = 0$, then, as is well known,

$$x = \frac{-b \pm \sqrt{b^2 - 4ac}}{2a}.$$

Reconcile this with problem 3, which says that if x is rational, then the denominator of x divides a.

5. Suppose that b and c are integers and that

$$r = -b + \sqrt{b^2 - c}$$

is rational. Since r is a root of the equation

$$x^2 + 2bx + c = 0,$$

it follows from problem 3 that r is an integer and $r|c$. Prove this directly.

6. Prove that $(a^2, b^2) = (a,b)^2$.

7. Factor the numbers 1 456 813, 1 468 823, 1 476 221, and 1 488 391 into primes (and prove that you have primes when you are done), given only that none of these numbers have any prime factors less than 35 and

$$1\ 456\ 813 \cdot 1\ 488\ 391 = 1\ 468\ 823 \cdot 1\ 476\ 221.$$

8. Show that if there are integral solutions to the equation

$$ax + by + cz = e,$$

then $(a,b,c)|e$. Suppose that $(a,b,c)|e$. Show that there are integers w and z such that

$$(a,b)w + cz = e$$

(see problem 1). Then show that there are integers x and y such that

$$ax + by = (a,b)w.$$

(This same technique works for one equation in n unknowns. Solutions may be found—when they exist—by the analogous process of converting the equation to $n - 1$ successive equations in two unknowns.)

9. Use the method of problem 8 to find all the integral solutions of the equation

$$323x + 391y + 437z = 10473.$$

Your answer should have two integer variables in it. Find all positive solutions.

10. When Mr. Smith returned from Europe in 1966, he found that he had in his possession 35 British sixpence coins, 55 French ten-centime pieces, and 77 Greek drachmas. Mr. Smith converted each of these coins to its value in American money (rounded off to the nearest cent) and found that the total was worth $5.86. How much was each coin worth in 1966 (to the nearest cent)? If the phrase "rounded off to the nearest cent" were dropped from the problem, would your answer above necessarily be near the correct coin values?

11. For $n > 0$, show that if $2^n + 1$ is a prime, then n is a power of 2.

12. We may use the unique factorization theorem to give another proof (due to Euler) that there are infinitely many primes. Assume that there are only finitely many primes, $p_1, p_2, \ldots, p_k$. Prove that

$$\left(\sum_{a_1=0}^{\infty} \frac{1}{p_1^{a_1}} \right) \left(\sum_{a_2=0}^{\infty} \frac{1}{p_2^{a_2}} \right) \cdots \left(\sum_{a_k=0}^{\infty} \frac{1}{p_k^{a_k}} \right)$$

$$= \left[\left(1 - \frac{1}{p_1} \right) \left(1 - \frac{1}{p_2} \right) \cdots \left(1 - \frac{1}{p_k} \right) \right]^{-1}.$$

The product of the finite number of series on the left will converge absolutely since each series is absolutely convergent. Show that the product of the series on the left is

$$\sum_{n=1}^{\infty} \frac{1}{n},$$

which diverges. Hence there are infinitely many primes.

13. Prove that n is a common divisor of a, b, and c if and only if n is a divisor of (a,b,c).

Chapter 3

CONGRUENCES

3.1. Introduction

We may thank Gauss for the exceedingly useful concept of congruences. Some of the results of this chapter were known earlier, but Gauss was the first to systematically develop the subject.

Definition. Let a and b be integers and n a positive integer. If $n|(a - b)$, then we say that a is **congruent** to b **modulo** n, and we write

$$a \equiv b(\text{mod } n).$$

We also write $a \not\equiv b(\text{mod } n)$ when we wish to say that a is not congruent to b modulo n. This is equivalent to saying that $n \nmid (a - b)$.

Thus by the definition of divisibility, $a \equiv b(\text{mod } n)$ if and only if there exists k such that $a = b + kn$. For example,

$$37 \equiv 25(\text{mod } 12),$$

$$-9 \equiv 31(\text{mod } 10),$$

$$7216 \equiv 29\,216(\text{mod } 1000),$$

$$5 \not\equiv 7(\text{mod } 3).$$

Congruences occur in everyday life. Ordinary clocks and wrist watches measure hours (mod 12). Days of the week measure days (mod 7). A car speedometer (technically an odometer) measures mileage (mod 100 000). A speedometer that reads 51 937 does not say (even if it has not been tampered with) that the car has driven 51 937 miles; the fact that this is the unanimous interpretation is a comment on today's low-quality production methods, which virtually ensure that no car will last 151 937 miles.

51

The congruence sign, $\equiv$, resembles an equal sign. This is particularly appropriate since congruences possess many of the properties of ordinary equality. As an illustration, suppose that a and b are positive integers and

$$a \equiv b(\text{mod } 1000).$$

This says that

$$1000|(a - b),$$

or, in other words, the last three digits of $(a - b)$ are zeros. Thus the last three digits of a and b must be the same. Conversely, if the last three digits of a and b are the same, then

$$1000|(a - b)$$

and hence

$$a \equiv b(\text{mod } 1000).$$

Therefore, two positive integers are congruent modulo 1000 if and only if their last three digits agree. If you think about it for a minute, you will realize that the last three digits of the sum and product of two positive integers depends only on the last three digits of the two integers. Thus, if $a, b, c,$ and d are positive integers and

$$a \equiv b(\text{mod } 1000), \qquad c \equiv d(\text{mod } 1000),$$

then

$$a + c \equiv b + d(\text{mod } 1000), \qquad ac \equiv bd(\text{mod } 1000).$$

We will shortly prove these rules for all moduli n.

The above result on products modulo 1000 has an amusing application. A beginning mind reader asks a person to think of a number from 1 to 999, multiply it by 143, and state the last three digits of the answer. Once this is done, the mind reader promptly states the original number used and explains that being a beginner, he needed to make the person concentrate on his original number and hence the multiplication by 143. On the other hand, he did not want the audience to think that he was merely able to rapidly divide by 143, and hence he asked for only the last three digits of the answer.

The "mind reader" is of course no such thing. He simply takes the three-digit number given to him and multiplies by 7. The last three digits of the answer gives the original number. For instance, if 492 is the number thought of originally, then the last three digits of $492 \cdot 143 (= 70\,356)$ are 356. The product of 7 and 356 is 2492, the last three digits of which give the original number. The only remaining question is: Why does this work?

The trick is based on the fact that

$$7 \cdot 143 = 1001.$$

If x is any three-digit integer, then $1001x$ is simply two copies of x, for instance $1001 \cdot 492 = 492\,492$. The whole process of multiplying by 143, taking the last three digits and multiplying by 7, and taking the last three digits reduces to the process of multiplying by 1001 and taking the last three digits, thus getting the original number. In terms of congruences, we wish to find a three-, two-, or one-digit number, x. We are told only that a given number b consists of the last three digits of $143x$. Thus we are given the congruence

$$143x \equiv b(\text{mod } 1000).$$

Clearly,

$$7 \equiv 7(\text{mod } 1000);$$

we may multiply these congruences together and get

$$7 \cdot 143x \equiv 7b(\text{mod } 1000),$$

or

$$1001x \equiv 7b(\text{mod } 1000).$$

Since

$$1001 \equiv 1(\text{mod } 1000)$$

and

$$x \equiv x(\text{mod } 1000),$$

we see that

$$1001x \equiv x(\text{mod } 1000)$$

and therefore

$$x \equiv 7b(\text{mod } 1000).$$

This says that the last three digits of $7b$ and the last three digits of x agree. Since x has at most three digits, the last three digits of $7b$ give x completely.

What the "mind-reading" trick really boils down to is solving the congruence equation

$$143x \equiv b(\text{mod } 1000)$$

for x. We will look into such problems later in the chapter.

EXERCISES

1. True or false? $17 \equiv 2(\text{mod } 5)$, $14 \equiv -6(\text{mod } 10)$, and $97 \equiv 5(\text{mod } 13)$.
2. Verify that $3 \cdot 5 \equiv 3 \cdot 13(\text{mod } 4)$, $7 \cdot 18 \equiv 7(-2)(\text{mod } 10)$, and $3 \cdot 4 \equiv 3 \cdot 14$ (mod 6).
3. Which of the following are valid? $5 \equiv 13(\text{mod } 4)$, $18 \equiv -2(\text{mod } 10)$, and $4 \equiv 14(\text{mod } 6)$.

Problems 2 and 3 combined show that the assertion "If $ab \equiv ac(\text{mod } n)$, then $b \equiv c(\text{mod } n)$" is not always correct, even if $a \not\equiv 0$.

4. Find a if $a \equiv 97(\text{mod } 7)$ and $1 \leq a \leq 7$.
5. Find a if $a \equiv 32(\text{mod } 19)$ and $52 \leq a \leq 70$.
6. Show that, modulo 1000, adding 999 to a number is the same as subtracting 1.
7. Prove that if $a \equiv b(\text{mod } n)$, then $a + c \equiv b + c(\text{mod } n)$.
8. Prove that if $a \equiv b(\text{mod } n)$, then $ac \equiv bc(\text{mod } n)$.
9. Why did we restrict a and b to be positive integers when we said that "$a \equiv b(\text{mod } 1000)$ if and only if the last three digits of a and b are the same"?
10. At 5 P.M. (Eastern Standard Time), Dec. 7, 1967, how many hours had passed (mod 24) in New York City since the beginning of the century?
11. If n is positive, show that $n|a$ if and only if $a \equiv 0(\text{mod } n)$.

3.2. Fundamental Properties of Congruences

Theorem 3.1. Let n be a positive integer. For all integers a,

$$a \equiv a(\text{mod } n).$$

If $a \equiv b(\text{mod } n)$, then $b \equiv a(\text{mod } n)$. If $a \equiv b(\text{mod } n)$ and $b \equiv c(\text{mod } n)$, then $a \equiv c(\text{mod } n)$.

Proof. Since $n|0$, $a \equiv a(\text{mod } n)$ by definition. If $a \equiv b(\text{mod } n)$, then $n|(a - b)$. Thus n divides $(-1)(a - b) = (b - a)$ and therefore $b \equiv a(\text{mod } n)$. Finally, if $a \equiv b(\text{mod } n)$ and $b \equiv c(\text{mod } n)$, then $n|(a - b)$ and $n|(b - c)$. Therefore, n divides the number

$$(a - b) + (b - c) = a - c$$

and hence $a \equiv c(\text{mod } n)$. ▲

Theorem 3.1 is analogous to the corresponding result for equalities. It will be used in many ways usually without specific mention. For example, because of Theorem 3.1, we may write something like

$$a \equiv b \equiv c \equiv d \equiv e \equiv f(\text{mod } n)$$

and immediately infer that $a \equiv f(\text{mod } n)$. As another example, this theorem is the justification for the inference

"$1001x \equiv 7b(\text{mod } 1000)$ and $1001x \equiv x(\text{mod } 1000)$;

therefore, $x \equiv 7b(\text{mod } 1000)$,"

which was used in Section 3.1.

Theorem 3.2. If $a \equiv b(\text{mod } n)$ and $c \equiv d(\text{mod } n)$, then

$$a + c \equiv b + d(\text{mod } n), \qquad a - c \equiv b - d(\text{mod } n), \qquad ac \equiv bd(\text{mod } n).$$

If $a \equiv b(\text{mod } n)$, then for all c,

$$a + c \equiv b + c(\text{mod } n), \qquad a - c \equiv b - c(\text{mod } n), \qquad ac \equiv bc(\text{mod } n).$$

Proof. Since

$$(a + c) - (b + d) = (a - b) + (c - d),$$
$$(a - c) - (b - d) = (a - b) - (c - d),$$
$$ac - bd = c(a - b) + b(c - d),$$

the first part of the theorem follows from Theorem 1.2 and the definition of congruence. By Theorem 3.1, $c \equiv c(\text{mod } n)$ for all c, and thus the second part of the theorem is a special case of the first part.   ▲

We illustrate Theorem 3.2 by solving the following word problem. The town of Anyplace, U.S.A., derives its principal income from fines paid by nonresidents cited for speeding while passing through town. Mr. Storer, who as a boy was cited for bicycling down Main Street at 100 miles an hour, finds to his horror that he must go to Anyplace every seven months on business starting in October. Thus it occurred that every seven months beginning in October, Mr. Storer received a speeding citation. The first citation came in October, the second in the following May, the third in December, and so on. Which were the first two citations in the series that were received in January?

We assign the numbers $1, 2, 3, \ldots, 12$ to the months January, February, March, $\ldots$, December, respectively. October is assigned the number 10 and January the number 1. Thus we wish to find x, where

$$10 + 7(x - 1) \equiv 1(\text{mod } 12)$$

(the expression on the left is not merely $10 + 7x$, since the first citation, rather than the zeroth, occurred in the tenth month of the year). Hence

(1) $$7x \equiv -2(\text{mod } 12).$$

If we subtract (1) from the obviously true congruence,

$$12x \equiv 0(\text{mod } 12),$$

we get

(2) $5x \equiv 2(\text{mod } 12).$

Subtracting (2) from (1) gives

(3) $2x \equiv -4 \equiv 8(\text{mod } 12).$

Doubling (3) gives

$$4x \equiv 16 \equiv 4(\text{mod } 12),$$

and if we subtract this from (2) we see that

$$x \equiv -2 \equiv 10(\text{mod } 12).$$

Therefore,

$$x = 10 + 12k;$$

the first two positive values of x that result are $x = 10$ and 22. These values satisfy the original equation and hence the first two citations received by Mr. Storer in the month of January were the tenth and twenty-second of the series.

   The reader has possibly noticed that we had the opportunity to divide both sides of (3) by 2. Such a division would have given us the incorrect result

$$x \equiv 4(\text{mod } 12).$$

The conditions under which one can divide both sides of a congruence are given in the next theorem.

Theorem 3.3. If $(a,n) = 1$ and $ab \equiv ac(\text{mod } n)$, then $b \equiv c(\text{mod } n)$. More generally, if $(a,n) = d$ and $ab \equiv ac(\text{mod } n)$, then $b \equiv c(\text{mod } n/d)$.

Proof. Suppose that $(a,n) = d$ and $ab \equiv ac(\text{mod } n)$. Then there is an integer k such that

(4) $ab = ac + kn.$

Let

$$a_1 = \frac{a}{d}, \qquad n_1 = \frac{n}{d};$$

these numbers are integers and

$$(a_1, n_1) = \left(\frac{a}{d}, \frac{n}{d}\right) = 1.$$

We divide (4) by d and get

(5) $a_1(b - c) = kn_1$

and thus $a_1 | kn_1$. Since $(a_1, n_1) = 1$, $a_1 | k$ by Theorem 2.6. Thus there is an integer k_1 such that $k = a_1 k_1$. It follows from (5) that

$$b - c = k_1 n_1$$

or, in other words, $n_1 | (b - c)$. Therefore, by definition,

$$b \equiv c(\mathrm{mod}\ n_1). \qquad \blacktriangle$$

There is another very important fact about congruences that we shall prove here. We first prove that any integer is congruent modulo n to exactly one of the numbers $0, 1, 2, \ldots, n - 1$. We show this as follows. Given an integer a, the division algorithm says that we may write it in the form

$$a = qn + r, \qquad 0 \leq r < n.$$

Thus

$$a \equiv qn + r \equiv q \cdot 0 + r \equiv r(\mathrm{mod}\ n)$$

and hence a is congruent, modulo n, to at least one of the numbers $0, 1, 2, \ldots, n - 1$. Suppose that r_1 and r_2 are two different integers in the range $0, 1, \ldots, n - 1$ and that

$$a \equiv r_1(\mathrm{mod}\ n), \qquad a \equiv r_2(\mathrm{mod}\ n).$$

We may as well assume that $r_1 > r_2$. We see that

$$r_1 \equiv r_2(\mathrm{mod}\ n)$$

and hence

$$n | (r_1 - r_2).$$

But

$$0 < r_1 - r_2 \leq (n - 1) - (0) < n;$$

that is, $r_1 - r_2$ is a positive number less than n. As such, it cannot be divisible by n and therefore a cannot be congruent to two different numbers in the range 0 to $n - 1$. Thus each integer is congruent (mod n) to exactly one of the numbers $0, 1, \ldots, n - 1$, as stated. The numbers $0, 1, \ldots, n - 1$ give one example of a complete system of residues (mod n).

Definition. A set of n integers, $a_1, a_2, \ldots, a_n$, is called a **complete system of residues** (or a **complete residue system**) (mod n) if every integer is congruent (mod n) to exactly one of the a_j's.

The reason for restricting the definition to n integers is that if the set $a_1, a_2, \ldots, a_r$ has the properties of the definition, then $n = r$. This is easy to show and is left to problems 17 and 18 at the end of the section. Note that for a complete residue system (a_j), if $j \neq k$, then $a_j \not\equiv a_k \pmod{n}$, as otherwise a_j would be congruent (mod n) to two members of the system: itself and a_k.

Theorem 3.4. Any set of n consecutive integers is a complete residue system (mod n).

Proof. We have already seen that the set $0, 1, 2, \ldots, n - 1$ is a complete residue system (mod n). Let b be the first of n consecutive integers which are then given by $b, b + 1, b + 2, \ldots, b + n - 1$. Given an integer a, there is, by the definition of a complete residue system, an integer j in the range $0 \leq j \leq n - 1$ such that

$$a - b \equiv j \pmod{n}.$$

Therefore,

$$a \equiv b + j \pmod{n}$$

and hence any integer is congruent to at least one of the numbers $b, b + 1, \ldots, b + n - 1$. Suppose that j_1 and j_2 are different integers in the range $0 \leq j \leq n - 1$ and there is an integer a such that

$$a \equiv b + j_1 \pmod{n}, \qquad a \equiv b + j_2 \pmod{n}.$$

Then

$$a - b \equiv j_1 \pmod{n}, \qquad a - b \equiv j_2 \pmod{n}$$

and thus $a - b$ is congruent (mod n) to two distinct members of the complete residue system (mod n), $0, 1, 2, \ldots, n - 1$, which is impossible. Hence every integer is congruent (mod n) to exactly one of the n integers $b, b + 1, \ldots, b + n - 1$; this set is therefore a complete system of residues (mod n). ▲

The most commonly used complete systems of residues (mod n) are the sets

$$0, 1, 2, \ldots, n - 1;$$

$$1, 2, 3, \ldots, n;$$

and when n is odd,

$$-\frac{n-1}{2}, -\frac{n-1}{2} + 1, \ldots, -1, 0, 1, \ldots, \frac{n-1}{2} - 1, \frac{n-1}{2}.$$

This last set is often given (for odd n) as the set of all integers j with the property that

$$|j| < \frac{n}{2}$$

(for example, when $n = 5$, the numbers j such that $|j| < \frac{5}{2}$ are $-2, -1, 0, 1, 2$).

The fact that the numbers $0, 1, \ldots, n - 1$ give a complete system of residues (mod n) says that any combination of sums, differences, and products of these numbers is congruent (mod n) to a unique integer of the system. This leads to the concept of arithmetic (mod n) or, as it is sometimes called, modular arithmetic. Figures 3.1 and 3.2 show the tables for addition and multiplication (mod 5) and (mod 6), respectively.

It follows from Theorem 3.2 that many of the usual laws of arithmetic are valid for arithmetic (mod n). For instance, the law

$$a(b + c) = ab + ac$$

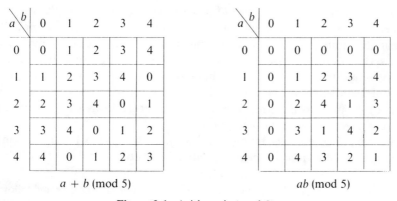

$a \backslash b$	0	1	2	3	4
0	0	1	2	3	4
1	1	2	3	4	0
2	2	3	4	0	1
3	3	4	0	1	2
4	4	0	1	2	3

$a + b$ (mod 5)

$a \backslash b$	0	1	2	3	4
0	0	0	0	0	0
1	0	1	2	3	4
2	0	2	4	1	3
3	0	3	1	4	2
4	0	4	3	2	1

ab (mod 5)

Figure 3.1. Arithmetic (mod 5).

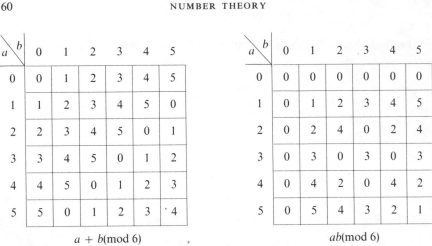

Figure 3.2. Arithmetic (mod 6).

becomes

(6) $a(b + c) \equiv ab + ac(\text{mod } n)$.

The reason that this is not an utter triviality in arithmetic (mod n) is that the numbers $b + c$, ab, ac found by the arithmetic (mod n) tables are, as likely as not, different from the usual sums and products. Thus (6) becomes a statement of the fact that when $a(b + c)$ is reduced (mod n) in two different ways to the complete system of residues $0, 1, \ldots, n - 1$, the results are the same. For example, by Figure 3.1,

$$3(4 + 4) \equiv 3(3) \equiv 4(\text{mod } 5),$$

$$3 \cdot 4 + 3 \cdot 4 \equiv 2 + 2 \equiv 4(\text{mod } 5).$$

We may put arithmetic (mod n) to another use. For example, we see in Figure 3.1 that the numbers $0^2, 1^2, 2^2, 3^2$, and 4^2 are congruent (mod 5) to one of the numbers 0, 1, and 4. Since the numbers 0, 1, 2, 3, and 4 give a complete system of residues (mod 5), every integer squared is congruent to either 0, 1, or 4 (mod 5). Thus, although there are infinitely many perfect squares, none of them leave the remainder 2 or 3 when divided by 5![1] It is not uncommon to find people who think that five verifications surely proves the theorem. Here it has actually happened.

The next theorem has many applications. We give a few of the more interesting applications immediately after its proof.

_____

[1] This is not to be interpreted as five factorial (there would then be a period following the exclamation mark).

Theorem 3.5. Let $f(x)$ be a polynomial with integral coefficients. If $a \equiv b(\text{mod } n)$, then

$$f(a) \equiv f(b)(\text{mod } n).$$

Proof. Let

$$f(x) = a_k x^k + a_{k-1} x^{k-1} + \cdots + a_1 x + a_0,$$

where $a_0, a_1, \ldots, a_k$ are integers. Then by Theorem 3.2,

$$a_k a^k + a_{k-1} a^{k-1} + \cdots + a_0 \equiv a_k b^k + a_{k-1} b^{k-1} + \cdots + a_0(\text{mod } n);$$

that is,

$$f(a) \equiv f(b)(\text{mod } n). \qquad \blacktriangle$$

As our first application of Theorem 3.5, we show that there is no polynomial $f(x)$ with integral coefficients and degree ≥ 1 such that $f(a)$ is a prime for all integers a. Let

$$f(x) = a_k x^k + a_{k-1} x^{k-1} + \cdots + a_0,$$

where $a_0, a_1, \ldots, a_k$ are integers and $a_k \neq 0$. If $a_k > 0$, then $f(x) \to +\infty$ as $x \to +\infty$, while if $a_k < 0$, then $f(x) \to -\infty$ as $x \to +\infty$. Since the word prime has been defined to mean positive, we let $a_k > 0$. The same proof would show that if $a_k < 0$, we do not always get negative primes. Since $f(x) \to \infty$ as $x \to \infty$, we can take the integer a sufficiently large to ensure that

$$n = f(a) > 1.$$

Just because we have used the letter n does not mean that n is not a prime. Let j be so large that

$$f(a + jn) > n;$$

this is again possible since $f(x) \to \infty$ as $x \to \infty$. But

$$a + jn \equiv a(\text{mod } n)$$

and hence, by Theorem 3.5,

$$f(a + jn) \equiv f(a) \equiv n \equiv 0(\text{mod } n).$$

Thus

$$n \mid f(a + jn),$$

and since

$$1 < n < f(a + jn),$$

$f(a + jn)$ is not a prime.

As another illustration, we show that every positive integer is congruent to the sum of its digits (mod 9). Let $n > 0$ have the decimal representation

$$n = a_k \cdot 10^k + a_{k-1} \cdot 10^{k-1} + \cdots + a_1 \cdot 10 + a_0,$$

where for all j, $0 \le a_j \le 9$; the numbers $a_0, \ldots, a_k$ are thus the digits of the number n. Let

$$f(x) = a_k x^k + a_{k-1} x^{k-1} + \cdots + a_1 x + a_0.$$

Theorem 3.5 says that

$$f(10) \equiv f(1)(\text{mod } 9);$$

that is,

$$n \equiv a_k + a_{k-1} + \cdots + a_0 (\text{mod } 9),$$

which was to be shown. For example,

$$139\,854\,872 \equiv 1 + 3 + 9 + 8 + 5 + 4 + 8 + 7 + 2$$

$$\equiv 47 \equiv 4 + 7 \equiv 11 \equiv 1 + 1 \equiv 2(\text{mod } 9).$$

Usually one does not actually sum the digits of n as we did above, but rather one sums them (mod 9). In particular, one may neglect any 9's that show up or any combination of numbers that add up to 9. In the above calculation, for example, one may ignore the 9, the 8 and 1, the 5 and 4, the 7 and 2; only the digits 3 and 8 are left, their sum is 11 which is congruent to 2(mod 9). It is because we may ignore 9's that this result goes by the name "casting out nines."

The process of casting out nines serves as a partial check on the arithmetic operations of addition, subtraction, and multiplication. For example, if we wished to check the claim that

$$147^2 = 21\,509$$

we would cast out the nines in 147 and 21 509:

$$147 \equiv 1 + 4 + 7 \equiv 3(\text{mod } 9),$$

$$147^2 \equiv 3^2 \equiv 0(\text{mod } 9),$$

but

$$21\,509 \equiv 2 + 1 + 5 + 0 + 9 \equiv 8(\text{mod } 9).$$

Therefore, $147^2 \ne 21\,509$. The method is not foolproof; casting out nines would not have disproved the absurd claim that $147^2 = 18$. In fact, one out of every nine integers is congruent to $147^2(\text{mod } 9)$. Thus, speaking very loosely, casting out nines will find 8 of 9 errors.

As a last illustration of Theorem 3.5, we present a process that is sometimes called "casting out elevens." Let the decimal representation of a positive integer n be

$$n = a_k \cdot 10^k + a_{k-1} \cdot 10^{k-1} + \cdots + a_1 \cdot 10 + a_0,$$

where the numbers $a_0, \ldots, a_k$ are the digits of n and are in the range 0 to 9. Let

$$f(x) = a_k x^k + a_{k-1} x^{k-1} + \cdots + a_1 x + a_0.$$

Then

$$f(10) \equiv f(-1)(\bmod 11);$$

that is,

$$n \equiv (-1)^k a_k + (-1)^{k-1} a_{k-1} + \cdots + a_4 - a_3 + a_2 - a_1 + a_0 (\bmod 11).$$

In words, we add the units, hundreds, ten thousands, ..., digits of n and subtract from this the sum of the tens, thousands, hundred thousands, ..., digits of n. The result is congruent to $n(\bmod 11)$. For example,

$$37\ 147\ 289 \equiv (9 + 2 + 4 + 7) - (8 + 7 + 1 + 3) \equiv 3(\bmod 11).$$

The casting-out-elevens process also serves as a partial check on the arithmetic operations of addition, subtraction, and multiplication. For example, in the supposed equality

$$147^2 = 21\ 509,$$

we find that

$$147 \equiv 7 - 4 + 1 \equiv 4(\bmod 11),$$

$$147^2 \equiv 4^2 \equiv 16 \equiv 6 - 1 \equiv 5(\bmod 11),$$

while

$$21\ 509 \equiv (9 + 5 + 2) - (0 + 1) \equiv 4(\bmod 11).$$

Thus

$$147^2 \neq 21\ 509.$$

Here again the method of casting out elevens is not foolproof, but, loosely speaking, it will discover 10 of 11 errors.

EXERCISES
1. Show that if $a \equiv b(\bmod n)$ and $d|n$, then $a \equiv b(\bmod d)$.
2. Show that if $a \equiv b(\bmod n)$ and $c > 0$, then $ac \equiv bc(\bmod nc)$.

3. Use Figure 3.2 to calculate $4 \cdot (2 \cdot 5)$ and $(4 \cdot 2) \cdot 5 \pmod 6$.

4. Construct the tables for addition and multiplication (mod 7) corresponding to Figures 3.1 and 3.2.

5. Show that a perfect square is congruent to either 0 or 1 (mod 4).

6. Show that a perfect square is congruent to either 0, 1, or 4 (mod 8).

7. Show that for all n, $n^3 \equiv n \pmod 3$.

8. Show that if $5 \nmid n$, then $n^4 \equiv 1 \pmod 5$.

9. We did not prove that if $a \equiv b \pmod n$, then $c^a \equiv c^b \pmod n$. Let $c = 2$, $n = 7, a = 2, b = 9$, and show that for these values, $c^a \not\equiv c^b \pmod n$, thus disproving such a result.

10. Show that the prime numbers split up into the three classes: 2, those primes congruent to 1(mod 4), and those primes $\equiv$ 3(mod 4).

11. Show that every prime number is in one of the six classes: $2, 3, p \equiv 1 \pmod{12}, p \equiv 5 \pmod{12}, p \equiv 7 \pmod{12}$, and $p \equiv 11 \pmod{12}$.

12. In 1825 Gauss gave the following construction for writing a prime congruent to 1(mod 4) as the sum of two squares: Let $p = 4k + 1$ be a prime number. Determine x (this is uniquely possible by Theorem 3.4) so that

$$ x \equiv \frac{(2k)!}{2(k!)^2} \pmod p, \qquad |x| < \frac{p}{2}. $$

Now determine y so that

$$ y \equiv x \cdot (2k)! \pmod p, \qquad |y| < \frac{p}{2}. $$

Gauss showed that $x^2 + y^2 = p$. Verify Gauss's result for $p = 5$ and $p = 13$.

13. Find all the possible values of the sum of two squares (mod 4). Use your result to show that 4 926 834 923 is not the sum of two squares.

14. There is reason to believe (but it has never been proved) that there are infinitely many primes which are the sum of the squares of three different prime numbers (the smallest example is $83 = 3^2 + 5^2 + 7^2$). Let $p = p_1^2 + p_2^2 + p_3^2$, where p, p_1, p_2, and p_3 are primes. Use congruences (mod 3) to show that one of the three primes p_1, p_2, and p_3 is, in fact, 3.

15. Suppose that $m \geq 0$. Show that $17|(3 \cdot 5^{2m+1} + 2^{3m+1})$. [Hint: Use congruences (mod 17). A further hint is given in the answers at the back of the book.]

16. Suppose that $m \geq 0$. Show that $49|(5 \cdot 3^{4m+2} + 53 \cdot 2^{5m})$.

17. Given m integers where $m > n$, show that two of these integers must be congruent (mod n). (Hint: Any integer is congruent to one of the numbers

$0, 1, \ldots, n - 1$; show that two of the m integers are congruent to the same thing and hence to each other.)

18. Given m integers where $m < n$, show that there is an integer in the range $0, 1, \ldots, n - 1$ which is congruent to none of the given integers.

19. Show that if n is a positive integer with two or more digits, then the sum of the digits of n is less than n. This shows that the process of casting out nines ultimately leads to a single digit.

20. Show that an integer is divisible by 9 if and only if the process of casting out nines leads ultimately to 0 or 9.

21. Show that an integer is divisible by 3 if and only if the process of casting out nines leads ultimately to 0, 3, 6, or 9.

22. Show that an integer is divisible by 11 if and only if the process of casting out elevens leads ultimately to 0.

23. Let $n = a_2 \cdot 10^2 + a_1 \cdot 10 + a_0$. Show that $n \equiv a_0 + 3a_1 + 2a_2 (\text{mod } 7)$ and use this result to find criteria for divisibility of a three (or less)-digit number by 7.

24. It is a fact that $23\,538 \equiv 38 + 35 + 02 \equiv 75 (\text{mod } 99)$. Prove the result that this suggests. The result could well be called "casting out ninety nines." [*Hint:* In the case of 23 538, the relevant polynomial is $f(x) = 2x^2 + 35x + 38$ with $f(100) \equiv f(1) \ (\text{mod } 99)$.]

25. Show that a number is divisible by 11 if and only if the casting-out-ninety-nines method of the previous problem ultimately gives either 0 or a two-digit number with both digits equal.

26. It is a fact that $4\,176\,204\,105 \equiv 105 - 204 + 176 - 4 \equiv 73 (\text{mod } 1001)$. Prove the result that this suggests; we shall call it "casting out one thousand and ones."

27. Suppose that the method of casting out one thousand and ones of the previous problem ultimately reduces n to the three (or less)-digit number m. Prove that $7|n$ if and only if $7|m$. Prove that $11|n$ if and only if $11|m$. Finally, show that $13|n$ if and only if $13|m$. It may be useful to know that $1001 = 7 \cdot 11 \cdot 13$.

28. In the proof that a polynomial of degree greater than or equal to one never gives only primes, where did we use the fact that the degree is greater than or equal to one? [We had better have used it someplace; the result is not true for polynomials of degree zero, for example, $f(x) = 3$.]

29. Show that for infinitely many n, $43|(n^2 + n + 41)$.

30. What is involved in checking an arithmetic operation (mod 10)?

31. Show that if $a_1, \ldots, a_n$ have the property that no two of them are congruent (mod n), then they form a complete residue system (mod n).

32. Find $(a, 26)$ given that $a^{10} \equiv 10 (\text{mod } 26)$.

33. Show that if $n^2 + 2$ and $n^2 - 2$ are both primes, then $3|n$.
34. The polynomial $x^2 + 1$ cannot be factored. Does this contradict the fact that not all numbers of the form $n^2 + 1$ are primes?
35. Under the present calendar system, every fourth year is a leap year. There are three exceptions to this rule every 400 years. If a year number is divisible by 100, then it is a leap year if and only if it is divisible by 400. Thus 1800, 1900, 2100 are not leap years, but 2000 is a leap year. The beginning of the twentieth century, January 1, 1900, was a Monday. Show that although Sunday begins every week, it will never begin a a century.
36. Show that anybody born between 1901 and 2071 will celebrate his twenty-eighth birthday on the same day of the week as the day he was born.

3.3. Linear Congruence Equations

An equation of the form

(7) $$a_1 x_1 + a_2 x_2 + \cdots + a_k x_k \equiv b(\mathrm{mod}\ n),$$

with unknowns $x_1, \ldots, x_k$, is a linear congruence equation in k variables. A solution to this equation is a set of *integers* which satisfies the equation. The definition of congruence shows that equation (7) is equivalent to the Diophantine equation

(8) $$a_1 x_1 + a_2 x_2 + \cdots + a_k x_k - n x_{k+1} = b$$

with $k + 1$ unknowns. Equation (8) either has no solutions or it has infinitely many. Thus the same is true of (7). In the case that $k = 1$, we know exactly how to find the solutions to (8) (when they exist) and hence (7). In dealing with (7), we wish to know how many solutions there are (mod n). By this we mean that two different solutions of (7) are the same (mod n) if the different values of x_j are congruent (mod n) for all j. Thus we say that the solution $x = 1, y = 2, z = 3$ to

$$x + y + z \equiv -1(\mathrm{mod}\ 7)$$

is the same (mod 7) as the solution $x = 8$, $y = -5$, $z = 17$ but different (mod 7) from the solution $x = 1$, $y = 3$, $z = 2$. In particular, when there is only one solution to (7) (mod n), we say that the solution is unique (mod n).

Theorem 3.6. The equation

(9) $$ax \equiv b(\mathrm{mod}\ n)$$

has solutions if and only if $d|b$, where $d = (a,n)$. If $d|b$, then the solution is unique (mod n/d). If $(a,n) = 1$, then (9) always has a solution and it is unique (mod n).

Proof. If $x = x_0$ is a solution to (9), then there is an integer y_0 such that

$$ax_0 = b + ny_0;$$

that is, the equation

(10) $$ax - ny = b$$

has a solution. If $x = x_0$, $y = y_0$ is a solution to (10), then

$$ax_0 \equiv ax_0 - ny_0 \equiv b(\bmod n),$$

and thus (9) has a solution. Therefore, (9) has solutions if and only if (10) has solutions and further, any solution for x in (10) gives a solution for x in (9). By Theorem 2.18, (10) has solutions if and only if $d|b$. Thus (9) has solutions if and only if $d|b$. Suppose that $d|b$, so that (10) has a solution. Let $x = x_0$, $y = y_0$ be a solution to (10). By Theorem 2.18, every solution of (10) is then of the form

$$x = x_0 + t\frac{n}{d}, \qquad y = y_0 + t\frac{a}{d},$$

where t is an integer. Thus every solution to (9) is of the form

$$x = x_0 + t\frac{n}{d}.$$

Since

$$x_0 + t\frac{n}{d} \equiv x_0 \left(\bmod \frac{n}{d}\right),$$

we see that all solutions to (9) are congruent to $x_0(\bmod n/d)$ and hence the solution to (9) is unique (mod n/d). The last statement of the theorem follows from the first two. ▲

We developed a systematic process in Chapter 2 for solving (10) and as a result (9). Usually, one can shortcut the Euclidean algorithm by taking advantage of situations as they arise. If in (9), $d = (a,n) > 1$, then it is best to divide everything through by d using Theorem 3.3. We are then left with an equation of the same type as (9) but with $(a,n) = 1$. We give several illustrations. The equation

$$14x \equiv 13(\bmod 21)$$

has no solutions since $(14,21) = 7$ and $7 \nmid 13$. We now solve the equation

(11) $9x \equiv 15(\text{mod } 21)$.

Here $(9,21) = 3$ and $3|15$. Thus the equation will have a unique solution (mod 7). We first divide everything by 3, by Theorem 3.3,

$$3x \equiv 5(\text{mod } 7).$$

Therefore,

$$3x \equiv 5 + 7 \equiv 12(\text{mod } 7)$$

and since $(3,7) = 1$, Theorem 3.3 says that

(12) $x \equiv 4(\text{mod } 7)$.

The original equation was (mod 21); we may wish to know the solutions (mod 21) also. This is easily done. In any complete residue system (mod 7), there is a unique solution to (11) and it can be found from (12). Thus in the set $0, 1, 2, 3, 4, 5, 6$, $x = 4$ is the unique solution to (11); in the set $7, 8, 9, 10, 11, 12, 13$, $x = 11$ is the unique solution to (11); and in the set $14, 15, 16, 17, 18, 19, 20$, $x = 18$ is the unique solution to (11). These three sets combined give a complete residue system (mod 21). Thus there are 3 solutions to (11) (mod 21). They are

(13) $x \equiv 4, 11, 18(\text{mod } 21)$.

Equations (12) and (13) are two ways of saying the same thing. In like manner, the equations

$$x \equiv 7(\text{mod } 8), \qquad x \equiv 7, 15, 23, 31, 39(\text{mod } 40)$$

are equivalent. In general, the congruence

$$x \equiv a(\text{mod } n)$$

has the m solutions (mod mn) given by

$$x \equiv a, a + n, a + 2n, \ldots, a + (m - 1)n(\text{mod } mn).$$

Let us illustrate the systematic and nonsystematic ways of solving the equation

(14) $8x \equiv 7(\text{mod } 13)$.

The systematic method involves using the Euclidean algorithm to find $(8, 13)$. All x satisfy

$$13x \equiv 13(\text{mod } 13).$$

Subtracting (14) from this gives

(15) $$5x \equiv 6(\text{mod } 13).$$

Subtracting this from (14) gives

(16) $$3x \equiv 1(\text{mod } 13).$$

Subtracting this from (15) gives

$$2x \equiv 5(\text{mod } 13).$$

Subtracting this from (16) gives

$$x \equiv -4 \equiv 9(\text{mod } 13).$$

We already know that (14) has a unique solution (mod 13); this must be it. In contrast to the systematic method, the nonsystematic methods usually take advantage of the possibility of dividing both sides by common factors. When the coefficient of x is small, it is usually possible to arrange for a common factor by inspection. This is what we did in deriving (12). As another example, (14) may be written

$$8x \equiv 7 + 13 \equiv 20(\text{mod } 13)$$

and, since $(4,13) = 1$,

$$2x \equiv 5 \equiv 5 + 13 \equiv 18(\text{mod } 13)$$

and thus, since $(2,13) = 1$,

$$x \equiv 9(\text{mod } 13).$$

The unexpected may naturally occur when one proceeds in a nonsystematic manner. The following example is particularly instructive. Since $(7,39) = 1$, the equation

(17) $$7x \equiv 22(\text{mod } 39)$$

has a unique solution (mod 39). Subtracting equation (17) five times from

$$39x \equiv 0(\text{mod } 39)$$

gives

$$4x \equiv -110 \equiv -110 + 3 \cdot 39 \equiv 7(\text{mod } 39).$$

We subtract this from (17) and get

(18) $$3x \equiv 15(\text{mod } 39).$$

Here we have the opportunity to divide both sides by 3. But since $(3,39) = 3$,

Theorem 3.3 says that the result is *not necessarily*

$$x \equiv 5 (\text{mod } 39)$$

but only

$$x \equiv 5 (\text{mod } 13).$$

This is equivalent to

(19) $x \equiv 5, 18, 31 (\text{mod } 39).$

We seem to be claiming that (17) has three solutions (mod 39) instead of a unique solution as guaranteed by Theorem 3.6. Of course, this is not true. We have merely shown that if x is a solution to (17), then x satisfies (19) also. Thus two of the "solutions" in (19) will be extraneous solutions that will not satisfy (17); the other will be our desired solution. In this instance,

$$x \equiv 31 (\text{mod } 39)$$

is the correct solution; the obvious $x \equiv 5 (\text{mod } 39)$ is not a solution and neither is $x \equiv 18 (\text{mod } 39)$.

What has happened above is not that unusual in mathematics. It is very easy to get extraneous solutions to equations. For example, we may solve the equation

(20) $\sqrt{x + \sqrt{x - 2}} = 2$

by successive squarings:

$$x + \sqrt{x - 2} = 4,$$

$$\sqrt{x - 2} = 4 - x,$$

$$x - 2 = 16 - 8x + x^2,$$

$$x^2 - 9x + 18 = 0,$$

$$(x - 3)(x - 6) = 0,$$

(21) $x = 3, 6.$

Again, we have not shown that $x = 3$ and $x = 6$ are solutions to (20) but only that if x is a solution to (20), then $x = 3$ or $x = 6$. In fact, $x = 3$ is a solution to (20) while $x = 6$ leads to $\sqrt{8} = 2$, which is absurd. Thus $x = 6$ is an extraneous solution to (20) (which is a kind way of saying that it is not a solution at all). One should always check one's answer with the original problem. There are times when it is legal (but not advisable) to sidestep the checking procedure. Sometimes an existence theorem (such as Theorem 3.6) will tell you that a certain equation has a unique solution; then if you find only one possibility, you know that you have indeed found the unique solution—assuming you made no arithmetic mistakes.

We will now discuss the subject of several equations in one unknown. Sometimes they are inconsistent, as in the two equations

$$x \equiv 2(\text{mod } 4), \qquad x \equiv 1(\text{mod } 6).$$

The first equation requires x to be even while the second requires x to be odd, and thus there is no common solution to both equations. Another way we can see the inconsistency of the two equations is to look at them both (mod 2) [a congruence (mod mn) is also a valid congruence (mod n)]. Then we see that a common solution satisfies both

$$x \equiv 2(\text{mod } 2), \qquad x \equiv 1(\text{mod } 2),$$

which is false since $2 \not\equiv 1(\text{mod } 2)$.

There are other times when two such equations do have a common solution. For example, consider the equations

(22)
$$x \equiv 2(\text{mod } 4),$$
$$x \equiv 3(\text{mod } 5).$$

which have the common solution $x = -2$. Let us find all the solutions of (22). The first equation is satisfied by x if and only if

$$x = 2 + 4t,$$

where t is an integer. We put this in the second equation and get

(23)
$$2 + 4t \equiv 3(\text{mod } 5),$$
$$-t \equiv 4t \equiv 1(\text{mod } 5),$$
$$t \equiv -1 \equiv 4(\text{mod } 5).$$

Equation (23) has a unique solution (mod 5), this is the only possibility and hence is the unique solution to (23). Thus t is a solution to (23) if and only if t can be written

$$t = 4 + 5k,$$

where k is an integer. Thus x satisfies both of equations (22) if and only if there is an integer k such that

$$x = 4(4 + 5k) + 2 = 20k + 18.$$

[It is easily seen that this really is a solution to both of equations (22).] The common solution to equations (22) is unique (mod 20). The situation in (22) is perfectly general.

Theorem 3.7. If $(m,n) = 1$, then the equations

(24)
$$x \equiv a(\text{mod } m),$$
$$x \equiv b(\text{mod } n)$$

have a unique common solution (mod mn).

Proof. An integer x satisfies the first equation if and only if there is an integer t such that

(25) $x = a + mt$.

This satisfies the second equation if and only if

(26) $mt \equiv b - a(\text{mod } n)$.

Since $(m,n) = 1$, this last equation has a unique solution (mod n), say,

$$t \equiv c(\text{mod } n).$$

Thus t satisfies (26) if and only if there is an integer k such that

$$t = c + nk.$$

We put this in (25) and find that x is a common solution to equations (24) if and only if

$$x = a + m(c + nk)$$
$$= (a + mc) + mnk.$$

Hence equations (24) have common solutions, $a + mc$ is one, and all solutions are congruent to $a + mc(\text{mod } mn)$. Thus the common solutions to (24) are unique (mod mn).   ▲

Theorem 3.7 is a special case of a more general result which was known to to the ancient Chinese.

Theorem 3.8 (Chinese Remainder Theorem[2]). Let $m_1, \ldots, m_k$ be positive integers which are relatively prime in pairs. Then the k equations

(27)
$$x \equiv a_1(\text{mod } m_1)$$
$$\vdots$$
$$x \equiv a_k(\text{mod } m_k)$$

have a unique solution (mod $m_1 m_2 \cdots m_k$).

[2] More rarely known as the Formosa Theorem.

Proof. We use Theorem 3.7 $(k-1)$ times. By Theorem 3.7, the first two equations have a unique solution (mod $m_1 m_2$); let this solution be given by

(28) $$x \equiv b_2 (\bmod \ m_1 m_2).$$

The third equation is

(29) $$x \equiv a_3 (\bmod \ m_3).$$

By hypothesis,

$$(m_1, m_3) = (m_2, m_3) = 1$$

and therefore, by Theorem 2.7,

$$(m_1 m_2, m_3) = 1.$$

We can now apply Theorem 3.7 to (28) and (29). There is a unique solution to (28) and (29) (mod $m_1 m_2 m_3$); in other words, there is a unique solution to the first three equations of (27). Let that unique solution be

(30) $$x \equiv b_3 (\bmod \ m_1 m_2 m_3).$$

Consider the fourth equation of (27),

(31) $$x \equiv a_4 (\bmod \ m_4).$$

Since

$$(m_1, m_4) = (m_2, m_4) = (m_3, m_4) = 1,$$

we see that

$$(m_1 m_2 m_3, m_4) = 1.$$

Thus there is a unique solution to (30) and (31) (mod $m_1 m_2 m_3 m_4$) and it is the unique solution to the first four equations of (27). We continue in this manner. After $(k-1)$ applications of Theorem 3.7, we will arrive at the fact that there is a unique solution (mod $m_1 m_2 \cdots m_k$) to

$$x \equiv b_{k-1} (\bmod \ m_1 m_2 \cdots m_{k-1})$$

and

$$x \equiv a_k (\bmod \ m_k);$$

this solution also provides the unique solution (mod $m_1 m_2 \cdots m_k$) to the k equations in (27). ▲

If for each j, we restrict a_j to the range from 0 to $m_j - 1$, then the existence part of the Chinese remainder theorem may be put in the form; if $m_1, \ldots, m_k$ are pairwise relatively prime, then there exists an integer x such that for each

$j(j = 1, 2, \ldots, k)$, x leaves the remainder a_j when divided by m_j. This explains the name of the theorem. (It also explains how a theorem about congruences could be known before congruences were invented; we have merely put an old theorem in modern form.)

The reader is no doubt wondering why we restricted our attention in Theorems 3.7 and 3.8 to equations having the coefficient of x equal to one. Consider the set of equations

(32) $a_1 x \equiv b_1 (\mod n_1), \ldots, a_k x \equiv b_k (\mod n_k)$.

If there is a common solution, then each equation is individually solvable. This means that $d_i | b_i$, where $d_i = (a_i, n_i)$, and this is true for $i = 1, 2, \ldots, k$. We know what the solutions to the individual equations of (32) look like,

$$x \equiv c_1 (\mod m_1), \ldots, x \equiv c_k (\mod m_k)$$

where the m_i are given by the formulas

$$m_1 = \frac{n_1}{d_1}, \ldots, m_k = \frac{n_k}{d_k}.$$

Thus if for all i, $d_i | b_i$ and if the numbers $m_1, \ldots, m_k$ are relatively prime in pairs, then equations (32) have a unique common solution (mod $m_1 m_2 \cdots m_k$). Thus Theorem 3.8 is sufficient for the theoretical purpose of proving that under certain conditions, (32) has a unique solution (mod $m_1 m_2 \cdots m_k$). When actually solving a specific problem, it is a complete waste of time to first solve each of (32) individually and then find the common solution. It takes only half the work to simply solve the first equation of (32), put the solution of the first equation into the second equation of (32) and solve, put the common solution of the first two equations in the third, and so on.

We now touch briefly on the subject of linear congruence equations with more than one unknown. We will content ourselves with examining the simplest case of two equations and two unknowns; the result, however, will be very useful in Chapter 4.

Theorem 3.9. Suppose $(cf - de, n) = 1$. Then the equations

(33)
$$cx + ey \equiv a (\mod n),$$
$$dx + fy \equiv b (\mod n)$$

have a unique common solution for x and y (mod n).

Proof. Let us suppose that there is a solution to (33) and attempt to find it. We multiply the first equation by f, the second equation by e, and subtract;

the result is

(34) $(cf - de)x \equiv af - be(\text{mod } n).$

We multiply the first equation by d, the second equation by c, and subtract the first from the second; the result is

(35) $(cf - de)y \equiv bc - ad(\text{mod } n).$

Since $(cf - de, n) = 1$, Theorem 3.6 says that there is a unique $x(\text{mod } n)$ satisfying (34) and a unique $y(\text{mod } n)$ satisfying (35). Hence if there is a solution to (33), it is unique (mod n).

Unfortunately, the fact that (34) and (35) have solutions does not mean that the original equations (33) have solutions. It is clear how we should proceed, however. We take the solutions to (34) and (35) and show that they do satisfy (33). There is a slight hitch in this. We cannot just divide both sides of (34) and (35) by $(cf - de)$, because we have not defined congruences for rational numbers.[3] But we can find an integer which does the same job (mod n) as $1/(cf - de)$. Since $(cf - de, n) = 1$, Theorem 3.6 says that there is an integer z such that

(36) $z(cf - de) \equiv 1(\text{mod } n).$

We use the integer z to solve (34) and (35). If x and y satisfy (34) and (35), then

(37)
$$x \equiv z(cf - de)x \equiv z(af - be)(\text{mod } n),$$
$$y \equiv z(cf - de)y \equiv z(bc - ad)(\text{mod } n).$$

Now we can verify that (33) does have solutions. Let x, y, and z be given by (36) and (37). Then

$$cx + ey \equiv cz(af - be) + ez(bc - ad)(\text{mod } n)$$
$$\equiv czaf - ezad \qquad (\text{mod } n)$$
$$\equiv az(cf - de) \qquad (\text{mod } n)$$
$$\equiv a \qquad (\text{mod } n)$$

and

$$dx + fy \equiv dz(af - be) + fz(bc - ad)(\text{mod } n)$$
$$\equiv -dzbe + fzbc \qquad (\text{mod } n)$$
$$\equiv bz(cf - de) \qquad (\text{mod } n)$$
$$\equiv b \qquad (\text{mod } n).$$

---

[3] It may be useful to remind the reader that the solutions to two equations in two unknowns with integral coefficients are usually not integers but only rational numbers. Here we are looking for integers which satisfy the equations (mod n).

Thus there are integral solutions to (33); we have already shown that they must be unique (mod n).   ▲

EXERCISES

In problems 1–13, either find all integral solutions (all common integral solutions, if more than one equation) or show that there are none.

1. Solve: $5x \equiv 1 \pmod{7}$.
2. Solve: $14x \equiv 5 \pmod{45}$.
3. Solve: $14x \equiv 35 \pmod{87}$.
4. Solve: $3x \equiv 2 \pmod{78}$.
5. Solve: $6x \equiv 10 \pmod{14}$.
6. Solve: $9x \equiv 21 \pmod{12}$.
7. Solve: $x \equiv 2 \pmod{3}$, $x \equiv 3 \pmod{4}$.
8. Solve: $x \equiv 7 \pmod{9}$, $x \equiv 13 \pmod{23}$, $x \equiv 1 \pmod{2}$.
9. Solve: $2x \equiv 3 \pmod{5}$, $4x \equiv 3 \pmod{7}$.
10. Solve: $6x \equiv 8 \pmod{10}$, $15x \equiv 30 \pmod{55}$.
11. Solve: $5x + 4y \equiv 6 \pmod{7}$, $3x - 2y \equiv 6 \pmod{7}$.
12. Solve: $2x + 7y \equiv 8 \pmod{13}$, $5x + 10y \equiv 7 \pmod{13}$.
13. Solve: $x + 2y + 3z \equiv 1 \pmod{11}$, $x + 5y + 6z \equiv 3 \pmod{11}$, $x + 4y + 7z \equiv 5 \pmod{11}$.
14. Find all solutions (mod 7): $3x + 4y \equiv 1 \pmod{7}$.
15. Find all solutions (mod 8): $3x + 7y \equiv 2 \pmod{8}$.
*16. Show that if $(a,n) = (b,n) = 1$, then the equation

$$ax + by \equiv c \pmod{n}$$

has exactly n different solutions (mod n).
17. Find all solutions (mod 6): $2x + 3y \equiv 1 \pmod{6}$.
18. Find all common solutions (mod 12) (or show that there are none) to

$$4x + y \equiv 6 \pmod{12}, \qquad x + 4y \equiv 2 \pmod{12}.$$

19. Find all common solutions (mod 12) (or show that there are none) to

$$4x + y \equiv 6 \pmod{12}, \qquad x + 4y \equiv 9 \pmod{12}.$$

20. Find all positive integers less than 1000 which leave the remainder 1 when divided by 2, 3, 5, and 7.
21. A multiplication has been performed incorrectly, but the answer is correct (mod 9), (mod 10), and (mod 11). What is the closest that the incorrect result can possibly be to the correct result?
22. The following multiplication was correct, but unfortunately the printer inserted an x in place of a digit in the answer

$$172\ 195 \cdot 572\ 167 = 98\ 524\ 2x6\ 565.$$

Determine x without redoing the multiplication.

23. Show that an integer is divisible by 4 if and only if the number left when all digits other than the last two are eliminated is divisible by 4. Use this rule to find conditions for divisibility by 12.

24. Show that every integer satisfies at least one of the following six congruences: $x \equiv 0 \pmod 2$, $x \equiv 0 \pmod 3$, $x \equiv 1 \pmod 4$, $x \equiv 1 \pmod 6$, $x \equiv 3 \pmod 8$, and $x \equiv 11 \pmod{12}$.

25. Prove Theorem 3.8 by induction.

3.4. Reduced Residue Systems and Euler's ϕ Function

We see from Theorem 3.6 that the equation

$$ax \equiv 1 \pmod n$$

is solvable (in integers) if and only if $(a,n) = 1$. The numbers that are relatively prime to n have other interesting congruence properties. In this section we single these numbers out for special attention.

Definition. Let S be a complete residue system (mod n). The set S' consisting of those members of S which are relatively prime to n is called a **reduced residue system** (mod n).

If $b \equiv a \pmod n$, then we may write b in the form $b = a + kn$. By Theorem 2.4,

$$(b,n) = (a + kn, n) = (a,n).$$

Thus if $a \equiv b \pmod n$, then a is relatively prime to n if and only if b is relatively prime to n.

Theorem 3.10. Let S' be a reduced residue system (mod n). If $(a,n) = 1$, then a is congruent (mod n) to a unique member of S'. If S'' is another reduced residue system (mod n), then S' and S'' have exactly the same number of members and, in fact, if $a_1, \ldots, a_k$ are the members of S', then the members of S'' can be listed in such an order, say $b_1, \ldots, b_k$, that $a_1 \equiv b_1, \ldots, a_k \equiv b_k \pmod n$ [that is, (mod n), S'' is simply a rearrangement of S'].

Proof. Let S be a complete residue system containing S'. By the definition of S, there is a unique integer b in S such that

$$a \equiv b \pmod n.$$

Since $(a,n) = 1$, $(b,n) = 1$ also. Therefore, by definition, b is in S'. Since b is unique in S, it is certainly unique in S'. This takes care of the first part of the theorem.

It follows from the first part of the theorem that each element of S'' is congruent to a unique member of S'. No two members of a complete residue system (mod n) are congruent (mod n). Therefore, no two members of S'' are congruent (mod n) to the same member of S', and since different members of S'' are congruent (mod n) to different members of S', we see that S' has at least as many members as S''. But everything done thus far is equally valid with S' and S'' interchanged. Thus S'' has at least as many members as S'. This combined with the previous statement says that S'' and S' have exactly the same number of members. Let k be the number of integers in each set. The last part of the theorem is now clear: Since different members of S'' are congruent (mod n) to different members of S' and since each has k members, the sets S' and S'' are the same (mod n) except for the order of the elements. ▲

Definition. For $n \geq 1$, let $\phi(n)$ denote the number of integers in a reduced residue system (mod n). [By Theorem 3.10, $\phi(n)$ does not depend on which reduced residue system (mod n) is chosen.] This function of n is called **Euler's ϕ function.** (It is also sometimes called **Euler's totient function.**)

It is interesting to note that although the function was invented by Euler, the present notation was given by Gauss. The following theorem is often used as the definition of $\phi(n)$.

Theorem 3.11. If $n \geq 1$, then the number of positive integers which are less than or equal to n and relatively prime to n is $\phi(n)$.

Proof. The numbers $1, 2, \ldots, n$ form a complete residue system (mod n). Thus the positive integers which are less than or equal to n and relatively prime to n form a reduced residue system (mod n). Their number is thus $\phi(n)$. ▲

A few examples are shown in Figure 3.3 (we take as our complete residue system (mod n), the numbers $1, 2, 3, \ldots, n$). Since $(n,n) = n$, we see that n is in a reduced residue system (mod n) if and only if $n = 1$. Thus for $n > 1$, $\phi(n)$ is the number of positive integers which are less than n and relatively prime to n. Every positive integer less than a prime p is relatively prime to p and hence

$$\phi(p) = p - 1.$$

n	1	2	3	4	5	6	7	8	9	10
A reduced residue system mod (n)	1	1	1, 2	1, 3	1, 2, 3, 4	1, 5	1, 2, 3, 4, 5, 6	1, 3, 5, 7	1, 2, 4, 5, 7, 8	1, 3, 7, 9
$\phi(n)$	1	1	2	2	4	2	6	4	6	4

Figure 3.3

The next theorem gives another simple property of numbers relatively prime to n.

Theorem 3.12. Suppose $(a,n) = 1$. If the numbers $a_1, \ldots, a_n$ form a complete residue system (mod n), then for all b the numbers $aa_1 + b, \ldots, aa_n + b$ also form a complete residue system (mod n). If the numbers $a_1, \ldots, a_{\phi(n)}$ form a reduced residue system (mod n), then so do the numbers $aa_1, \ldots, aa_{\phi(n)}$.

Proof. Since $(a,n) = 1$, by Theorem 3.6, there is an integer c such that

$$ac \equiv 1 \ (\text{mod } n).$$

Suppose that $a_1, \ldots, a_n$ gives a complete residue system (mod n). If d is an integer, then there is a (unique) k such that

$$c(d - b) \equiv a_k (\text{mod } n).$$

But then

$$d - b \equiv ac(d - b) \equiv aa_k (\text{mod } n) \qquad \text{or} \qquad d \equiv aa_k + b (\text{mod } n).$$

If

$$d \equiv aa_j + b (\text{mod } n) \qquad \text{and} \qquad d \equiv aa_k + b (\text{mod } n),$$

then

$$c(d - b) \equiv aca_j \equiv a_j (\text{mod } n), \qquad c(d - b) \equiv aca_k \equiv a_k (\text{mod } n),$$

which is false if $j \neq k$. Thus every integer is congruent to exactly one of the n integers $aa_1 + b, \ldots, aa_n + b$, and thus this set is a complete system of residues (mod n). If $a_1, \ldots, a_{\phi(n)}$ form a reduced residue system (mod n), then they are distinct elements of a complete residue system, and thus by the first part of the theorem with $b = 0$, $aa_1, \ldots, aa_{\phi(n)}$ are distinct elements of a complete residue system (mod n). A reduced residue system has exactly $\phi(n)$ elements; therefore, we need only show that each of the numbers

$aa_1, \ldots, aa_{\phi(n)}$ is relatively prime to n and we will have found all $\phi(n)$ elements of a reduced residue system (mod n). Since $(a_k, n) = 1$ for all k and $(a, n) = 1$, we see from Theorem 2.7 that $(aa_k, n) = 1$ for all k, $k = 1, \ldots, \phi(n)$. ▲

The preceding theorem has a remarkable consequence.

Theorem 3.13 (Euler's Theorem; also known as the Euler–Fermat Theorem). If $(a, n) = 1$, then $a^{\phi(n)} \equiv 1(\mathrm{mod}\ n)$.

Proof. Let $a_1, \ldots, a_{\phi(n)}$ be a reduced residue system (mod n). Then the numbers $aa_1, \ldots, aa_{\phi(n)}$ are also a reduced residue system (mod n). By Theorem 3.10, the numbers $aa_1, \ldots, aa_{\phi(n)}$ are just a rearrangement (mod n) of the numbers $a_1, \ldots, a_{\phi(n)}$, and thus the products of all the numbers in the two systems are the same (mod n):

$$(38) \qquad (aa_1)(aa_2) \cdots (aa_{\phi(n)}) \equiv a_1 a_2 \cdots a_{\phi(n)}(\mathrm{mod}\ n).$$

Since each a_k is relatively prime to n, it may be canceled from both sides of (38). When we do this for each k, we are left with

$$a^{\phi(n)} \equiv 1(\mathrm{mod}\ n). \qquad ▲$$

Theorem 3.14 (Fermat's Theorem; also known as the Little Fermat Theorem). Suppose p is a prime. Then for all a,

$$a^p \equiv a(\mathrm{mod}\ p).$$

If $p \nmid a$, then

$$a^{p-1} \equiv 1(\mathrm{mod}\ p).$$

Proof. When $p | a$, both sides of the first congruence are congruent to $0(\mathrm{mod}\ p)$ and hence congruent. When $p \nmid a$, then the first congruence is a times the second, and thus it suffices to prove that the second congruence is valid. Since p is a prime, $p \nmid a$ is equivalent to $(a, p) = 1$. This combined with the fact that $\phi(p) = p - 1$ makes this theorem a corollary of the previous theorem. ▲

Historically, Fermat's theorem was stated in 1640, and it was generalized by Euler in 1760 to the form that if $(a, n) = 1$, then $n|(a^{\phi(n)} - 1)$. A special case of Fermat's theorem is that if p is a prime, then

$$p|(2^p - 2).$$

The ancient Chinese knew this fact and also believed that the converse is true. The converse states that if $n > 1$ and

$$n|(2^n - 2),$$

then n is a prime. It is probable that the Chinese observed this experimentally and never thought of proving their conjecture. In any event, they were wrong. For example,

$$341|(2^{341} - 2),$$

even though $341 = 11 \cdot 31$. The theory of congruences makes it possible to check this without having to find 2^{341} (a number of 103 digits). Since 11 is a prime, Fermat's theorem says that

$$2^{10} \equiv 1 (\bmod\ 11).$$

Therefore,

$$2^{341} \equiv 2(2^{10})^{34} \equiv 2(1)^{34} \equiv 2(\bmod\ 11)$$

and hence

$$11|(2^{341} - 2).$$

Also,

$$2^5 \equiv 32 \equiv 1(\bmod\ 31)$$

and therefore

$$2^{341} \equiv 2(2^5)^{68} \equiv 2(1)^{68} \equiv 2(\bmod\ 31)$$

and therefore

$$31|(2^{341} - 2).$$

Since 11 and 31 are relatively prime, their product divides $(2^{341} - 2)$; that is,

$$341|(2^{341} - 2).$$

Thus the ancient Chinese were wrong. In honor of their mistake, we say today that a composite integer n such that

$$n|(2^n - 2)$$

is a **pseudoprime.** The first two pseudoprimes are $341 = 11 \cdot 31$ and $561 = 3 \cdot 11 \cdot 17$. There are infinitely many pseudoprimes. In fact, there are infinitely many even pseudoprimes, but they are harder to find. The first known example of an even pseudoprime is

$$161\ 038 = 2 \cdot 73 \cdot 1103,$$

which was found in 1950 by D. H. Lehmer.

EXERCISES

1. Find a reduced residue system (mod 20) and give $\phi(20)$.
2. Find a reduced residue system (mod 30) and give $\phi(30)$.
3. Give two examples to show that when 2 is added to every member of a reduced residue system (mod n), the result may or may not be a reduced residue system (mod n).
4. Show that $561 = 3 \cdot 11 \cdot 17$ is a pseudoprime.
5. Show that, in fact, for all integers a, $a^{561} \equiv a(\text{mod } 561)$.
6. Show that $3^3 \equiv -4(\text{mod } 31)$ and use this to show that $3^{10} \equiv -6(\text{mod } 31)$. Use this result and Euler's theorem to show that

$$3^{341} \not\equiv 3(\text{mod } 31)$$

   and therefore

$$3^{341} \not\equiv 3(\text{mod } 341).$$

7. Find a composite number n such that $n|(3^n - 3)$. (*Hint:* There is such a number less than 10.)
8. Show that if $(a,561) = 1$, then $a^{80} \equiv 1(\text{mod } 561)$. [*Note:* Do not try to find $\phi(561)$, as it is greater than 300.]
9. Show that if p is a prime, a is an integer, and k is a nonnegative integer, then

$$a^{1+k(p-1)} \equiv a(\text{mod } p).$$

10. Show that if n is odd and a is an integer,

$$a^n \equiv a(\text{mod } 3).$$

3.5. More on Euler's ϕ Function

We see that if we are going to apply Euler's theorem in a particular problem, then we must be able to calculate $\phi(n)$ from n. If n is small, it is fairly easy to find all the numbers less than or equal to n which are relatively prime to n, and then we immediately have $\phi(n)$. But what if we wish something like $\phi(1776)$? In this section we develop a formula for $\phi(n)$ in terms of the prime factorization of n. The key to this result will be a proof of the fact that $\phi(n)$ is multiplicative.

We first illustrate the proof that $\phi(n)$ is multiplicative by showing that $\phi(30) = \phi(5) \cdot \phi(6)$. We first write the numbers from 1 to 30 in a rectangular array as in Figure 3.4. We note that a number is relatively prime to 30 if and only if it is relatively prime to both 5 and 6. Thus $n(1 \leq n \leq 30)$ is relatively prime to 30 if and only if it is one of $\phi(5)$ numbers in $\phi(6)$ columns

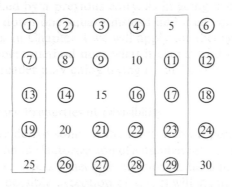

Figure 3.4. Numbers relatively prime to 6 are in the large vertical rectangles, numbers relatively prime to 5 are in circles. Numbers relatively prime to 30 are in both the large vertical rectangles and circles. The numbers from 1 to 30 that are relatively prime to 6 are in the first and fifth columns [$\phi(6)$ columns in all] and each column contains exactly four [which equals $\phi(5)$] numbers relatively prime to 5.

(the first and fifth). There are $\phi(5) \cdot \phi(6)$ such numbers and thus $\phi(30) = \phi(5)\phi(6)$. The situation here is perfectly general and leads to a proof that $\phi(n)$ is multiplicative.

Theorem 3.15. $\phi(n)$ is multiplicative.

Proof. Suppose m and n are positive integers with $(m,n) = 1$. We put the first mn positive integers in a rectangular array with m columns and n rows as in Figure 3.5. The numbers in the jth column are $m \cdot 0 + j$, $m \cdot 1 + j$, $m \cdot 2 + j, \ldots, m(n-1) + j$. By Theorem 2.4,

$$(ma + j, m) = (j,m),$$

1	2	$\cdots$	m
$m + 1$	$m + 2$	$\cdots$	$m + m$
$2m + 1$	$2m + 2$	$\cdots$	$2m + m$
.	.	$\cdots$	.
.	.	$\cdots$	.
.	.	$\cdots$	.
$(n-1)m + 1$	$(n-1)m + 2$	$\cdots$	$(n-1)m + m$

Figure 3.5

and thus either every element of the jth column is relatively prime to m or no element of the jth column is relatively prime to m. Therefore, exactly $\phi(m)$ columns contain numbers relatively prime to m and every entry in these $\phi(m)$ columns is relatively prime to m.

Since $(m,n) = 1$, by Theorem 3.12 the n numbers in the jth column form a complete residue system (mod n). Thus by definition, the jth column contains exactly $\phi(n)$ numbers relatively prime to n. Hence in each of the $\phi(m)$ columns containing the numbers relatively prime to m, there are $\phi(n)$ numbers relatively prime to n, and these are the only numbers relatively prime to both m and n. That is, there are exactly $\phi(m)\phi(n)$ numbers in the array of Figure 3.5 that are relatively prime to both m and n. But an integer is relatively prime to mn if and only if it is relatively prime to both m and n. Therefore,

$$\phi(mn) = \phi(m)\phi(n). \qquad \blacktriangle$$

After Theorem 3.15, we can evaluate $\phi(n)$ if we can find $\phi(p^a)$, where p is a prime. This is an easy task and the result is

Theorem 3.16. Let the prime factorization of n be

$$n = p_1^{a_1} p_2^{a_2} \cdots p_k^{a_k}.$$

Then

$$\phi(n) = n(1 - 1/p_1)(1 - 1/p_2) \cdots (1 - 1/p_k).$$

Proof. We begin by finding $\phi(p^a)$, where p is a prime and $a \geq 1$. A number is relatively prime to p^a unless it is divisible by p. The numbers from 1 to p^a which are divisible by p are $1 \cdot p, 2 \cdot p, \ldots, p^{a-1} \cdot p$. Thus exactly p^{a-1} positive integers less than or equal to p^a are divisible by p, and therefore there are exactly $p^a - p^{a-1}$ positive integers less than or equal to p^a which are not divisible by p. Hence

$$\phi(p^a) = p^a - p^{a-1} = p^a(1 - 1/p).$$

But now, by Theorem 3.15,

$$\phi(n) = \phi(p_1^{a_1})\phi(p_2^{a_2}) \cdots \phi(p_k^{a_k})$$

$$= p_1^{a_1}\left(1 - \frac{1}{p_1}\right) p_2^{a_2}\left(1 - \frac{1}{p_2}\right) \cdots p_k^{a_k}\left(1 - \frac{1}{p_k}\right)$$

$$= p_1^{a_1} p_2^{a_2} \cdots p_k^{a_k}\left(1 - \frac{1}{p_1}\right)\left(1 - \frac{1}{p_2}\right) \cdots \left(1 - \frac{1}{p_k}\right)$$

$$= n\left(1 - \frac{1}{p_1}\right)\left(1 - \frac{1}{p_2}\right) \cdots \left(1 - \frac{1}{p_k}\right). \qquad \blacktriangle$$

As an example, let us answer the question that started all this, and find $\phi(1776)$. We need to factor 1776 first; fortunately, we see that there is a factor of 2 and 3 (casting out nines reduces 1776 to 3) to get us started. We then find without much trouble that

$$1776 = 2^4 \cdot 3 \cdot 37.$$

Thus

$$\begin{aligned}
\phi(1776) &= 1776(1 - \tfrac{1}{2})(1 - \tfrac{1}{3})(1 - \tfrac{1}{37}) \\
&= 2^4 \cdot 3 \cdot 37 \cdot \tfrac{1}{2} \cdot \tfrac{2}{3} \cdot \tfrac{36}{37} \\
&= 2^4 \cdot 36 \\
&= 576.
\end{aligned}$$

This is considerably simpler than examining the first 1776 positive integers to see which have factors of 2, 3, and 37 and which do not.

There are other methods of deriving the crucial Theorem 3.15. Two of them are given in the miscellaneous exercises. We have now been able to evaluate three different functions from the knowledge that they are multiplicative $[d(n), \sigma(n), \text{and } \phi(n)]$. In the next theorem, we again make use of the knowledge that a function is multiplicative in order to find it.

Theorem 3.17. If $n \geq 1$, then

$$\sum_{d|n} \phi(d) = n.$$

Proof. Let

$$f(n) = \sum_{d|n} \phi(d).$$

Since $\phi(n)$ is multiplicative, Theorem 2.15 says that $f(n)$ is also multiplicative. Thus we first wish to find $f(p^a)$, where p is a prime and $a \geq 1$. Here we have

$$\begin{aligned}
f(p^a) &= \sum_{d|p^a} \phi(d) \\
&= \phi(1) + \phi(p) + \phi(p^2) + \cdots + \phi(p^a) \\
&= 1 + p\left(1 - \frac{1}{p}\right) + p^2\left(1 - \frac{1}{p}\right) + \cdots + p^a\left(1 - \frac{1}{p}\right) \\
&= 1 + (p - 1) + (p^2 - p) + (p^3 - p^2) + \cdots + (p^a - p^{a-1}) \\
&= p^a,
\end{aligned}$$

with all the other terms canceling each other. Therefore, if n factors as

$$n = p_1^{a_1} p_2^{a_2} \cdots p_k^{a_k},$$

then

$$f(n) = f(p_1^{a_1}) f(p_2^{a_2}) \cdots f(p_k^{a_k})$$
$$= p_1^{a_1} p_2^{a_2} \cdots p_k^{a_k}$$
$$= n. \qquad \blacktriangle$$

EXERCISES
1. Find $\phi(n)$ for $n = 20, 60, 63, 341,$ and 561.
2. Show that if n is odd, $\phi(2n) = \phi(n)$.
3. Show that if n is even, $\phi(2n) = 2\phi(n)$.
4. What goes wrong with the proof of Theorem 3.15 if $(n, m) > 1$?
5. Does Theorem 3.16 give $\phi(1)$?
6. Verify Theorem 3.17 for $n = 30$.
7. Let x be the smallest positive integer such that $2^x \equiv 1 (\text{mod } 63)$. Find x and verify that $x | \phi(63)$.
*8. Repeat problem 7 with 63 replaced by 105.
*9. Find the last three digits of 7^{9999}.

3.6. Polynomial Congruences

Definition. Let $f(x)$ and $g(x)$ be polynomials with integral coefficients. If the coefficients of each power of x are congruent (mod n), then we say that $f(x)$ and $g(x)$ are **congruent as polynomials** (mod n) and we write

$$f(x) \equiv g(x)(\text{poly mod } n).$$

For example

$$5x^2 - 6x + 3 \equiv x^2 + 2x - 1(\text{poly mod } 4),$$

$$x^3 + 4x^2 - 2x + 1 \equiv 3x^4 + 4x^3 + x^2 - 8x + 22(\text{poly mod } 3),$$

$$x^3 - 1 \equiv (x - 1)^3(\text{poly mod } 3),$$

$$x^2 + 2x + 1 \not\equiv 6x^2 + 3x + 6(\text{poly mod } 5),$$

$$x^5 \not\equiv x(\text{poly mod } 5).$$

Missing powers of x are assumed to have the coefficient 0. Thus in the second example, we may assume that the polynomial on the left is

$$0x^4 + x^3 + 4x^2 - 2x + 1$$

and in the third example the polynomial on the left is

$$x^3 + 0x^2 + 0x - 1.$$

The fifth example is extremely instructive. The polynomials x^5 and x are not congruent as polynomials (mod 5) since the coefficients of x are incongruent (mod 5). This is true in spite of the fact that by Fermat's theorem (3.14)

$$x^5 \equiv x(\mathrm{mod}\ 5)$$

for all integers x. Thus it may happen that two polynomials may be incongruent as polynomials (mod n) and still be congruent (mod n) for all integral values of the variable. This is a situation which does not occur with equalities. Two polynomials $f(x)$ and $g(x)$ are the same if and only if $f(x) = g(x)$ for all integers x. Thus when we write $f(x) = g(x)$, it does not matter whether we think of the equality as saying f and g are the same polynomials or f and g give the same values for all values of x; the two concepts are the same.

This is our reason for distinguishing between $f(x)$ and $g(x)$ being congruent as polynomials (mod n) and just being congruent (mod n). When we write

(39) $$f(x) \equiv g(x)(\mathrm{mod}\ n)$$

we simply mean that (39) is true for all integral values of x. There is yet a second meaning of (39) and that is the meaning of solving an equation. This is a situation that occurs with equalities. For instance,

$$(x + 1)^2 = x^2 + 2x + 1$$

is an identity, true for all values of x, while

$$(x + 1)^2 = 2x^2 + 2x$$

is true only for certain values of x which can be found by solving the equation. The reader should be able to distinguish between the two meanings of (39), particularly since they will usually be accompanied by a phrase such as "for all integers, x," or "solve."[4]

The following result, although trivial, is sufficiently fundamental to be called a theorem.

---

[4] It should be noted that the notation

$$f(x) \equiv g(x)(\mathrm{poly}\ \mathrm{mod}\ n)$$

is unique with this book and is presented as a public service to help minimize confusion. Other books and articles customarily use (39) for this purpose also and leave it to the reader to figure out which meaning is being used. (The kind author will attach a phrase such as "congruent as polynomials," but he is under no compulsion to do so and usually does not.)

Theorem 3.18. If

$$f(x) \equiv g(x)(\text{poly mod } n)$$

then for all integers x,

$$f(x) \equiv g(x)(\text{mod } n).$$

Proof. The result follows from the definition of

$$f(x) \equiv g(x)(\text{poly mod } n)$$

and Theorem 3.2 (the theorem on sums and products of congruent numbers being congruent).   ▲

As the example

$$x^5 \equiv x(\text{mod } 5)$$

shows, the converse of Theorem 3.18 is not necessarily true. Thus the statement

$$f(x) \equiv g(x)(\text{poly mod } n)$$

contains more information than the statement

$$f(x) \equiv g(x)(\text{mod } n).$$

Let us give an example of how much more information the first statement can carry. It is easy to show that if p is a prime, then

(40) $x^p - x \equiv x(x + 1)(x + 2) \cdots (x + p - 1)(\text{mod } p)$

for all integers x. By Fermat's theorem (3.14), the left side of (40) is congruent to $0(\text{mod } p)$ for all integers x. Since the numbers $1, 2, \ldots, p$ form a complete residue system $(\text{mod } p)$, given an integer x, there is an integer j in the range from 1 to p such that

$$x \equiv j(\text{mod } p).$$

But then the factor $[x + (p - j)]$ in the right side of (40) is congruent to 0:

$$x + p - j \equiv j + p - j \equiv 0(\text{mod } p),$$

and thus the right side of (40) is congruent to $0(\text{mod } p)$ for all integers x. This proves the relation (40). Equation (40) is not very useful as it stands, but it so happens that (40) is also true as a congruence of polynomials; that is,

(41) $x^p - x \equiv x(x + 1)(x + 2) \cdots (x + p - 1)(\text{poly mod } p).$

We will prove this later in this section. If we accept this as true, then the coefficients on each side of (41) are congruent (mod p). [This is precisely the difference between (40) and (41).] If we equate the coefficient of x on both sides of (41), we get

(42) $-1 \equiv (p - 1)!(\mathrm{mod}\ p).$

Thus from (40) we get nothing, but (41) is full of information [and (42) is just the result of equating one coefficient of (41)]. As examples of (42),

$$2|(1! + 1), \quad 5|(4! + 1), \quad 7|(6! + 1), \quad 101|(100! + 1).$$

The reader can verify the first three of these easily enough, but the fourth might take a few days (100! has 158 digits).

By Fermat's theorem, the congruence

$$x^p - x \equiv 0(\mathrm{mod}\ p)$$

has the roots $0, -1, -2, \ldots, -(p - 1)$ (among others). With ordinary equality, roots lead to factors and thus, by analogy, (41) seems quite reasonable. This is, in fact, how we will eventually derive (41). In the meantime, we prove several preliminary theorems on polynomial congruences familiar to the reader as equalities.

Theorem 3.19. If

$$f_1(x) \equiv f_2(x)(\mathrm{poly}\ \mathrm{mod}\ n),$$

$$g_1(x) \equiv g_2(x)(\mathrm{poly}\ \mathrm{mod}\ n),$$

then

$$f_1(x) + g_1(x) \equiv f_2(x) + g_2(x)(\mathrm{poly}\ \mathrm{mod}\ n),$$

$$f_1(x)g_1(x) \equiv f_2(x)g_2(x)(\mathrm{poly}\ \mathrm{mod}\ n).$$

Proof. Coefficients of sums and products of polynomials are determined as combinations of sums and products of the coefficients of the original polynomials. Therefore, changing the coefficients of the original polynomials to congruent numbers (mod n) changes the coefficients of the answer to congruent numbers (mod n).   ▲

Theorem 3.20. If $f(x)$ is a polynomial with integral coefficients and $f(a) \equiv 0(\mathrm{mod}\ n)$, then there is a polynomial $g(x)$ with integral coefficients such that

$$f(x) \equiv (x - a)g(x)(\mathrm{poly}\ \mathrm{mod}\ n).$$

Proof. Let us divide $f(x)$ by $(x - a)$. The result is a quotient $g(x)$ with integral coefficients and a remainder b:

$$f(x) = (x - a)g(x) + b.$$

Therefore,

$$f(a) = (a - a)g(a) + b = b$$

and hence

$$b \equiv f(a) \equiv 0 (\mathrm{mod}\ n).$$

Therefore,

$$f(x) \equiv f(x) - b \equiv (x - a)g(x)(\text{poly mod } n). \qquad \blacktriangle$$

For example, let $f(x) = x^2 + x + 1$. Then

$$f(1) \equiv 0 (\mathrm{mod}\ 3).$$

When we divide $f(x)$ by $x - 1$, we get the result

$$
\begin{array}{r}
x + 2 \\
x - 1 \overline{)\ x^2 + x + 1} \\
\underline{x^2 - x} \\
2x + 1 \\
\underline{2x - 2} \\
3
\end{array}
$$

$$x^2 + x + 1 = (x - 1)(x + 2) + 3.$$

(The reader acquainted with synthetic division may perform the calculations quicker and easier.) Hence

$$x^2 + x + 1 \equiv (x - 1)(x + 2)(\text{poly mod } 3).$$

Since

$$x + 2 \equiv x - 1 (\text{poly mod } 3),$$

this may be written

$$x^2 + x + 1 \equiv (x - 1)^2 (\text{poly mod } 3).$$

If we were told that

$$f(x) = (x - a)g(x),$$

$$f(b) = 0, \qquad b \neq a,$$

then we would deduce that

$$(b - a)g(b) = 0, \qquad b - a \neq 0$$

and hence

$$g(b) = 0.$$

From this we see that there is a factor of $x - b$ in $g(x)$ and this gives factors of $(x - a)(x - b)$ in $f(x)$. This is the process used in showing that successive roots of the equation

$$f(x) = 0$$

correspond to successive factors of f. We wish to imitate this process for congruences, but we immediately find that there are difficulties. Consider the example

(43) $x^2 - 1 \equiv 0 (\mathrm{mod}\ 8).$

Since

$$x \equiv 1 (\mathrm{mod}\ 8)$$

is a solution, there is a factor of $(x - 1)$ in $x^2 - 1$:

(44) $x^2 - 1 \equiv (x - 1)(x + 1) (\mathrm{poly\ mod}\ 8)$

But now

$$x \equiv 3 (\mathrm{mod}\ 8)$$

is another solution to (43), in spite of the fact that $x + 1$ does not have a factor of $x - 3 (\mathrm{mod}\ 8)$. If we substitute $x = 3$ into (44), we find the trouble,

$$0 \equiv 2 \cdot 4 (\mathrm{mod}\ 8).$$

The proof above that two roots of an equation correspond to two factors depends on the fact that if the product of two numbers is zero, then one of the numbers is zero. This does not always happen for congruences, as is seen above. Arithmetic (mod n) does not behave sufficiently like ordinary arithmetic to enable us to show that two distinct roots of a congruence equation correspond to two distinct factors of the polynomial.[5]

[5] The question at issue is whether the polynomial can have both factors simultaneously. For example, $x^2 - 1$ does have a factor of $(x - 1)$ (mod 8), and it also has a factor of $(x - 3)(\mathrm{mod}\ 8)$:

$$x^2 - 1 \equiv (x - 3)(x + 3) (\mathrm{poly\ mod}\ 8),$$

but $x^2 - 1$ does not have both $(x - 1)$ and $(x - 3)$ as factors simultaneously.

There is an instance that the "product of two numbers is zero implies one of them is zero" theorem carries over to congruences. If p is a prime and

$$a_1 a_2 \cdots a_k \equiv 0 (\text{mod } p),$$

then for some j,

$$a_j \equiv 0 (\text{mod } p).$$

This follows immediately from the definition of congruence and Theorem 2.8. Thus when we have a polynomial congruence (mod p), there is hope that we can proceed as with equalities. This is, in fact, true, and for the rest of this section we will restrict ourselves to prime moduli.

Theorem 3.21. If $f(x)$ is a polynomial with integral coefficients and if $a_1, a_2, \ldots, a_k$ are pairwise incongruent integers (mod p) (where p is a prime) which are solutions of the congruence equation

$$f(x) \equiv 0 (\text{mod } p),$$

then there is a polynomial $g(x)$ with integral coefficients such that

$$f(x) \equiv (x - a_1)(x - a_2) \cdots (x - a_k) g(x) (\text{poly mod } p).$$

Proof. By Theorem 3.20, there is a polynomial $g_1(x)$ with integral coefficients such that

$$f(x) \equiv (x - a_1) g_1(x) (\text{poly mod } p).$$

We now proceed to give a proof by induction. Suppose that there is a polynomial $g_j(x)$ with integral coefficients such that

(45) $f(x) \equiv (x - a_1)(x - a_2) \cdots (x - a_j) g_j(x) (\text{poly mod } p).$

(As we have just seen, this is true for $j = 1$.) Then

$$0 \equiv f(a_{j+1}) \equiv (a_{j+1} - a_1)(a_{j+1} - a_2) \cdots (a_{j+1} - a_j) g_j(a_{j+1}) (\text{mod } p).$$

By hypothesis, none of the numbers $a_{j+1} - a_1, a_{j+1} - a_2, \ldots, a_{j+1} - a_j$ is congruent to 0(mod p) and therefore

$$g_j(a_{j+1}) \equiv 0 (\text{mod } p).$$

Thus, by Theorem 3.20, there is a polynomial $g_{j+1}(x)$ with integral coefficients such that

$$g_j(x) \equiv (x - a_{j+1}) g_{j+1}(x) (\text{poly mod } p).$$

By Theorem 3.19, we may insert this into (45) and get

$$f(x) \equiv (x - a_1)(x - a_2) \cdots (x - a_{j+1}) g_{j+1}(x) (\text{poly mod } p).$$

This completes the inductive step. Since (45) is possible for $j = 1$, it is possible for $j = 2$ and then $j = 3, \ldots$, and finally, $j = k$.   ▲

Thus far, we have said nothing about the degree of a polynomial. It will be necessary to use this concept in the near future.

Definition. Let $f(x) = a_k x^k + a_{k-1} x^{k-1} + \cdots + a_1 x + a_0$ be a polynomial with integral coefficients. We define the **degree (mod n)** of $f(x)$ to be the largest number d such that

$$a_d \not\equiv 0 (\bmod n).$$

If $d < k$ this means that

$$a_{d+1} \equiv a_{d+2} \equiv \cdots \equiv a_k \equiv 0 (\bmod n).$$

It is possible that every coefficient of $f(x)$ is congruent to $0 (\bmod n)$, and then we say that the degree (mod n) of f is **undefined.**

Thus the degree (mod n) of a polynomial $f(x)$ is undefined if and only if

$$f(x) \equiv 0 (\text{poly mod } n).$$

For example, if

$$f(x) = 30x^4 - 60x^3 + 12x^2 - 6x + 3,$$

then

the degree (mod 12) of $f(x) = 4$,

the degree (mod 7) of $f(x) = 4$,

the degree (mod 5) of $f(x) = 2$,

the degree (mod 6) of $f(x) = 0$,

the degree (mod 3) of $f(x)$ is undefined.

Theorem 3.22. Let p be a prime and let $f(x)$ be a polynomial with integral coefficients and let the degree (mod p) of $f(x)$ be defined and equal to n. Then the equation

(46) $$f(x) \equiv 0 (\bmod p)$$

has at most n distinct roots (mod p).

Proof. Suppose the integers $a_1, a_2, \ldots, a_{n+1}$ are pairwise incongruent (mod p) and are also solutions to (46). By Theorem 3.21, there is a polynomial

$g(x)$ with integral coefficients such that

(47) $f(x) \equiv (x - a_1)(x - a_2) \cdots (x - a_{n+1})g(x)(\text{poly mod } p)$.

It is clear from the definition of degree (mod p) that two polynomials, which are congruent as polynomials (mod p), have the same degree (mod p). Thus, since the degree (mod p) of $f(x)$ is n, the degree (mod p) of the right side of (47) is defined and equal to n. If the degree (mod p) of g is not defined, then

$$g(x) \equiv 0(\text{poly mod } p),$$

and it follows from (47) that

$$f(x) \equiv 0(\text{poly mod } p).$$

This is false, since it would imply that the degree (mod p) of $f(x)$ is undefined, whereas it is part of the hypothesis that the degree (mod p) of f is defined. Thus the degree (mod p) of $g(x)$ is also defined. Let k be the degree (mod p) of $g(x)$. By the definition of degree (mod p),

$$k \geq 0.$$

Let b_j be the coefficient of x^j in $g(x)$. Thus

$$b_k \not\equiv 0(\text{mod } p).$$

It is possible that there are powers of x higher than the kth in $g(x)$, but all their coefficients are congruent to $0(\text{mod } p)$ by the definition of k. Therefore,

$$g(x) \equiv b_k x^k + b_{k-1} x^{k-1} + \cdots + b_0(\text{poly mod } p)$$

and thus

(48) $(x - a_1)(x - a_2) \cdots (x - a_{n+1})g(x) \equiv (x - a_1)(x - a_2) \cdots$
$$\cdots (x - a_{n+1})(b_k x^k + \cdots + b_0)(\text{poly mod } p).$$

Hence the right side of (48) also has n as its degree (mod p). But this is absurd, since the coefficient of x^{n+1+k} on the right side is

$$b_k \not\equiv 0(\text{mod } p).$$

This contradicts our supposition that (46) can have $n + 1$ distinct roots (mod p). It follows that (46) cannot have more than $n + 1$ distinct roots (mod p) and hence must have at most n distinct roots (mod p).   ▲

As the example

$$x^2 - 1 \equiv 0(\text{mod } 8)$$

[with distinct roots (mod 8) of 1, 3, 5, 7] shows, the restriction to primes in Theorem 3.22 cannot be eliminated.

Theorem 3.23. Let p be a prime, $f(x)$ be a polynomial with integral co-efficients and degree (mod p) defined and equal to n. Suppose that b is the coefficient of x^n in $f(x)$ and suppose that the n integers $a_1, a_2, \ldots, a_n$ are distinct (mod p) solutions to the equation

$$f(x) \equiv 0 \pmod{p}.$$

Then

$$f(x) \equiv b(x - a_1)(x - a_2) \cdots (x - a_n)(\text{poly mod } p).$$

Proof. We could prove this substantially the same as we proved Theorem 3.22. But instead, we will illustrate a typical application of Theorem 3.22 in this proof. Let

$$g(x) = b(x - a_1)(x - a_2) \cdots (x - a_n),$$

$$h(x) = f(x) - g(x).$$

We will use Theorem 3.22 to show that

$$h(x) \equiv 0(\text{poly mod } p),$$

and this is obviously equivalent to the result of the theorem. Clearly $g(a_j) = 0$ for each a_j and by hypothesis $f(a_j) \equiv 0 \pmod{p}$ for each a_j. Therefore, the distinct (mod p) integers $a_1, a_2, \ldots, a_n$ are solutions to the congruence equation

$$h(x) \equiv 0 \pmod{p}.$$

Thus by Theorem 3.22, either the degree (mod p) of $h(x)$ is defined and greater than or equal to n or the degree (mod p) of $h(x)$ is undefined. But $g(x)$ has no terms involving powers of x higher than the nth power and hence, if $k > n$, the coefficient of x^k in $h(x)$ is the same as the coefficient of x^k in $f(x)$, which is congruent to $0 \pmod{p}$ by the definition of n. Also, by the definition of b, the coefficient of x^n in $h(x)$ is 0. Hence the degree (mod p) of $h(x)$ cannot possibly be defined and greater than or equal to n. Thus, by Theorem 3.22, the congruence

$$h(x) \equiv 0 \pmod{p}$$

has too many solutions to allow the degree (mod p) of $h(x)$ to be defined. Therefore, the degree (mod p) of $h(x)$ is not defined and therefore

$$h(x) \equiv 0(\text{poly mod } p). \qquad \blacktriangle$$

We illustrate the application of Theorem 3.23 by proving the result in (42). It is more convenient to derive this from the factorization of $x^{p-1} - 1$ than from (41).

Theorem 3.24 (Wilson's Theorem). If p is a prime, then

$$(p - 1)! \equiv -1(\text{mod } p).$$

Proof. By Fermat's theorem (3.14), the $p - 1$ pairwise incongruent (mod p) numbers $-1, -2, -3, \ldots, -(p - 1)$ satisfy the congruence

$$x^{p-1} - 1 \equiv 0(\text{mod } p).$$

Clearly the degree (mod p) of $x^{p-1} - 1$ is $p - 1$ and the coefficient of x^{p-1} in $x^{p-1} - 1$ is 1. Therefore, by Theorem 3.23,

(49) $\qquad x^{p-1} - 1 \equiv (x + 1)(x + 2) \cdots (x + p - 1)(\text{poly mod } p).$

Hence the constant terms (the coefficient of x^0) on both sides are congruent,

$$-1 \equiv 1 \cdot 2 \cdot 3 \cdots (p - 1)(\text{mod } p). \qquad \blacktriangle$$

EXERCISES
1. Factor $x^2 - 3x - 3$ into linear factors (poly mod 5).
2. Factor $2x^2 - 3x - 2$ into linear factors (poly mod 7).
3. Factor $x^2 + 1$ into linear factors (poly mod 13).
4. Factor $x^2 + 1$ into linear factors (poly mod 17).
5. Factor $x^3 + x^2 + 4x + 1$ into linear factors (poly mod 7).
6. Factor $x^3 + 4x^2 + 3x + 6$ into linear factors (poly mod 7).
7. Is it allowable to divide both sides of (40) by x and then substitute $x = 0$ in the resulting congruence to prove Theorem 3.24? Explain.
8. Use the fact that

$$p - j \equiv -j(\text{mod } p)$$

to show that if p is an odd prime, $p = 2k + 1$, then

$$(p - 1)! \equiv (-1)^k \left[\left(\frac{p - 1}{2} \right)! \right]^2 (\text{mod } p).$$

9. Use the result of problem 8 to show that if p is a prime, $p \equiv 1(\text{mod } 4)$, then $[(p - 1)/2]!$ is a solution to the congruence equation

$$x^2 + 1 \equiv 0(\text{mod } p).$$

10. If $n > 4$ is a composite number, show that $n|(n - 1)!$ Conclude that

$$(n - 1)! \not\equiv -1(\text{mod } n).$$

(This shows that Wilson's theorem can be used as a proof of primality. It is unfortunately not practical for large numbers.)

11. What is the result when the coefficients of x^{p-2} are equated on both sides of (49)? Is your result valid for $p = 2$? Explain.

*12. What is the result when the coefficients of x^{p-3} are equated on both sides of (49)? Separate the cases of $p = 2$, $p = 3$, $p > 3$. Prove the result for $p > 3$ without the use of polynomials. The formulas

$$\sum_{j=1}^{n} j = \frac{n(n+1)}{2}, \qquad \sum_{j=1}^{n} j^2 = \frac{n(n+1)(2n+1)}{6}, \qquad \sum_{j=1}^{n} j^3 = \left[\frac{n(n+1)}{2}\right]^2$$

may be helpful.

13. Find all distinct solutions (mod 5) or show there are none:

$$x^2 + x + 3 \equiv 0 \pmod{5}.$$

14. Find all distinct solutions (mod 5) or show there are none:

$$x^2 + 2x + 3 \equiv 0 \pmod{5}.$$

15. Find all distinct solutions (mod 15) or show there are none:

$$x^2 \equiv 4 \pmod{15}.$$

16. Find all distinct solutions (mod 8) or show there are none:

$$x^2 + x + 4 \equiv 0 \pmod{8}.$$

17. Find all distinct solutions (mod 11) or show there are none:

$$x^3 + 2x^2 + 5x + 6 \equiv 0 \pmod{11}.$$

*18. Use the coefficient of x on both sides of (49) to prove that if $p \geq 3$, then $p|a$, where

$$\frac{a}{b} = 1 + \frac{1}{2} + \frac{1}{3} + \cdots + \frac{1}{p-1}.$$

19. Use the congruence equation $x^2 - 1 \equiv 0 \pmod{p}$ to show that if $(a,p) = 1$, then

$$a^{(p-1)/2} \equiv \pm 1 \pmod{p}.$$

3.7. Primitive Roots

We have seen that if $(a,n) = 1$, then

$$a^{\phi(n)} \equiv 1 \pmod{n}.$$

In this section we investigate the twin problems of finding the smallest positive power such that a to that power is congruent to 1(mod n) and finding those a for which this power is actually $\phi(n)$.

Definition. Suppose that $(a,n) = 1$. We define the **order** of $a \pmod n$ to be the smallest positive integer, call it b, such that

$$a^b \equiv 1 \pmod n$$

and we write

$$b = \text{ord}_n(a).$$

For example, $\text{ord}_{13}(5) = 4$, since

$$5^1 \equiv 5 \pmod{13}, \qquad 5^2 \equiv 25 \equiv -1 \pmod{13},$$

$$5^3 \equiv -5 \pmod{13}, \qquad 5^4 \equiv -25 \equiv 1 \pmod{13}.$$

Thanks to Euler's theorem (3.13), if $(a,n) = 1$, we are guaranteed that there is a power to which a can be raised to be congruent to $1 \pmod n$ and thus there really is such a thing as $\text{ord}_n(a)$. It also follows from Euler's theorem that if $(a,n) = 1$,

$$\text{ord}_n(a) \leq \phi(n).$$

On the other hand, if $(a,n) > 1$, then the equation

(50) $$ax \equiv 1 \pmod n$$

has no solutions, by Theorem 3.6. Thus for $b \geq 1$,

$$a^b \equiv 1 \pmod n$$

is impossible, since it would provide the solution $x = a^{b-1}$ to (50). Thus if $(a,n) > 1$, it is impossible to define the order of $a \pmod n$.

Definition. If $(a,n) = 1$ and $\text{ord}_n(a) = \phi(n)$, then we say that a is a **primitive root** of n.

Unfortunately, the analogous situation in equalities is worded differently: If $\alpha^k = 1$, but $\alpha^m \neq 1$ for $0 < m < k$, then it is said that α is a **primitive kth root of unity**. By analogy, if $\text{ord}_n(a) = \phi(n)$, then

$$a^{\phi(n)} \equiv 1 \pmod n, \qquad \text{but } a^m \not\equiv 1 \pmod n \qquad \text{for } 0 < m < \phi(n),$$

and at the very least we would expect something like "a is a primitive root of unity $\pmod n$." But this is never said. As an example, 3 is a primitive root of 10, since $\phi(10) = 4$ and

$$3^1 \equiv 3 \pmod{10}, \qquad 3^2 \equiv 9 \equiv -1 \pmod{10},$$

$$3^3 \equiv -3 \pmod{10}, \qquad 3^4 \equiv -9 \equiv 1 \pmod{10}.$$

As another example, there are no primitive roots of 8, since $\phi(8) = 4$ and a reduced residue system is given by the numbers 1, 3, 5, and 7 with

$$1^1 \equiv 1(\text{mod } 8), \qquad 3^2 \equiv 1(\text{mod } 8),$$
$$5^2 \equiv 1(\text{mod } 8), \qquad 7^2 \equiv 1(\text{mod } 8).$$

We will shortly show that there are primitive roots mod p if p is a prime. In the meantime, we prove a result that will help us limit the possible values of $\text{ord}_n(a)$.

Theorem 3.25. If $b > 0$, $c > 0$, $d = (b,c)$, and

$$a^b \equiv 1(\text{mod } n), \qquad a^c \equiv 1(\text{mod } n),$$

then

$$a^d \equiv 1(\text{mod } n).$$

Proof. By Theorem 2.2, there are integers r and s such that

$$d = br - cs.$$

This means that for all integers t,

$$d = b(r + ct) - c(s + bt),$$

and since if we take t sufficiently large, both $r + ct$ and $s + bt$ will be positive, we may as well assume that $r > 0$, $s > 0$. But then

$$a^d \equiv 1^s a^d \equiv (a^c)^s a^d \equiv a^{cs+d} \equiv a^{br} \equiv (a^b)^r \equiv 1^r \equiv 1(\text{mod } n). \qquad \blacktriangle$$

We worried about making r and s positive because otherwise the multiplications in the last step would be divisions and we have not even defined $a/b(\text{mod } n)$, let alone prove any results about it. As an example of the last theorem, we show that $x \equiv 1(\text{mod } 29)$ is the only solution to the congruence

(51) $x^{13} \equiv 1(\text{mod } 29).$

Suppose $x = a$ is a solution to (51). Then $(a,29) = 1$. If follows from Fermat's theorem (3.14) that

$$a^{28} \equiv 1(\text{mod } 29).$$

But since $(13,28) = 1$, Theorem 3.25 says that

$$a^1 \equiv 1(\text{mod } 29)$$

also. [It is easily seen that this is a solution to (51).] As another example of Theorem 3.25 at work, we have the result,

Theorem 3.26. If $(a,n) = 1$ and for some $b > 0$,

$$a^b \equiv 1(\text{mod } n),$$

then $\text{ord}_n(a)|b$. In particular, $\text{ord}_n(a)|\phi(n)$. Conversely, if $\text{ord}_n(a)|b$, then

$$a^b \equiv 1(\text{mod } n).$$

Proof. Let

$$d = (\text{ord}_n(a),b).$$

Then

(52) $\qquad\qquad\qquad\qquad d \leq \text{ord}_n(a)$

and also, by Theorem 3.25,

$$a^d \equiv 1 \ (\text{mod } n).$$

But by the definition of $\text{ord}_n(a)$, this means that

$$d \geq \text{ord}_n(a).$$

This, combined with (52), says that

$$\text{ord}_n(a) = d.$$

By the definition of d, $d|b$, and hence $\text{ord}_n(a)|b$. Conversely, if $\text{ord}_n(a)|b$, then for some k,

$$b = k \cdot \text{ord}_n(a)$$

and hence

$$a^b \equiv [a^{\text{ord}_n(a)}]^k \equiv 1(\text{mod } p). \qquad\qquad \blacktriangle$$

Theorem 3.26 lightens the labor when we attempt to find $\text{ord}_n(a)$, since it is now only necessary to check the divisors of $\phi(n)$. As an example, let us find $\text{ord}_{23}(2)$. Since $\phi(23) = 22$, we see that $\text{ord}_{23}(2)$ is one of the four divisors of $22: 1, 2, 11, 22$. The numbers 1 and 2 are obvious failures. [In fact, $\text{ord}_n(a) = 1$ if and only if $a \equiv 1(\text{mod } n)$.] Thus we need only look at $2^{11}(\text{mod } 23)$. If we make no mistakes, we will come up with $2^{11} \equiv \pm 1(\text{mod } 23)$ (see problem 19 of Section 3.6), but this remark is meant to serve only as a check. The useful thing is that we do not have to look at each power of 2 in getting up to 2^{11}.

$$2^2 \equiv 4(\text{mod } 23),$$

$$2^4 \equiv 4^2 \equiv 16 \equiv -7(\text{mod } 23),$$

$$2^8 \equiv (-7)^2 \equiv 49 \equiv 3(\text{mod } 23),$$

$$2^{10} \equiv 2^8 \cdot 2^2 \equiv 3 \cdot 4 \equiv 12 (\text{mod } 23),$$

$$2^{11} \equiv 2^{10} \cdot 2 \equiv 12 \cdot 2 \equiv 1 (\text{mod } 23).$$

Thus $\text{ord}_{23}(2) = 11$.

At this point we will restrict ourselves to prime moduli. As the example of $n = 8$ showed, not all n have primitive roots. The next theorem will aid us in showing that all primes have primitive roots.

Theorem 3.27. If p is a prime and $d|(p - 1)$, then the congruence equation

$$x^d \equiv 1 (\text{mod } p)$$

has exactly d distinct solutions.

Proof. We already know that the equation

$$x^{p-1} - 1 \equiv 0 (\text{mod } p)$$

has exactly $p - 1$ distinct solutions (mod p) (this is Fermat's theorem, 3.14). Since $d|(p - 1)$, let $kd = p - 1$. Then

$$x^{p-1} - 1 = x^{kd} - 1$$

$$= (x^d - 1)[x^{d(k-1)} + x^{d(k-2)} + x^{d(k-3)} + \cdots + x^d + 1].$$

Therefore, there are $(p - 1)$ distinct solutions to the congruence

(53) $(x^d - 1)[x^{d(k-1)} + x^{d(k-2)} + \cdots + x^d + 1] \equiv 0 (\text{mod } p),$

and since p is a prime, every solution to (53) is a solution of at least one of the two congruences

(54) $x^d - 1 \equiv 0 (\text{mod } p),$

(55) $x^{d(k-1)} + x^{d(k-2)} + \cdots + x^d + 1 \equiv 0 (\text{mod } p).$

By Theorem 3.22, (55) has at most $d(k - 1)$ distinct solutions (mod p). Since there are $p - 1$ distinct numbers (mod p) that satisfy at least one of (54) and (55), and at most $d(k - 1)$ of them satisfy (55), at least $(p - 1) - d(k - 1)$ distinct integers (mod p) satisfy (54); that is, (54) has at least

$$(p - 1) - d(k - 1) = [(p - 1) - dk] + d = d$$

distinct solutions (mod p). By Theorem 3.22, (54) can have no more than d solutions and thus has exactly d solutions. ▲

Theorem 3.28. Let p be a prime and let $d|(p - 1)$. Then there are exactly $\phi(d)$ distinct integers (mod p) whose order (mod p) is d. In particular, there are exactly $\phi(p - 1)$ primitive roots of p.

Proof. Since any solution of the equation

$$x^d \equiv 1 \pmod{p}$$

must be relatively prime to p [recall that the definition of order (mod p) had this condition], we need only show that there are $\phi(d)$ solutions to this equation that do not satisfy similar equations with smaller d. We will give a proof by contradiction (which in this case is a means of avoiding a proof by induction). Let us suppose that the theorem is false. Then there is a smallest d for which the theorem is false and we will let n be that smallest value. Thus if $d < n$ and $d|(p-1)$, there are exactly $\phi(d)$ distinct numbers (mod p) whose order is d. Let us look at the solutions to

(56) $$\qquad\qquad\qquad x^n \equiv 1 \pmod{p}.$$

By Theorem 3.26, $x = a$ is a solution of (56) if and only if $\operatorname{ord}_p(a)|n$. Thus the number of solutions to (56) with order (mod p) less than n is

$$\sum_{\substack{d|n \\ d < n}} (\text{number of distinct integers (mod } p) \text{ whose order (mod } p) \text{ is } d)$$

$$= \sum_{\substack{d|n \\ d < n}} \phi(d)$$

$$\sum_{d|n} \phi(d) - \phi(n)$$

$$= n - \phi(n),$$

by Theorem 3.17. Thus the number of solutions of (56) with order (mod p) equal to n is

$$n - [n - \phi(n)] = \phi(n).$$

This contradicts the definition of n, thereby proving the theorem.   ▲

Shown in Table 2 at the end of the book are the smallest primitive roots of each $p < 500$. As we noted earlier, there are no primitive roots of 8. To satisfy the reader's curiosity, an integer n has primitive roots if and only if n is one of the five categories

$$n = 1, \qquad n = 2, \qquad n = 4, \qquad n = p^k, \qquad n = 2p^k,$$

where p is an odd prime and k is a positive integer. Thus 7^{47}, $2 \cdot 101^2$, and 10 have primitive roots (the last was verified earlier) while 8, 16, 15, and 20 do not. This result is left to miscellaneous exercises 3–7 at the end of the chapter. The result that there are $\phi(p-1)$ primitive roots of p is sometimes written: There are $\phi(\phi(p))$ primitive roots of p. This is the more general form, since it can be shown that if there is a primitive root of n, then there are exactly

$\phi(\phi(n))$ primitive roots of n which are distinct (mod n) (problem 11 at the end of the section).

We conclude the chapter with two applications of the above theorems. Another application to decimal expansions of rational numbers will be given in Chapter 6. As our first example, we investigate the question as to when the equation

$$x^2 + 1 \equiv 0(\text{mod } p)$$

has solutions. This is the equation for equality that resulted in the invention of complex numbers. Here the situation is different, as there are sometimes already integral solutions to the congruence equation.

Theorem 3.29. Let p be a prime. The equation

(57) $$x^2 \equiv -1(\text{mod } p)$$

has solutions if $p = 2$ or if $p \equiv 1(\text{mod } 4)$ but does not have any solutions if $p \equiv 3(\text{mod } 4)$.

Proof. When $p = 2$, $x = 1$ is a solution. Suppose that $p \equiv 1(\text{mod } 4)$. Then $4|(p - 1)$. By Theorem 3.28, there is an integer a such that

$$\text{ord}_p(a) = 4.$$

In particular,

$$(a^2 - 1)(a^2 + 1) \equiv a^4 - 1 \equiv 0(\text{mod } p)$$

and hence either

$$a^2 - 1 \equiv 0(\text{mod } p)$$

or

$$a^2 + 1 \equiv 0(\text{mod } p).$$

The first case cannot happen since $\text{ord}_p(a) = 4$ and therefore

$$a^2 + 1 \equiv 0(\text{mod } p).$$

Hence a is a solution to (57).

Suppose that $p \equiv 3(\text{mod } 4)$. Then

$$2|(p - 1), \qquad \text{but } 4 \nmid (p - 1)$$

and hence

(58) $$(p - 1, 4) = 2$$

Suppose that

(59) $$a^2 \equiv -1(\text{mod } p).$$

Then

(60) $$a^4 \equiv (a^2)^2 \equiv (-1)^2 \equiv 1(\text{mod } p)$$

[it follows that $(a,p) = 1$] and by Fermat's theorem (3.14),

(61) $$a^{p-1} \equiv 1(\text{mod } p)$$

also. Hence, by (58), (60), (61) and Theorem 3.25,

(62) $$a^2 \equiv 1(\text{mod } p).$$

It follows from (59) and (62) that

$$1 \equiv -1(\text{mod } p)$$

and hence

$$2 \equiv 0(\text{mod } p).$$

This is impossible, since it says that an odd prime divides 2. Therefore, (59) is untenable and hence (57) has no solutions when $p \equiv 3 \pmod 4$. ▲

The next theorem is a simple corollary of Theorem 3.29, but it will be used in Chapters 5 and 8.

Theorem 3.30. If p is a prime $\equiv 3(\text{mod } 4)$ and a and b are integers such that

$$a^2 + b^2 \equiv 0(\text{mod } p),$$

then

$$a \equiv b \equiv 0(\text{mod } p).$$

Proof. We first show that $b \equiv 0(\text{mod } p)$. If $b \not\equiv 0(\text{mod } p)$, then there exists an integer c such that

$$bc \equiv 1(\text{mod } p),$$

and then

$$(ac)^2 \equiv a^2 c^2 \equiv -b^2 c^2 \equiv -(bc)^2 \equiv -1(\text{mod } p).$$

But this is impossible by Theorem 3.29. Therefore,

$$b \equiv 0(\text{mod } p)$$

and hence

$$a^2 \equiv a^2 + 0^2 \equiv a^2 + b^2 \equiv 0 \pmod{p}$$

so that $p|a^2$, and thus $p|a$; that is,

$$a \equiv 0 \pmod{p}. \quad \blacktriangle$$

We turn now to a completely different area. Gauss showed that if α is a primitive kth root of unity (that is, $\alpha^k = 1$ but $\alpha^m \neq 1$ if $0 < m < k$), then

$$\left(\sum_{j=0}^{k-1} \alpha^{j^2}\right)^2 = \begin{cases} k & \text{if } k \equiv 1 \pmod 4, \\ 0 & \text{if } k \equiv 2 \pmod 4, \\ -k & \text{if } k \equiv 3 \pmod 4, \\ 2k\alpha^{k/4} & \text{if } k \equiv 4 \pmod 4. \end{cases}$$

This is difficult to check in particular cases because of the nature of α (a complex number with irrational real and imaginary parts when $k > 12$). Owing to the great similarities between the theory of congruences and equalities, we might suspect that a similar result is true for congruences. We have noted one major difference between arithmetic (mod n) and equalities: If n is composite, then 0 is congruent (mod n) to the product of two nonzero (mod n) integers. This made a great difference when we dealt with polynomial equations (mod n). Since Gauss's result certainly deals with polynomial equations, it seems best to investigate Gauss's result for congruences (mod p), where p is a prime. The analogue of a primitive kth root of unity is an integer whose order (mod p) is k. Thus we may conjecture that if p is a prime and

$$\text{ord}_p(a) = k,$$

then

(63) $$\left(\sum_{j=0}^{k-1} a^{j^2}\right)^2 \equiv \begin{cases} k \pmod p & \text{if } k \equiv 1 \pmod 4, \\ 0 \pmod p & \text{if } k \equiv 2 \pmod 4, \\ -k \pmod p & \text{if } k \equiv 3 \pmod 4, \\ 2ka^{k/4} \pmod p & \text{if } k \equiv 4 \pmod 4. \end{cases}$$

Let us check this conjecture for $k = 3$. We cannot choose any prime that comes to mind; by Theorem 3.26,

$$\text{ord}_p(a)|(p-1),$$

in this case, $3|(p-1)$. If $3|(p-1)$, then by Theorem 3.28, there will be integers whose order (mod p) is 3. Let us pick $p = 13$ for our example. The next question is: Can we find a number whose order (mod 13) is 3 without proceeding by trial and error? The answer is yes, if there is a table of primitive roots available. We see from Table two at the end of the book that 2 is a primitive root of 13. Thus

$$2^{12} \equiv 1(\text{mod } 13), \qquad \text{but} \quad 2^n \not\equiv 1(\text{mod } 13) \qquad \text{for} \quad 1 \le n \le 11.$$

It follows that

$$(2^4)^3 \equiv 1(\text{mod } 13), \qquad \text{but} \quad (2^4)^m \not\equiv 1(\text{mod } 13) \qquad \text{for} \quad m = 1, 2.$$

Thus by definition,

$$\text{ord}_{13}(16) = 3.$$

For our calculations, it will be best to replace 16 by the congruent number $3(\text{mod } 13)$ and thus we let $a = 3$. Then

$$
\begin{aligned}
\left(\sum_{j=0}^{2} 3^{j^2} \right)^2 &\equiv (3^0 + 3^1 + 3^4)^2 \ (\text{mod } 13) \\
&\equiv (1 + 3 + 3^3 \cdot 3)^2 (\text{mod } 13) \\
&\equiv (1 + 3 + 3)^2 \qquad (\text{mod } 13) \\
&\equiv 7^2 \qquad\qquad\quad (\text{mod } 13) \\
&\equiv 49 - 4 \cdot 13 \qquad (\text{mod } 13) \\
&\equiv -3 \qquad\qquad\quad (\text{mod } 13),
\end{aligned}
$$

as predicted by (63).

    This example is instructive for two reasons. First it has illustrated several of the theorems of this chapter at work in a numerical example. Second, the conjecture itself (assuming it is true) is likely to have a proof that parallels the proof of Gauss's result for equalities. Here, as in much of this chapter, we have taken proofs from results on equalities and adapted them to congruences. In other circumstances, we might have to readapt these proofs once again. This illustrates the great advantage of the modern axiomatic method. Starting from a certain set of axioms [in this case, the so-called field axioms, satisfied by the rational numbers, the real numbers, the complex numbers, congruences (mod p), and other systems] one derives certain theorems. The resulting theorems are then true for everything that satisfies the axioms, which results in a great saving of needless duplication of proofs.

EXERCISES

1. Did this section prove the conjecture in (63) for $k = 3$?
2. How many distinct solutions (mod 102) are there to the equation

$$x^{85} \equiv 1(\text{mod } 102)?$$

3. Suppose that $(a,n) = 1$. Prove that

$$a^b \equiv a^c (\text{mod } n)$$

if and only if

$$b \equiv c(\text{mod ord}_n(a)).$$

4. Show that if $(a,15) = 1$, then

$$a^{\phi(15)/2} \equiv 1(\text{mod } 15)$$

and hence 15 has no primitive roots. [*Hint:* Examine the congruence (mod 3) and (mod 5).]
5. Show that 21 has no primitive roots (see problem 4).
6. Show that 35 has no primitive roots (see problem 4).
7. Show that if g is a primitive root of n, then the numbers

$$g, g^2, g^3, \ldots, g^{\phi(n)}$$

form a reduced residue system (mod n).
8. Show that if p is a prime, $p \equiv 1(\text{mod } 4)$ and g is a primitive root of p, then $g^{(p-1)/4}$ is a solution to the equation

$$x^2 \equiv -1(\text{mod } p).$$

9. There are four solutions to the equation

$$x^2 + 1 \equiv 0(\text{mod } 65).$$

Find them by solving this equation (mod 5) and (mod 13) and then using the Chinese remainder theorem.
10. Given that 3 is a primitive root of 31, show that $3^5, 3^{10}, 3^{15}, 3^{20}, 3^{25}$, and 3^{30} are the six distinct roots of the equation

$$x^6 \equiv 1(\text{mod } 31).$$

Since

$$x^6 - 1 = (x^3 - 1)(x^3 + 1) = (x - 1)(x^2 + x + 1)(x + 1)(x^2 - x + 1),$$

the six numbers above satisfy the equation

$$(x - 1)(x^2 + x + 1)(x + 1)(x^2 - x + 1) \equiv 0(\text{mod } 31).$$

Which solution goes with which factor? (Do this without substituting the solutions into the factors, if possible.)
11. Show that if n has a primitive root, then n has exactly $\phi(\phi(n))$ primitive roots. (*Hint:* Use the result of problem 7 and decide which powers of g give the primitive roots of n.)

12. Find all solutions to the equation

$$x^{10} \equiv 1 (\bmod\ 14).$$

13. Verify the conjecture in (63) for $k = 5$ and whatever values of p and a that you find convenient.
14. Repeat problem 13 with $k = 6$.
15. Repeat problem 13 with $k = 7$.
16. Repeat problem 13 with $k = 8$.

MISCELLANEOUS EXERCISES

1. Show that if p and q are different odd primes, and if $(a,pq) = 1$, then

$$a^{\phi(pq)/2} \equiv 1 (\bmod\ pq).$$

2. Show that if $n > 2$, then $2 | \phi(n)$.
3. Use the ideas in problems 1 and 2 to show that if $n = ab$, where $a > 2$, $b > 2$, $(a,b) = 1$, then n has no primitive roots. Show that the only numbers which can possibly have primitive roots are those of the form $1, 2, 4, p^k$, and $2p^k$, where p is an odd prime.
4. Suppose p is an odd prime, k and g are positive integers. Use the binomial theorem to show that

$$(g + p)^{\phi(p^k)} \equiv g^{\phi(p^k)} - p^k g^{\phi(p^k)-1}(\bmod\ p^{k+1}).$$

This result is false if $p = 2$ (try $g = 1$, $k = 2$). Where does your proof use the fact that p is odd?
5. Show that if p is an odd prime and n is a positive integer then there is a primitive root of p^n. [*Hint:* Suppose g is a primitive root of p^k. Use problem 4 to show that either g or $g + p$ (or both) is a primitive root of p^{k+1}.]
6. Show that if p is an odd prime, $n > 0$, then there is a primitive root of $2p^n$. [*Hint:* Let g be a primitive root of p^n (such a number exists by problem 5). Show that either g or $g + p^n$ is a primitive root of $2p^n$.]
7. Show that if g is a primitive root of p^2, then g is a primitive root of p^n for all $n \geq 2$. [*Hint:* By problem 5, there are primitive roots of p^n and problem 11, page 107, there are exactly $\phi(\phi(p^n))$ such primitive roots. Investigate how these are related to the $\phi(\phi(p^2))$ primitive roots of p^2.]
8. The Fermat numbers are defined as

$$F_n = 2^{2^n} + 1.$$

Raise the congruence

$$2^{2^n} \equiv -1 (\bmod\ F_n)$$

to the (2^{2^n-n})th power and use your result to show that every F_n is either a prime or a pseudoprime. The Polish astronomer Banachiewicz has conjectured that Fermat knew this fact and that this is one of the reasons Fermat conjectured that every F_n is prime after having verified only that F_0, F_1, F_2, F_3, and F_4 are primes (recall that in Fermat's time, the Chinese conjecture that there is no such thing as a pseudoprime was still believed). Thus when Euler showed in 1732 that F_5 was not a prime, he had actually produced a pseudoprime 87 years before the first announced example of one.

9. Let

$$\Gamma_n = 2^{2^n} + 1$$

and suppose that $p|F_n$, where p is a prime (possibly F_n itself). Show that

$$2^{2^{n+1}} \equiv 1(\text{mod } p)$$

so that $\text{ord}_p(2)|2^{n+1}$. Use this to show that

$$\text{ord}_p(2) = 2^{n+1}$$

[first show that $\text{ord}_p(2)$ is a power of 2]. Since $\text{ord}_p(2)|(p-1)$, show that there is an integer k such that

$$p = k \cdot 2^{n+1} + 1.$$

In the next two problems, we will see that if $n > 1$, then k is even.

10. Suppose $p = 8m + 1$ is a prime. Show that

$$2^{4m} \cdot (4m)! = 2 \cdot 4 \cdot 6 \cdots 4m[p - (4m - 1)][p - (4m - 3)] \cdots [p - 1]$$

and use this to show that

$$
\begin{aligned}
2 \cdot 4 \cdot 6 \cdots 8m &\equiv 2 \cdot 4 \cdot 6 \cdots 4m[-(4m - 1)][-(4m - 3)] \cdots [-1](\text{mod } p) \\
&\equiv (-1)^{2m} 2 \cdot 4 \cdot 6 \cdots 4m(4m - 1)(4m - 3) \cdots (1)(\text{mod } p) \\
&\equiv (4m)!(\text{mod } p).
\end{aligned}
$$

Use this to prove that

$$2^{(p-1)/2} \equiv 1(\text{mod } p).$$

11. Use the results of problems 9 and 10 to show that if $n > 1$ and p is a prime divisor of F_n, then

$$\text{ord}_p(2)\left|\left(\frac{p-1}{2}\right)\right.$$

and thus there is an integer t such that

$$p = t \cdot 2^{n+2} + 1.$$

As an example, if $n = 5$, then any prime divisor of F_5 must be of the form

$$p = 128t + 1.$$

When $t = 5$, we get the prime $p = 641$, which does divide F_5:

$$F_5 = 641 \cdot 6\,700\,417.$$

This was done in 1732 by Euler, who, in addition, announced that $6\,700\,417$ was a prime so that F_5 could be factored no further. If p is a prime divisor of $6\,700\,417$, show that there is an integer t such that

$$p = 128t + 1.$$

Given that $(128 \cdot 21 + 1)^2 > 6\,700\,417$, show that if $6\,700\,417$ is composite, then it is divisible by a prime among one of the twenty numbers

$$128t + 1, \qquad 1 \le t \le 20.$$

The primes in this list are $257(t = 2)$, $641(t = 5)$, $769(t = 6)$, $1153(t = 9)$, and $1409(t = 11)$, none of which divide $6\,700\,417$. Thus $6\,700\,417$ is a prime.

12. Show that every integer of the form

$$4 \cdot 14^k + 1, \qquad k \ge 1$$

is composite. (*Hint:* Show that there is a factor of 3 when k is odd and a factor of 5 when k is even.)

13. Show that every integer of the form

$$521 \cdot 12^k + 1, \qquad k \ge 1,$$

is composite. [*Hint:* Show that there is a factor of 13 when k is odd, a factor of 5 when $k \equiv 2(\mathrm{mod}\ 4)$, and a factor of 29 when $4|k$.]

14. Show that every integer of the form $a \cdot 2^k + 1$, where $k \ge 1$ and

$$a = 2\,935\,363\,331\,541\,925\,531,$$

is composite. You may assume that

$$a \equiv 1(\mathrm{mod}\ F_0 F_1 F_2 F_3 F_4 p_1), \qquad a \equiv -1(\mathrm{mod}\ p_2),$$

where

$$p_1 = 6\,700\,417, \qquad p_2 = 641,$$

and

$$p_1 p_2 = F_5.$$

[*Hint:* Consider separately the cases $k \equiv 1(\mathrm{mod}\ 2)$, $k \equiv 2(\mathrm{mod}\ 4)$, $k \equiv 4(\mathrm{mod}\ 8)$, $k \equiv 8(\mathrm{mod}\ 16)$, $k \equiv 16(\mathrm{mod}\ 32)$, $k \equiv 32(\mathrm{mod}\ 64)$, and $k \equiv 64(\mathrm{mod}\ 64)$.] This result was first proved in 1960 by Sierpinski, who compared it with the unsolved problem of whether or not there are infinitely many primes of the form $1 \cdot 2^k + 1$ (when $k = 2^n$ we have F_n). Sierpinski also noted that if $1 \le a \le 100$, then there is at least one prime of the form $a \cdot 2^k + 1$.

15. The Mersenne numbers, M_m, defined by

$$M_m = 2^m - 1$$

have been well known since 1644 when Mersenne made an incorrect conjecture on the primality (or lack of primality) of all M_m with $m \le 257$. Show that if m is a prime, then M_m is either a prime or a pseudoprime. Show further that if m is a pseudoprime, then M_m is a pseudoprime. Conclude that there are infinitely many pseudoprimes. The number $M_{11} = 23 \cdot 89$ is a pseudoprime and thus there was a number well known to Fermat available as an example of a pseudoprime 175 years before the first announced example of one.

16. Suppose p is a prime $\equiv 1(\mathrm{mod}\ 8)$. Use the result of problem 10 to show that 2 is congruent $(\mathrm{mod}\ p)$ to an even power of a primitive root of p and hence show that the equation

$$x^2 \equiv 2(\mathrm{mod}\ p)$$

has solutions.

17. Suppose that p is a prime and $(a,p) = 1$. Show that the equation

$$x^2 \equiv a(\mathrm{mod}\ p)$$

has solutions if

$$a^{(p-1)/2} \equiv 1(\mathrm{mod}\ p)$$

and does not have solutions if

$$a^{(p-1)/2} \equiv -1(\mathrm{mod}\ p).$$

18. Determine whether or not 945 827 is a prime, given that

$$a \equiv 149\ 762(\mathrm{mod}\ 945\ 827),$$

where a is the product of all the primes less than 1000.

19. Show that if p is a prime and $\text{ord}_p(a) = 3$, then

$$\left(\sum_{j=0}^{2} a^{j^2} \right)^2 \equiv -3 \pmod{p}.$$

20. Show that if p is a prime and $\text{ord}_p(a) = 4$, then

$$\left(\sum_{j=0}^{3} a^{j^2} \right)^2 \equiv 8a \pmod{p}.$$

21. Show that if p is a prime and $\text{ord}_p(a) = 6$, then

$$\sum_{j=0}^{5} a^{j^2} \equiv 0 \pmod{p}.$$

22. Show that for all integers a and b,

$$ab(a^2 - b^2)(a^2 + b^2)$$

is divisible by 30. When showing that 2, 3, or 5 divides this number, do not break the problem up into cases (such as, for example, case 1 : one of a and b even; case 2 : both of a and b odd).

23. Show that

$$\lim_{k \to \infty} \frac{n^{k+1}}{\sigma(n^k)} = \phi(n)$$

and use this equation to prove that $\phi(n)$ is multiplicative. (Of course, the way we have done things in this chapter, this equation could not be derived without the knowledge that $\phi(n)$ is multiplicative.)

24. The Möbius function is defined as

$$\mu(n) = \begin{cases} 1 & \text{if } n = 1, \\ (-1)^k & \text{if } n = p_1^{a_1} p_2^{a_2} \cdots p_k^{a_k}, \quad a_1 = a_2 = \cdots = a_k = 1, \\ 0 & \text{if } n = p_1^{a_1} p_2^{a_2} \cdots p_k^{a_k}, \quad \text{some } a_j \geq 2. \end{cases}$$

Show that $\mu(n)$ is multiplicative and that

$$\sum_{d|n} \mu(d) = \begin{cases} 1 & \text{if } n = 1, \\ 0 & \text{if } n > 1. \end{cases}$$

25. Since $d|(j, n)$ if and only if $d|j$ and $d|n$, use the result of problem 24 to show that

$$\phi(n) = \sum_{j=1}^{n} \sum_{\substack{d|j \\ d|n}} \mu(d) = \sum_{d|n} \sum_{\substack{j=1 \\ d|j}}^{n} \mu(d) = \sum_{d|n} \frac{n}{d} \mu(d) = n \sum_{d|n} \frac{\mu(d)}{d}$$

and use this to show that $\phi(n)$ is multiplicative. Show also that this result gives $\phi(p^a)$ when p is a prime and $a \geq 1$.

26. There is a general connection between Theorem 3.17 and the result of problem 25, which is known as the Möbius inversion formula. Use the result of problem 24 to show that

$$f(n) = \sum_{d|n} g(d)$$

for all n if and only if

$$g(n) = \sum_{d|n} f(n/d)\mu(d)$$

for all n.

27. After problem 26, the result of problem 25 follows immediately from Theorem 3.17. Prove Theorem 3.17 directly by showing that

$$n = \sum_{d|n} \sum_{\substack{j=1 \\ (j,n)=d}}^{n} 1 = \sum_{d|n} \phi(n/d).$$

28. From a given date we may calculate how many days have passed (or will pass) between then and today. This number of days (mod 7) will tell us the day of the week of the given date (assuming we know what day it is). This principle allows us to determine the day of the week of any date in history. Let the days of the week be represented by the numbers 0, 1, 2, 3, 4, 5, 6 (Sunday being 0 and Saturday being 6). Suppose that the 0th of a given month and year (that is, the last day of the previous month) falls on the weekday M. Show that the dth day of the month falls on the day w of the week given by

$$w \equiv M + d \pmod 7.$$

We can find M from the weekday Y which represents the 0th day of the year (that is, the last day of the previous year). Unfortunately, for March and later months, M will depend not only on Y but also on whether or not the year involved is a leap year. Most perpetual calendars get around this problem by defining the beginning of the year to be March 1. Thus in calculating the day of the week of February 28, 1970, we would use the year 1969 in our calculations. From this point on, we assume that January and February belong to the year containing the previous December. Thus Y is the day of the week of March 0 [that is, February 28 or 29, according as to whether the year before (ending in February) is a normal or a leap year].

Day	w		Month	m
Sunday	0		March	0
Monday	1		April	3
Tuesday	2		May	5
Wednesday	3		June	1
Thursday	4		July	3
Friday	5		August	6
Saturday	6		September	2
			October	4
			November	0
			December	2
			January	5
			February	1

Show that

$$M \equiv Y + m(\text{mod } 7),$$

where m is given in the accompanying table (30 days hath September, April, June, and so on). Let y be the last two digits in the year (for example, in January 1900, $y = 99$; in April 1914, $y = 14$). Let c be the weekday of March 0th of the first year of the century containing the date in question. Show that, (mod 7), a normal year has one day and a leap year two days and hence prove that

$$Y \equiv c + y + \left[\frac{y}{4}\right](\text{mod } 7),$$

where $[y/4]$ indicates that any remainder in $y/4$ ($\frac{1}{4}, \frac{2}{4}, \frac{3}{4}$) is thrown out (it is useful to note that the right-hand side usually increases by 1, but every fourth y, starting with 4, the right-hand side increases by 2). Combining all the equations, we have

$$w \equiv m + d + y + \left[\frac{y}{4}\right] + c(\text{mod } 7).$$

For example, on January 19, 1944, we have

$$w \equiv 5 + 19 + 43 + \left[\frac{43}{4}\right] + c(\text{mod } 7)$$

$$\equiv c(\text{mod } 7),$$

and assuming that in 1900, $c = 3$, we see that January 19, 1944 is a Wednesday. Use today's date and day to show that in 1900, $c = 3$.

Century	c	
1500	3	
1600	2	
1700	0	
1800	5	
1900	3	periodic
2000	2	
2100	0	
2200	5	

Since March 0, 1900 ($w = c = 3$) was the same day as February 28, 1900 ($m = 1, d = 28, y = 99$), show that in 1800, $c = 5$. Verify the values given of c in 1500, 1600, 1700, 2000, 2100, and 2200 (1600 and 2000 are the only leap years among these; a year divisible by 100 is a leap year only when it is divisible by 400).

29. Our present calendar, the Gregorian calendar, was introduced by Pope Gregory XIII in 1582 to correct a slight error in the Julian calendar (introduced by Julius Caesar in 46 B.C.) which was gradually accumulating into a significant error. The Julian calendar is the same as the Gregorian calendar, except that every year (such as 1900) divisible by 100 is a leap year. Thus the Julian calendar has three extra days every four centuries. In 1582, the Julian calendar was in error by 10 days; thus October 5, 1582 (Julian calendar) was converted to October 15, 1582 (Gregorian calendar). In the notation of the previous problem, show that the Julian calendar c of 1500 is $c = 6$. Show that in the Julian calendar, c decreases by one

Century	c	
1300	1	
1400	0	
1500	6	
1600	5	
1700	4	periodic
1800	3	
1900	2	
2000	1	

(mod 7) each century. The Gregorian calendar was adopted in 1582 by France and Spain, but England and her American colonies waited until 1752 to adopt it and Russia did not adopt it until after the revolution in 1917.

30. On what days of the week did the following events occur (see problems 28 and 29)?

August 6, 258 (A.D.)   (Julian calendar). The martyrdom of Pope Sixtus II.

August 6, 1637   (Julian calendar). Ben Jonson, dramatist, died.

August 6, 1644   Louise de la Valliere, mistress of Louis XIV, born.

August 6, 1660   Diego Velasquez, Spanish painter, died.

August 6, 1759   Eugene Aram, English scholar and murderer, hanged.

August 6, 1811   Peter Barlow readies for print his book, *An Elementary Investigation of the Theory of Numbers with Its Applications to the Indeterminant and Diophantine Analysis, the Analytical and Geometrical Division of the Circle and Several Other Curious Algebraical and Arithmetical Problems*, containing a proof of Fermat's last theorem which depends on an incorrect corollary on page 20.

August 6, 1848   *H.M.S. Daedalus* sights a sea serpent.

August 6, 1890   First successful operation of an electric chair, State Prison, Auburn, New York.

August 6, 1930   Judge Crater, justice on the New York Supreme Court, disappears.

August 6, 1939   The author's birthday.

August 6, 1945   The first wartime use of an atomic bomb, Hiroshima.

August 6, 1966   The President's daughter's marriage on an inauspicious day.

31. Show that if $f(x)$ is a polynomial with integral coefficients and

$$f(a) \equiv f'(a) \equiv f''(a) \equiv \cdots \equiv f^{(r)}(a) \equiv 0 (\text{mod } n),$$

$$(r!,n) = 1,$$

then there is a polynomial $g(x)$ with integral coefficients and degree (mod n) equaling the degree (mod n) of $f(x)$ minus $(r + 1)$ such that

$$f(x) \equiv (x - a)^{r+1} g(x)(\text{poly mod } n).$$

Apply this result with $a = 3$ to the polynomial equation

$$x^4 + 6x^3 + 2x^2 + 4x + 2 \equiv 0 (\text{mod } 11).$$

32. Suppose that $(r!,n) = 1$ and let the numbers $a_0, a_1, \ldots, a_r$ be defined by the equations

$$(j!)a_j \equiv 1(\text{mod } n), \qquad j = 0, 1, \ldots, r.$$

Suppose that $f(x)$ is a polynomial with integral coefficients and degree (mod n) equal to r. Show that if a is an integer, then

$$f(x) \equiv \sum_{j=0}^{r} a_j f^{(j)}(a)(x - a)^j (\text{poly mod } n).$$

There are times that this problem overlaps problem 31. Apply the result here with $a = 3$ to the polynomial of problem 31.

Chapter 4

MAGIC SQUARES

4.1. The Uniform Step Method

An $n \times n$ square with n^2 entries is said to be a **magic square** if the sum of the entries of any row or column is always the same. This common sum is called the **magic sum**. Figure 4.1 is a famous example of a 4×4 magic square. In 1693 De la Loubère gave a method (illustrated in Chapter 1) of finding magic squares for any odd n. In 1929 D. N. Lehmer investigated by means of congruences a generalization of Loubère's method called the uniform step method. Both Loubère and Lehmer gave a rule for placing the b^2 consecutive integers $1, 2, \ldots, n^2$ in an $n \times n$ square so as to make it magic. We will find it convenient to use the numbers $0, 1, 2, \ldots, n^2 - 1$; this simply involves subtracting 1 from every entry in the Loubère or Lehmer squares and thus subtracting n from each row and column sum.

Each of the n^2 possible locations for a number in our square will be called a **cell**. We may give the cells coordinates as illustrated in Figure 4.2. We let x_j and y_j be the X and Y coordinates of the cell containing j. For example, the cell containing 0 has coordinates (4,3) and the cell containing 7 has coordinates (2,5) and thus $(x_0, y_0) = (4,3)$, $(x_7, y_7) = (2,5)$. As another example,

$$y_1 = y_7 = y_{13} = y_{19} = y_{20} = 5.$$

We first develop the equations of the uniform step method for a particular example. We consider a 5×5 square. We place 0 in any cell whatsoever, say $(x_0, y_0) = (4,3)$. We no longer restrict ourselves to going one to the right and one up for the next entry. For our example this time, we go over one to the right and up two for consecutive entries. Thus

$$x_j = 4 + j, \qquad y_j = 3 + 2j.$$

As with the Loubère method, we soon leave our square if we follow these equations. The cure in the Loubère method was to subtract 5 as often as

118

16	3	2	13
5	10	11	8
9	6	7	12
4	15	14	1

Figure 4.1. This square is in the engraving "Melancolia" by Albrecht Dürer (1514). The sum of the four numbers in any row or column is 34, as is the sum of the four numbers on either of the two diagonals or the central four numbers or the corner four and others. Also the sum of two symmetrically placed numbers about the center is 17.

necessary so as to stay in the original square. In other words, x_j and y_j are given by congruences (mod 5) rather than equalities:

(1) $$x_j \equiv 4 + j \pmod 5, \qquad y_j \equiv 3 + 2j \pmod 5.$$

These congruences are enough to determine x_j and y_j in the range from 1 to 5, and this is all we desire. This gives us the positions of 0, 1, 2, 3, and 4 in the square of Figure 4.2.

When we come to inserting 5 in the square, however, we find that 5 is assigned the same cell as 0. This happened with Loubère's method also; his cure was to place 5 in another vacant cell. Loubère chose to place 5 in the cell that was one to the left and two down from the cell containing 0, but this is not necessary. This time, we arbitrarily put 5 in the cell one to the right and three up (mod 5) from the cell containing 0. Then we continue going one to the right and two up as before. Thus for $j \geq 5$, instead of (1), we use the congruences

(2) $$x_j \equiv 4 + j + 1 \pmod 5, \qquad y_j \equiv 3 + 2j + 3 \pmod 5.$$

Y					
5	19	7	20	13	1
4	10	3	16	9	22
3	6	24	12	0	18
2	2	15	8	21	14
1	23	11	4	17	5

| | 1 | 2 | 3 | 4 | 5 |
| | (6, | 7, | 8, | etc.) | |

Figure 4.2. $\begin{pmatrix} a & c & e \\ b & d & f \end{pmatrix} = \begin{pmatrix} 4 & 1 & 1 \\ 3 & 2 & 3 \end{pmatrix}$, $n = 5$. This square is filled, magic, diabolic, and symmetric. (This and later figures will be referred to several times in the course of the chapter. The strange words and notation will be defined before the last referrals.)

This suffices to place 5, 6, 7, 8, and 9 in the positions shown in Figure 4.2, but 10 is assigned to the same cell as 5. Again, we make use of the additional step, one further to the right and three up. Thus, for $j \geq 10$, instead of (2), we use the congruences

(3) $x_j \equiv 4 + j + 1 \cdot 2 (\text{mod } 5), \qquad y_j \equiv 3 + 2j + 3 \cdot 2 (\text{mod } 5).$

This same problem will arise when we come to placing 15 and 20. If we are to avoid having to deal with five separate sets of congruences, each valid for only five values of j, then we are going to have to find functions of j which remain constant for five consecutive values of j and then change, always by the same amount.

We define a function which will help solve the problem just raised.

Definition. If α is a real number, then we define the **greatest integer function** of α, $[\alpha]$, to be the greatest integer less than or equal to α. In other words, $[\alpha] = a$, where a is the unique integer satisfying the inequality,

$$a \leq \alpha < a + 1.$$

Thus, for example, $[7.2] = 7, [\pi] = 3, [-\pi] = -4, [2] = 2,$ and $[-8] = -8.$ The greatest integer function is just what we need; consider the function of j,

$$f(j) = \left[\frac{j}{5} \right].$$

For $j = 0, 1, 2, 3, 4, f(j) = 0.$ For $j = 5, 6, 7, 8, 9, f(j) = 1.$ For $j = 10, 11, 12, 13, 14, f(j) = 2$; and so on. Thus $f(j)$ has the property that for five consecutive values of j it remains the same and then it changes by one. This enables us to write all the x_j and y_j in one congruence equation: For $0 \leq j \leq 24,$

(4)
$$x_j \equiv 4 + j + \left[\frac{j}{5} \right] (\text{mod } 5),$$
$$y_j \equiv 3 + 2j + 3 \left[\frac{j}{5} \right] (\text{mod } 5).$$

These equations give the square shown in Figure 4.2. For most values of j (namely those j that are not one less than a multiple of 5), $[j/5]$ remains the same when we change j to $j + 1$, and thus equation (4) states that the cell containing $j + 1$ has x coordinate one greater than that containing j and y coordinate two greater than the cell containing j; that is, $j + 1$ is one cell to the right and two cells up from $j (\text{mod } 5)$. But for certain values of j

(namely those values of j that are one less than a multiple of 5: 4, 9, 14, 19), $[j/5]$ increases by one as we change from j to $j + 1$; without this addition change, $j + 1$ would go in the same cell as $j - 4$, but with it $j + 1$ is moved an additional cell to the right and three cells up (mod 5).

We now define the general uniform step method.

Definition. Let a, b, c, d, e, and f be integers and let n be a positive integer. The **uniform step method** for an $n \times n$ square puts the n^2 numbers $j = 0$, $1, 2, \ldots, n^2 - 1$ in the cells with coordinates (x_j, y_j), where

(5)
$$x_j \equiv a + cj + e\left[\frac{j}{n}\right] (\mathrm{mod}\ n),$$

$$y_j \equiv b + dj + f\left[\frac{j}{n}\right] (\mathrm{mod}\ n).$$

The uniform step method thus has the potential to furnish a great many $n \times n$ squares, depending on the parameters a, b, c, d, e, and f. When $n = 5$ and $a = 4$, $b = 3$, $c = 1$, $d = 2$, $e = 1$, and $f = 3$, we have the square of Figure 4.2.

The uniform step method places 0 in the cell with coordinates (a, b). Each successive integer is then placed in a cell c units to the right and d units up from the preceding integer except that after every n steps an additional step of e units to the right and f units up is called for. If any of the numbers c, d, e, and f are negative, then the words "right" and "up" should be replaced by "left" and "down" in the appropriate places (for example, $c = -1$ corresponds to one cell to the left). The square in Figure 4.2 is a magic square (and a few other things also) but this does not guarantee that the uniform step method always gives magic squares. Nor does one example guarantee that the uniform step method always actually fills the square. Rather than worry about it, we give two examples to show that the uniform step method is not perfect.

1,3,8		
	2,4,6	
		0,5,7

Figure 4.3. $\begin{pmatrix} a & c & e \\ b & d & f \end{pmatrix} = \begin{pmatrix} 3 & 1 & 1 \\ 1 & -1 & 2 \end{pmatrix}$, $n = 3$. Although not filled, this square is magic in the rows, columns, and negative diagonals. It is also symmetric.

As seen in Figure 4.3, there are times that the uniform step method does not completely fill the square but rather n^2 numbers are put in fewer than n^2 cells, with some cells thus receiving more than their share.

Definition. The uniform step method will be said to **fill** an $n \times n$ square if each of the n^2 cells in the square has exactly one entry.

The sum of the entries in any row or column in Figure 4.3 is always 12. We will find it convenient to call this square magic also. This is not the usual terminology; magic squares as usually defined will be in our terminology, both magic and filled. The two properties are separate and are best treated separately.

The square in Figure 4.4 is filled, but it is not magic since the sums of the numbers in the first and second rows are different. The reader may picture the square that results if we take $c = d = e = f = 0$. Every number goes in the cell with coordinates (a,b). Such a square is neither filled nor magic. Thus we clearly need some sort of conditions to be placed on $c, d, e,$ and f if the square is to be filled or magic. We shall give these conditions in Section 4.2.

$$
\begin{array}{cccc}
3 & 5 & 11 & 13 \\
4 & 10 & 12 & 2 \\
9 & 15 & 1 & 7 \\
14 & 0 & 6 & 8
\end{array}
$$

Figure 4.4. $\begin{pmatrix} a & c & e \\ b & d & f \end{pmatrix} = \begin{pmatrix} 2 & 1 & -1 \\ 1 & 1 & -2 \end{pmatrix}$, $n = 4$. This square is given by the Loubère method. It is not magic, but it is column magic.

EXERCISES
In problems 1–6, find the square given by the uniform step method from the given parameters and state whether it is filled, magic, both, or neither.

1. $\begin{pmatrix} a & c & e \\ b & d & f \end{pmatrix} = \begin{pmatrix} 2 & 1 & 2 \\ 3 & 1 & 1 \end{pmatrix}$, $n = 3$.

2. $\begin{pmatrix} a & c & e \\ b & d & f \end{pmatrix} = \begin{pmatrix} 1 & 1 & 1 \\ 1 & 1 & 3 \end{pmatrix}$, $n = 4$.

3. $\begin{pmatrix} a & c & e \\ b & d & f \end{pmatrix} = \begin{pmatrix} 2 & 1 & -1 \\ 3 & 1 & -2 \end{pmatrix}$, $n = 5$.

4. $\begin{pmatrix} a & c & e \\ b & d & f \end{pmatrix} = \begin{pmatrix} 1 & 1 & 2 \\ 3 & 0 & 1 \end{pmatrix}$, $\quad n = 3$.

5. $\begin{pmatrix} a & c & e \\ b & d & f \end{pmatrix} = \begin{pmatrix} 1 & 1 & 2 \\ 5 & -1 & 1 \end{pmatrix}$, $\quad n = 6$.

6. $\begin{pmatrix} a & c & e \\ b & d & f \end{pmatrix} = \begin{pmatrix} 2 & 1 & 2 \\ 2 & 2 & 1 \end{pmatrix}$, $\quad n = 5$.

7. The following square is given by the uniform step method. Find $a, b, c, d, e,$ and f.

5	17	4	11	23
21	8	15	2	14
12	24	6	18	0
3	10	22	9	16
19	1	13	20	7

8. Show that the Loubère method is always given by $c = 1, d = 1, e = -1$, and $f = -2$.

9. Can the numbers 0 through 5 be inserted in a 2×3 rectangle (two rows and three columns) in such a way that the sums of the entries of the two rows are equal?

10. The numbers 0 through $n^2 - 1$ are placed in an $n \times n$ square in such a way as to make it magic (the process used is not necessarily the uniform step method). What is the magic sum? (*Hint:* Consider the sum of every number in the square and how this total is related to the magic sum.)

11. Suppose m and n are positive integers, $m \neq n$. Can the numbers $1, 2, \ldots, mn$ be inserted into an $m \times n$ rectangle in such a way as to make the sums of the entries of all rows and columns the same?

12. Show that if $0 \leq j \leq n^2 - n - 1$, then the uniform step method places $j + n$ in the cell e units to the right and f units up (mod n) from the cell containing j.

4.2. Filled and Magic Squares

In this section we find conditions that ensure that the uniform step method gives filled and/or magic squares. We first present a preliminary theorem which will prove very useful. For the rest of the chapter, we assume that $n \geq 2$, since the case $n = 1$ is of no interest.

Theorem 4.1. Let j be an integer in the range $0 \leq j \leq n^2 - 1$. There are unique integers u and v such that

(6) $$0 \leq u \leq n - 1, \qquad 0 \leq v \leq n - 1,$$

and

$$j = vn + u.$$

These integers are completely determined by (6) and the conditions

$$j \equiv u(\text{mod } n), \qquad \left[\frac{j}{n}\right] = v.$$

Proof. There is a unique integer u in the range of n numbers $0, 1, \ldots, n - 1$ such that

$$j \equiv u(\text{mod } n).$$

By the definition of congruence, there is an integer v such that

$$j = u + nv.$$

Since

$$0 \leq j \leq n^2 - 1, \qquad 0 \leq u \leq n - 1,$$

we see that

$$-1 < \frac{0 - (n - 1)}{n} \leq \frac{j - u}{n} \leq \frac{(n^2 - 1) - 0}{n} < n,$$

and since $(j - u)/n = v$,

$$-1 < v < n.$$

But v is an integer and therefore

$$0 \leq v \leq n - 1.$$

This shows that the representation of j in the theorem is possible.
  Suppose that
$$j = vn + u,$$
where u and v satisfy inequalities (6). Therefore,

$$j \equiv u(\text{mod } n),$$

and this along with the inequality

$$0 \le u \le n - 1$$

completely determines u. Using this inequality again, we see that

$$v = \frac{j - u}{n} \le \frac{j}{n} < \frac{j + (n - u)}{n} = v + 1.$$

Thus, by the definition of the greatest integer function,

$$v = \left[\frac{j}{n} \right],$$

and thus v is also uniquely determined. We have also incidentally proved the last part of the theorem.   ▲

The inequality $0 \le v \le n - 1$ was verified but never actually used in the course of proving Theorem 4.1. Its use will be in the applications of the theorem. In the case that $n = 10$, the whole theorem is nothing more than an obvious statement about the representation of a number in the decimal notation. The condition $0 \le j \le 10^2 - 1$ says that j has a maximum of two digits. Inequalities (6) and the equation $j = v \cdot 10 + u$ say that u is the units' digit and v the tens' digit. Any positive integer is congruent to its unit digit (mod 10) and if $j > 0$, $[j/10]$ is clearly what is left when the units digit is erased (try all this out, for example, on $j = 97$). In like manner, this theorem is a statement about two digit numbers written in the base n.

If for j in the interval $0 \le j \le n^2 - 1$, we put j in the form given by Theorem 4.1; then the uniform step method puts j in the cell (x_j, y_j) given by

(7)
$$x_j \equiv a + cu + ev \pmod{n},$$

$$y_j \equiv b + du + fv \pmod{n}.$$

We will use this form of the uniform step method in our proofs.

Theorem 4.2. The uniform step method fills the $n \times n$ square if $(cf - de, n) = 1$.

[*Remark:* The converse is also true; if the square is filled, then $(cf - de, n) = 1$. However, the proof is more difficult and will not be given.]

Proof. The uniform step method places n^2 numbers in n^2 cells. Thus the only way that the square would not be filled is that some cell contains more than one entry. Thus if we can show that no cell contains two different entries, it will follow that the square is filled. We give a proof by contradiction.

Let j_1 and j_2 ($j_1 \neq j_2$ and both in the range from 0 to $n^2 - 1$) be placed in the same cell so that

$$x_{j_1} = x_{j_2}, \qquad y_{j_1} = y_{j_2}.$$

By Theorem 4.1, we may put j_1 and j_2 in the form

$$j_1 = v_1 n + u_1, \qquad j_2 = v_2 n + u_2,$$

where u_1, v_1, u_2, v_2 are in the range 0 through $n - 1$. Thus

$$a + cu_1 + ev_1 \equiv a + cu_2 + ev_2 (\bmod n),$$

$$b + du_1 + fv_1 \equiv b + du_2 + fv_2 (\bmod n)$$

or

(8)
$$cu_1 + ev_1 \equiv cu_2 + ev_2 (\bmod n),$$

$$du_1 + fv_1 \equiv du_2 + fv_2 (\bmod n).$$

Equations (8) may be thought of as two congruence equations in the two unknowns u_1 and v_1. By hypothesis, the determinant of the coefficients, $cf - de$, is relatively prime to n and thus, by Theorem 3.9, there is a unique solution to (8) for u_1 and v_1 (mod n). But clearly

$$u_1 \equiv u_2 (\bmod n), \qquad v_1 \equiv v_2 (\bmod n)$$

is a solution to (8) and hence must be the unique solution to (8). Since u_1, u_2, v_1, and v_2 are all in the range 0 through $n - 1$, it must be that

$$u_1 = u_2, \qquad v_1 = v_2$$

and thus

$$j_1 = j_2.$$

This is a contradiction and thus different numbers go in different cells. ▲

As an example, when $c = 1, d = 1, e = -1, f = -2$ (the Loubère method), we see that $(cf - de, n) = 1$ for all n and thus the Loubère method fills every $n \times n$ square. The following theorem gives the key to finding conditions that guarantee that a square is magic.

Theorem 4.3. Let q, r, and s be integers and

$$(q,n) = (r,n) = 1.$$

Then there are exactly n integers in the range

$$0 \leq j \leq n^2 - 1$$

that satisfy the congruence

$$qj + r\left[\frac{j}{n}\right] \equiv s(\text{mod } n),$$

and the sum of these n integers is $n(n^2 - 1)/2$.

Proof. For each j in the range $0 \leq j \leq n^2 - 1$, there are by Theorem 4.1 unique integers u and v such that

(9) $$0 \leq u \leq n - 1, \qquad 0 \leq v \leq n - 1,$$

and

$$j = vn + u.$$

Conversely, given integers u and v satisfying the above inequalities, the number

$$j = vn + u$$

satisfies the inequality

$$0 = 0 \cdot n + 0 \leq vn + u \leq (n - 1)n + (n - 1) = n^2 - 1.$$

Thus each j in the range 0 to $n^2 - 1$ determines unique u and v in the range from 0 to $n - 1$ and conversely. In terms of u and v, our congruence is

(10) $$qu + rv \equiv s(\text{mod } n).$$

Our first task is to show that there are exactly n pairs of numbers u, v which satisfy (9) and (10).

Since $(r,n) = 1$, Theorem 3.6 tells us that for a given u, v is the unique solution to

$$rv \equiv s - qu(\text{mod } n).$$

There are n values of u in the range $0 \leq u \leq n - 1$, and each of these determines a unique v in the same range such that the pair u, v satisfies (10). Hence there are n values of j in the range $0 \leq j \leq n^2 - 1$ satisfying

$$qj + r\left[\frac{j}{n}\right] \equiv s(\text{mod } n).$$

Let these n values of j be $j_0, j_1, \ldots, j_{n-1}$, and let the corresponding values of u and v be $u_0, v_0, u_1, v_1, \ldots, u_{n-1}, v_{n-1}$. We have already noted that the u_k assume each value in the range from 0 to $n - 1$ exactly once. The same can be said for the v_k. Given a value of v, there is a unique u satisfying the equation

$$qu \equiv s - rv(\text{mod } n),$$

since $(q, n) = 1$. Therefore, every number v in the range 0 to $n - 1$ has a corresponding u in that range such that the pair u, v satisfies (10). Thus the v_k also give each value in the range 0 to $n - 1$ exactly once. In other words, the numbers $u_0, u_1, \ldots, u_{n-1}$ are simply a rearrangement of the numbers $0, 1, \ldots, n - 1$ and the same is true of the numbers $v_0, v_1, \ldots, v_{n-1}$ (the rearrangement is undoubtedly a different rearrangement, but this does not matter). Therefore,

$$\sum_{k=0}^{n-1} j_k = \sum_{k=0}^{n-1} (nv_k + u_k) = n \sum_{k=0}^{n-1} v_k + \sum_{k=0}^{n-1} u_k$$

$$= n \sum_{k=0}^{n-1} k + \sum_{k=0}^{n-1} k$$

$$= (n + 1) \sum_{k=0}^{n-1} k$$

$$= (n + 1)\frac{(n - 1)(n)}{2}$$

$$= \frac{n(n^2 - 1)}{2}. \qquad \blacktriangle$$

Let us illustrate the last part of the proof of Theorem 4.3. Take $n = 5$ and let the congruence equation be

(11)
$$2j + 3\left[\frac{j}{5}\right] \equiv 2(\bmod 5)$$

or, in terms of u and v,

$$2u + 3v \equiv 2(\bmod 5).$$

The five solutions to (11) with $0 \le j \le 24$ are $j = 1, 7, 13, 19$, and 20. The u, v representations of these numbers are

$$1 = 5 \cdot 0 + 1,$$
$$7 = 5 \cdot 1 + 2,$$
$$13 = 5 \cdot 2 + 3,$$
$$19 = 5 \cdot 3 + 4,$$
$$20 = 5 \cdot 4 + 0.$$

Their sum is

$$
\begin{aligned}
1 + 7 + 13 + 19 + 20 &= 5 \cdot (0 + 1 + 2 + 3 + 4) + (1 + 2 + 3 + 4 + 0) \\
&= 5 \cdot (0 + 1 + 2 + 3 + 4) + (0 + 1 + 2 + 3 + 4) \\
&= 6(0 + 1 + 2 + 3 + 4) \\
&= 60.
\end{aligned}
$$

Equation (11) is actually the equation of the top row of the square in Figure 4.2. The top row there is given by the equation

$$y_j = 5,$$

and y_j is given by

$$y_j \equiv 3 + 2j + 3\left[\frac{j}{5}\right] (\text{mod } 5).$$

These two equations give equation (11). The sum of the elements in the first row in Figure 4.2 is 60. We see that this must be true of the other rows also; their equations are the same as (11) except that the number on the right side changes from row to row. The numbers change from row to row but their sum will always be

$$5(0 + 1 + 2 + 3 + 4) + (0 + 1 + 2 + 3 + 4) = 60.$$

We can perhaps better appreciate the proof of the previous theorem and its application to magic squares if we take the magic square of Figure 4.2 and write its entries not in the decimal system but in the base 5 (Figure 4.5). (Ordinarily, a number such as 4 would be written 4 and not 04; we add the 0 just to make a point.) The reader can see at a glance that in any row or column, the units' digits (the u's of Theorem 4.3) are always 0, 1, 2, 3, 4 (rearranged), and the fives' digits (the v's of Theorem 4.3) are also always 0, 1, 2, 3, 4 (rearranged). Thus the sum of the numbers in any row or column is always (base 10)

$$5(0 + 1 + 2 + 3 + 4) + (0 + 1 + 2 + 3 + 4) = 60.$$

As a result, the square is magic.

34	12	40	23	01
20	03	31	14	42
11	44	22	00	33
02	30	13	41	24
43	21	04	32	10

Figure 4.5. The square of Figure 4.2 with entries written base 5. [For example, the number in the upper left hand corner is (base 10) $3 \cdot 5 + 4 = 19$.]

We have not yet formally defined what we mean by magic; we remedy this oversight now.

Definition. Suppose n^2 different integers are put in various cells of an $n \times n$ square (not necessarily filling it). If the sum of the entries of each row is always the same, we say that the square is **row magic**. If the sum of the entries of each column is always the same, we say that the square is **column magic**. If the square is both row magic and column magic, then we simply call it a **magic square**. The sum that results in each of these cases is called the **magic sum**.

As we noted earlier, other authors will add to this definition the condition that the square be filled. The reader should be aware of this if he should ever read other books or articles on the subject. Almost all authors insist, as we do, that the n^2 numbers in the square be different. Some authors go even further and insist that the entries be consecutive integers; other magic squares are not quite legitimate to them. We have not made this extra requirement part of our definition, but of course when we deal with the uniform step method, we are satisfying such a requirement, like it or not.

Theorem 4.4. Suppose that the numbers $0, 1, \ldots, n^2 - 1$ are put in an $n \times n$ square by the uniform step method in equation (5).
If $(c,n) = (e,n) = 1$, then the square is column magic.
If $(d,n) = (f,n) = 1$, then the square is row magic.
If $(c,n) = (d,n) = (e,n) = (f,n) = 1$, then the square is magic. In each case the magic sum is $n(n^2 - 1)/2$.
(*Remark:* The conditions above are actually both necessary and sufficient in each case, but we prove the theorem as stated.)

Proof. Suppose that $(c,n) = (e,n) = 1$. The number j is inserted in the kth column if and only if $x_j = k$. Since

$$x_j \equiv a + cj + e\left[\frac{j}{n}\right](\bmod n),$$

we see that j is put in the kth column if and only if

$$cj + e\left[\frac{j}{n}\right] \equiv k - a(\bmod n).$$

By Theorem 4.3, the sum of such j in the range we are considering, $0 \le j \le n^2 - 1$, is

$$\frac{n(n^2 - 1)}{2}.$$

Thus the sum of the elements in every column is $n(n^2 - 1)/2$. Therefore, the square is column magic with magic sum $n(n^2 - 1)/2$.

Suppose now that $(d,n) = (f,n) = 1$. The number j is placed in the kth row if and only if $y_j = k$. Since

$$y_j \equiv b + dj + f\left[\frac{j}{n}\right] (\bmod\, n),$$

we see that j is put in the kth row if and only if

$$dj + f\left[\frac{j}{n}\right] \equiv k - b(\bmod\, n).$$

Thus, by Theorem 4.3, the sum of the numbers in the kth row is

$$\frac{n(n^2 - 1)}{2}.$$

Hence the square is row magic and the magic sum is $(n^2 - 1)/2$. The third part of the theorem follows from the first two parts. ▲

Examples of Theorems 4.2 and 4.4 at work can be seen in Figures 4.2, 4.3, and 4.4. Note that the converse of Theorem 4.2 does hold in Figure 4.3 and the converse of Theorem 4.4 holds in Figure 4.4.

EXERCISES

1. Find $[\alpha]$ if $\alpha = 9.76$, $\sqrt{2}$, -81, $-\sqrt{3}$, and 472.
2. Find $[2^{9/4}]$ without resorting to tables or slide rules.
3. Give an example of an unfilled 2×2 magic square using the numbers 0, 1, 2, and 3.
4. Show that there is no 2×2 filled magic square (this will depend upon a technicality in the definition of magic).

In problems 5–9, determine whether the square given by the uniform step method is filled, column magic, row magic, or magic. You may assume that the converses of Theorems 4.2 and 4.4 are true.

5. $\begin{pmatrix} a & c & e \\ b & d & f \end{pmatrix} = \begin{pmatrix} 7 & 4 & 1 \\ 3 & 2 & 5 \end{pmatrix}$, $n = 11$.

6. $\begin{pmatrix} a & c & e \\ b & d & f \end{pmatrix} = \begin{pmatrix} 7 & 4 & 1 \\ 3 & 2 & 4 \end{pmatrix}$, $n = 9$.

7. $\begin{pmatrix} a & c & e \\ b & d & f \end{pmatrix} = \begin{pmatrix} 1 & 4 & 1 \\ 2 & 1 & 10 \end{pmatrix}$, $n = 15$.

8. $\begin{pmatrix} a & c & e \\ b & d & f \end{pmatrix} = \begin{pmatrix} 37 & 8 & -5 \\ 51 & 1 & 10 \end{pmatrix}$, $n = 51$.

9. $\begin{pmatrix} a & c & e \\ b & d & f \end{pmatrix} = \begin{pmatrix} 1 & 1 & -1 \\ 1 & 1 & -2 \end{pmatrix}$, $n = 101$.

10. Show that the Loubère method gives a filled column magic square for all n and that if n is odd, the square is magic.

11. Assuming the converse of Theorems 4.2 and 4.4, show that the uniform step method never gives a filled magic square when n is even. (*Hint:* Show that c, d, e, f, and $cf - de$ must all be odd and derive a contradiction.)

12. A 6×6 square is found by the uniform step method. Find a, b, c, d, e, and f from the following facts. The square is filled and row magic. The numbers 10 and 11 are in the cells with coordinates (6,2) and (1,3), respectively. The x coordinate of the cell containing 12 is 6. You may use the converses of Theorems 4.2 and 4.4 if necessary.

4.3. Diabolic and Symmetric Squares

An $n \times n$ square has two main diagonals. The diagonal consisting of the cells running from the lower left corner to the upper right corner is given by the equation $y = x$. The diagonal consisting of the cells running from the upper left corner down to the lower right corner is given by the equation $x + y = n + 1$. These diagonals can be given equally well by the respective congruences,

$$y \equiv x(\bmod n), \qquad x + y \equiv 1(\bmod n).$$

But in addition to the two main diagonals, there are $2n - 2$ other diagonals, the broken diagonals. The n **positive diagonals** are given by the congruences

$$x + y \equiv k(\bmod n)$$

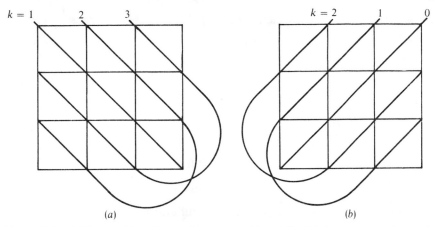

Figure 4.6. (a) The positive diagonals $x + y \equiv k(\text{mod } 3)$; (b) the negative diagonals $y \equiv x + k(\text{mod } 3)$.

and the n **negative diagonals** are given by the congruences

$$y \equiv x + k(\text{mod } n).$$

The six diagonals of a 3×3 square are shown in Figure 4.6.

Definition. Suppose that n^2 different integers are put in various cells of an $n \times n$ square (not necessarily filling it). If the sum of the entries of each positive diagonal is always the same, we say that the square is **magic in the positive diagonals**. If the sum of the entries of each negative diagonal is always the same, we say that the square is **magic in the negative diagonals**. If the square is magic in both the positive and negative diagonals, then we say that it is **diabolic**. The sum that results in each of these cases will be called the **diabolic sum**.

As was the case with our definition of magic, other authors require in their definitions that the square be filled.

Theorem 4.5. Suppose that the numbers $0, 1, \dots, n^2 - 1$ are put in an $n \times n$ square by the uniform step method in equation (5). If

$$(c + d, n) = (e + f, n) = 1,$$

then the square is magic in the positive diagonals. If

$$(c - d, n) = (e - f, n) = 1,$$

then the square is magic in the negative diagonals. If

$$(c + d, n) = (c - d, n) = (e + f, n) = (e - f, n) = 1,$$

then the square is diabolic. In each case, the diabolic sum is

$$\frac{n(n^2 - 1)}{2}.$$

(*Remark:* These conditions are actually both necessary and sufficient.)

Proof. Suppose that

$$(c + d, n) = (e + f, n) = 1.$$

The equation of a positive diagonal is of the form

$$x + y \equiv k(\text{mod } n).$$

The number j will be in this diagonal if and only if

$$x_j + y_j \equiv k(\text{mod } n);$$

that is,

$$(c + d)j + (e + f)\left[\frac{j}{n}\right] \equiv k - a - b(\text{mod } n).$$

Therefore, by Theorem 4.3, the sum of the numbers on this diagonal is

$$\frac{n(n^2 - 1)}{2}.$$

Thus the square is magic in the positive diagonals and $n(n^2 - 1)/2$ is the diabolic sum.

Suppose now that

$$(c - d, n) = (e - f, n) = 1.$$

The equation of a negative diagonal is of the form

$$x - y \equiv -k(\text{mod } n).$$

The number j will be in this diagonal if and only if

$$x_j - y_j \equiv -k(\text{mod } n);$$

that is,

$$(c - d)j + (e - f)\left[\frac{j}{n}\right] \equiv b - a - k(\text{mod } n).$$

Therefore, by Theorem 4.3, the sum of the entries in this diagonal is

$$\frac{n(n^2 - 1)}{2}.$$

Thus the square is magic in the negative diagonals and $n(n^2 - 1)/2$ is the diabolic sum. The third part of the theorem follows from the first two parts.   ▲

The reader may check this theorem against the examples in Figures 4.2, 4.3, and 4.4. Note that the converse of Theorem 4.5 holds in Figure 4.3, where the square is not magic in the positive diagonals, and in Figure 4.4, where the square is not magic in either the positive or negative diagonals.

The last type of square that we shall discuss here is the symmetric square.

Definition. Suppose that n^2 different integers are placed in an $n \times n$ square (not necessarily filling it). If there is an integer s such that for each j in the square, $s - j$ is also in the square and the numbers j and $s - j$ are symmetrically placed about the center of the square, then we say that the square is a **symmetric square** and s is its **symmetric sum.**

For example, the squares in Figures 4.1, 4.2, and 4.3 are symmetric squares with symmetric sums 17, 24, and 8, respectively. The usual definition of a symmetric square includes the condition that the square be filled. In this case, there is a unique number symmetrically placed about the center from any given entry and thus the definition is usually phrased: A square is symmetric when the sum of any two symmetrically placed elements about the center is a constant (which is called the symmetric sum). When n is odd so that there is a center cell and the square is filled, the integer in the center is symmetrically placed about the center with only itself and thus the symmetric sum is twice the central entry (see the square in Figure 4.2 as an example).

Theorem 4.6. Suppose that a symmetric square is made up of the n^2 numbers $0, 1, 2, \ldots, n^2 - 1$. Then the symmetric sum is $n^2 - 1$.

Proof. Let s be the symmetric sum. Since 0 is in the square, $(s - 0)$ is in the square (symmetrically located about the center from 0). Since $n^2 - 1$ is the largest entry in the square,

(12) $$s = s - 0 \leq n^2 - 1.$$

Likewise, $s - (n^2 - 1)$ is in the square and since the smallest entry is 0,

$$s - (n^2 - 1) \geq 0$$

or

(13) $$s \geq n^2 - 1.$$

There is only one number that satisfies both (12) and (13),

$$s = n^2 - 1. \qquad \blacktriangle$$

Theorem 4.7. The uniform step method of equation (5) gives a symmetric $n \times n$ square if and only if

$$2a \equiv c + e + 1 (\mathrm{mod}\ n),$$

$$2b \equiv d + f + 1 (\mathrm{mod}\ n).$$

The proof is fairly simple and is sketched in miscellaneous exercises 9, 10, and 11. The squares in Figures 4.2, 4.3, and 4.4 furnish examples of this theorem. If we combine Theorems 4.2, 4.4, 4.5, and 4.7, we see that if

$$(n,c) = (n,d) = (n,e) = (n,f) = (n,c + d) = (n,c - d)$$

$$= (n, e + f) = (n, e - f) = (n, cf - de) = 1,$$

$$2a \equiv c + e + 1 (\mathrm{mod}\ n), \qquad 2b \equiv d + f + 1 (\mathrm{mod}\ n),$$

then the square produced by the uniform step method is filled, magic, diabolic, and symmetric. We have given nonstandard definitions of these words so that the reader could better see which condition was responsible for which property.

EXERCISES

In Problems 1–5, which of the following properties do the squares possess: filled, magic (or just row or column magic), diabolic (or just magic in the positive or negative diagonals), symmetric? You may use the converses of Theorems 4.2, 4.4, and 4.5.

1. $\begin{pmatrix} a & c & e \\ b & d & f \end{pmatrix} = \begin{pmatrix} 3 & 4 & 1 \\ 3 & 1 & 4 \end{pmatrix}, \qquad n = 5.$

2. $\begin{pmatrix} a & c & e \\ b & d & f \end{pmatrix} = \begin{pmatrix} 3 & 1 & 5 \\ 3 & 2 & 3 \end{pmatrix}, \qquad n = 6.$

3. $\begin{pmatrix} a & c & e \\ b & d & f \end{pmatrix} = \begin{pmatrix} 4 & 3 & 4 \\ 3 & 2 & 3 \end{pmatrix}, \qquad n = 7.$

4. $\begin{pmatrix} a & c & e \\ b & d & f \end{pmatrix} = \begin{pmatrix} 6 & 7 & 4 \\ 4 & -2 & 11 \end{pmatrix}, \qquad n = 30.$

5. $\begin{pmatrix} a & c & e \\ b & d & f \end{pmatrix} = \begin{pmatrix} 11 & 11 & 10 \\ 12 & 12 & 11 \end{pmatrix}, \qquad n = 42.$

6. When n is odd, in what cell should 0 be inserted so that the Loubère method makes the resulting square symmetric?
7. Show that when n is even, the Loubère method never gives a symmetric square.
8. Give an example of a filled, diabolic, and symmetric 4 × 4 square.
9. Show that the uniform step method never gives a magic symmetric square when n is even. You may use the converse of Theorem 4.4.
10. For which values of $n > 1$ does the uniform step method give a symmetric square if

$$\begin{pmatrix} a & c & e \\ b & d & f \end{pmatrix} = \begin{pmatrix} 25 & -59 & -63 \\ 100 & -99 & -101 \end{pmatrix}.$$

11. Suppose that the uniform step method has produced a symmetric (not necessarily filled) square where n is odd. Show that one of the occupants of the central cell is $(n^2 - 1)/2$. Since there is no central cell when n is even, your proof must fail when n is even. Where does your proof use the condition that n is odd?
12. Given a symmetric square, not necessarily produced by the uniform step method and not necessarily consisting of consecutive integer entries, with r and t being the smallest and largest entries. What is the symmetric sum?
13. Show that the uniform step method cannot produce a magic diabolic square when n is either even or divisible by 3. [Thus, if the uniform step method gives a magic diabolic square, then $(n,6) = 1$.] You may use the converses of Theorems 4.4 and 4.5.

4.4. Historical Comments

The concept of a magic square is an extremely old one; the uniform step method square given by $a = 2$, $b = 1$, $c = 1$, $d = 2$, $e = 2$, $f = 2$, $n = 3$ (actually every element was increased by 1) was seen in 2200 B.C. by the Chinese emperor Yu on the back of a Divine Tortoise. Many of the modern methods of constructing squares were introduced to Europeans in the Middle Ages by travelers who had learned of them in Asia. For instance, De la Loubère brought his method back with him from Siam, where he had been the ambassador of Louis XIV in 1687–1688. As one may judge by the terminology, the subject was once shrouded in mysticism and astrology. For example, it was once thought that a magic square engraved on a silver plate and worn around the neck would ward off the plague. A diabolic square originally meant what we call here a filled magic diabolic square.

Other less interesting names still used for this type of square are perfect
squares and pandiagonal squares.

While magic and diabolic squares have been studied for several centuries,
McClintock in 1897 was one of the first authors to discuss symmetric
squares. McClintock gave the example of the 7×7 square given by the
uniform step method with

$$\begin{pmatrix} a & c & e \\ b & d & f \end{pmatrix} = \begin{pmatrix} 5 & 1 & 1 \\ 1 & -2 & 3 \end{pmatrix}.$$

In addition to being a filled, magic, diabolic, symmetric square, it has the
property that a knight placed on the cell containing 0 may make 48 consecutive moves, according to the rules of chess, in such a way as to land *in
order* on the cells containing $1, 2, 3, \ldots, 48$ (it is assumed that the first and
seventh rows and columns are adjacent in this knight's tour). The reader
should reconcile the values of e and f with a knight's move.

There is one other type of square worth mentioning.

Definition. Suppose that the numbers $0, 1, \ldots, n^2 - 1$ are inserted in an
$n \times n$ square in such a way as to fill it. Suppose further that the entries are
written, not in the decimal system, but in the base n notation. If the square
has the property that in each row and column, the n digits $0, 1, \ldots, n - 1$
each appear exactly once among the units' digits and among the n's
digits, then we say that the square is **regular**.

For example, the square of Figure 4.2 (written in base 5 in Figure 4.5) is
regular. When the numbers $1, 2, \ldots, n^2$ are used, the square is said to be
regular if the square that results when 1 is subtracted from every entry is
regular. A regular square is always a (filled) magic square. The square in
Figure 4.1 is an example of a filled magic square which is not regular. Whenever the uniform step method gives a filled magic square, that square is
regular. These are some of the simple facts about regular squares. For hundreds of years, methods have been known that would produce regular $n \times n$
squares whenever n is odd or even and divisible by 4. After extensive trials,
Euler conjectured in 1782 that there are no regular squares when
$n \equiv 2 \pmod 4$. This is trivial for $n = 2$. Euler's conjecture for $n = 6$ was
verified by G. Tarry in 1900 by the exhausting method of systematically
trying all possibilities. Tarry's "method" is out of the question when $n = 10$;
the number of possibilities is so great that 100 hours on a computer (which
failed to find any 10×10 regular squares) did not even scratch the surface
of the possibilities involved. It was not until 1959 that anything new was

00	47	18	76	29	93	85	34	61	52
86	11	57	28	70	39	94	45	02	63
95	80	22	67	38	71	49	56	13	04
59	96	81	33	07	48	72	60	24	15
73	69	90	82	44	17	58	01	35	26
68	74	09	91	83	55	27	12	46	30
37	08	75	19	92	84	66	23	50	41
14	25	36	40	51	62	03	77	88	99
21	32	43	54	65	06	10	89	97	78
42	53	64	05	16	20	31	98	79	87

Figure 4.7. A 10×10 regular square. Note the 3×3 magic subsquare in the lower-right-hand corner. The preceding seemingly minor comment is of great importance to projective geometers who are searching for finite projective planes of order 10!

learned. In that year R. C. Bose, S. S. Shrikhande, and E. T. Parker destroyed Euler's conjecture. Parker gave the 10×10 regular square ($n = 10$ is convenient, as then base n means base 10) in Figure 4.7. It was shown a year later that Euler was wrong for all $n > 6$; that is, $n \times n$ regular squares exist for all n with the exceptions $n = 2$ and $n = 6$.

In other terminology, a regular $n \times n$ square (written base n) is often called a **Greco-Latin square.** In this terminology, what is emphasized is not the fact that the entries are integers (and, in fact, usually the units' digits are replaced by n different letters and the n's digits are replaced by n other different letters) but the pattern of the entries. Greco-Latin squares have been found to have wide applications in controlled experiments in various fields. It is not at all improbable that the 5×5 square of Figure 4.5 has had agricultural applications. Perhaps the word "magic" is not too poor a description at that.

MISCELLANEOUS EXERCISES

1. Prove that a regular square is magic.
2. Prove that a filled magic square given by the uniform step method is regular.
3. Show that Theorems 4.2, 4.4, 4.5, 4.6, and 4.7 remain true if one is added to each entry after a uniform step method square is constructed, the only exceptions being that the new magic sum is $n(n^2 + 1)/2$ and the new symmetric sum is $n^2 + 1$.
4. A 9×9 square may be split up naturally into nine 3×3 squares as shown in Figure 4.8. Using the numbers from 0 to 80, construct a filled magic square such that each of the nine 3×3 squares is also magic.

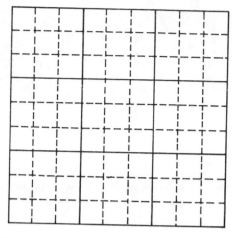

Figure 4.8

5. All the usual laws of arithmetic are satisfied by the set of four elements $\{0, 1, \alpha, \beta\}$ with the operations of plus and times shown in Figure 4.9. The mathematical name for this is a "field of four elements." One way this field arises is as follows: The equation

$$x^2 + x + 1 \equiv 0(\text{mod } 2)$$

has no solutions (mod 2); that is, $x = 0$ and $x = 1$ both fail. We therefore invent the solutions α and β:

$$(x - \alpha)(x - \beta) \equiv x^2 + x + 1(\text{mod } 2).$$

Then assuming that the usual laws of arithmetic hold,

$$\alpha + \beta \equiv -1 \equiv 1(\text{mod } 2),$$
$$\alpha\beta \equiv 1(\text{mod } 2).$$

Also

$$\alpha^2 \equiv -1(\alpha + 1) \equiv 1(\alpha + 1) \equiv \alpha + 1$$

and in like manner

$$\beta^2 \equiv \beta + 1.$$

X \ Y	0	1	α	β
0	0	1	α	β
1	1	0	β	α
α	α	β	0	1
β	β	α	1	0

$X + Y$

X \ Y	0	1	α	β
0	0	0	0	0
1	0	1	α	β
α	0	α	β	1
β	0	β	1	α

$X \cdot Y$

Figure 4.9

Further,

$$\alpha + 1 \equiv \alpha - 1 \equiv -\beta \equiv (-1)\beta \equiv (1)\beta \equiv \beta.$$

If we continue in this way, every entry in the above tables can be found.

(If all this seems unnatural and if it is unclear that such an invention of solutions to equations will result in the usual laws of arithmetic holding, then you can appreciate why the ancients gave the name "imaginary" to complex numbers. For the complex numbers are given by exactly the same process of inventing solutions to the equation $x^2 + 1 = 0$ where previously there were none.)

Use this field to construct a 4 × 4 regular square as follows. Let

$$x_{u,v} = u + v,$$

$$y_{u,v} = \alpha u + \beta v.$$

Using the plus and times tables of Figure 4.9, assign each of the 16 possible pairs u, v (u and v run independently through the four field elements $0, 1, \alpha, \beta$) to a position in the square shown in Figure 4.10.

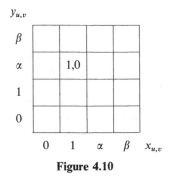

Figure 4.10

For example, $x_{1,0} = 1$, $y_{1,0} = \alpha$. Note that in each row and column, each of the four possible values of u and v occur exactly once. Replace α by 2, β by 3, and erase the commas; you should then have a 4×4 regular square written in base 4 notation. Convert your square to the decimal system; it should use each of the integers from 0 to 15. Notice that several (but not all) of the 2×2 subsquares have the same total sum.

6. A problem for the student who is familiar with fields and their properties. Problem 5 should be worked first. Let F be the field of four elements given in problem 5. Suppose the entries of the square in problem 5 are determined by the equations

$$x_{u,v} = a + cu + ev,$$

$$y_{u,v} = b + du + fv,$$

where a, b, c, d, e, and f are elements of F. Show that the square is filled by the sixteen possible pairs u, v if $cf - de \neq 0$. Show that in each row, the four possible values of u and v each occur exactly once if $d \neq 0$, $f \neq 0$. Show that if

$$cf - de \neq 0, \quad c \neq 0, \quad d \neq 0, \quad e \neq 0, \quad f \neq 0,$$

then we may replace α by 2 and β by 3 and get a regular magic square with entries written base 4. (This problem can be generalized to any finite field and to the even more general systems of finite commutative rings.)

7. Let

$$x_j \equiv 1 + 0j - \left[\frac{j}{3}\right] - \left[\frac{j}{9}\right] \pmod{3},$$

$$y_j \equiv 2 - j + \left[\frac{j}{3}\right] + 0\left[\frac{j}{9}\right] \pmod{3},$$

$$z_j \equiv 2 - j + 0\left[\frac{j}{3}\right] + \left[\frac{j}{9}\right] \pmod{3}.$$

Use these congruences to insert the numbers $0, 1, 2, \ldots, 26$ in a $3 \times 3 \times 3$ cube (this can be done by using three adjacent 3×3 squares, the first square representing $z = 1$, the second representing $z = 2$, and the third representing $z = 3$). You should have a filled, magic, symmetric cube; the row, column, and vertical sums are 39 and the symmetric sum is 26.

8. Let

$$x_r \equiv a + dr + g\left[\frac{r}{n}\right] + j\left[\frac{r}{n^2}\right] \pmod{n},$$

$$y_r \equiv b + er + h\left[\frac{r}{n}\right] + k\left[\frac{r}{n^2}\right] \pmod{n},$$

$$z_r \equiv c + fr + i\left[\frac{r}{n}\right] + l\left[\frac{r}{n^2}\right] \pmod{n}.$$

Suppose the numbers $r = 0, 1, \ldots, n^3 - 1$ are inserted in an $n \times n \times n$ cube according to these congruences. Let

$$m = \begin{vmatrix} d & g & j \\ e & h & k \\ f & i & l \end{vmatrix}$$

and let $D, E, F, G, H, I, J, K,$ and L be the 2×2 minors (determinants) corresponding to $d, e, f, g, h, i, j, k,$ and l, respectively. Show that if

$$(D,n) = (E,n) = (F,n) = (G,n) = (H,n) = (I,n)$$
$$= (J,n) = (K,n) = (L,n) = (m,n) = 1,$$

then the cube is filled and magic.

9. Show that j and k are symmetrically placed about the center of an $n \times n$ square if and only if $x_j + x_k = y_j + y_k = n + 1$. Since $2 \le x_j + x_k$ (or $y_j + y_k) \le 2n$, show that j and k are symmetrically placed about the center of an $n \times n$ square if and only if

$$x_j + x_k \equiv 1 \pmod{n}, \qquad y_j + y_k \equiv 1 \pmod{n}.$$

10. Let $0 \le j \le n^2 - 1$ and let the u-v representation of j given by Theorem 4.1 be $j = vn + u$. Show that $0 \le n^2 - 1 - j \le n^2 - 1$ and that the u-v representation of $n^2 - 1 - j$ is

$$n^2 - 1 - j = (n - 1 - v)n + (n - 1 - u).$$

11. Use problems 9, 10 and Theorem 4.6 to prove Theorem 4.7.

12. Use problems 9 and 10 to prove the following refinement of Theorem 4.7: The uniform step method gives a symmetric square if and only if there exists at least one j, $0 \le j \le n^2 - 1$ such that j and $n^2 - 1 - j$ are symmetrically placed about the center of the square.

13. Show that there are exactly eight different triplets of three distinct numbers between 1 and 9 that add to 15 (1,4,9 and 9,1,4 would be counted as the same triplet). Mr. X and Mr. Y play the following game: They alternately choose numbers from 1 to 9 (a number once chosen cannot be chosen again); the first person to get precisely three numbers adding to 15 is the winner. This game is known by you. What is it?

Chapter 5

DIOPHANTINE EQUATIONS

5.1. Introduction

The Greek Diophantus (somewhere in the period from the first to the third century) was the first to consider a whole class of problems having as their common feature the fact that there were more unknowns than equations. Diophantus was content with finding a solution, rather than all solutions, and he allowed the solution to be rational rather than integral. But still, he had the fundamental idea of restricting his solutions to particular types of numbers, and thus it is entirely fitting that the subject should be named for him. No great advancement was made in the subject during the 14 to 16 centuries between Diophantus and Fermat. Fermat's interest in the subject was aroused by reading Bachet's 1621 edition of what remained of Diophantus' work. It is with Fermat that the modern topic of Diophantine analysis (and, in fact, much of modern number theory) had its beginning.

We have already mentioned Fermat's last theorem (also called Fermat's great theorem). Fermat stated that he had a proof of the fact that if $n \geq 3$, there are no solutions to the equation

$$x^n + y^n = z^n$$

in nonzero integers. If a and b are positive integers and the equation

$$x^a + y^a = z^a$$

has no solutions in nonzero integers, then the equation

$$(x^b)^a + (y^b)^a = (z^b)^a,$$

being a special case of the first, has no solutions in nonzero integers. Since $(u^b)^a = u^{ab}$, this means that if Fermat's last theorem is true for $n = a$, then it is true for $n = ab$ also. Since every number greater than 2 is either divisible by an odd prime or is a power of 2 greater than the first power and hence is divisible by 4, it follows that if Fermat's last theorem is true in the cases that n is an odd prime and the case $n = 4$, then Fermat's last theorem is true for all $n \geq 3$. Fermat left a proof in the case of $n = 4$, and it is essentially this proof that we present in Section 5.4. It remains, then, to prove Fermat's last theorem for all odd primes. Euler was the first to publish anything; he established the theorem for the case that $n = 3$ (his proof had a gap, but it was filled later), but even this case is too complicated for inclusion in this book. In dealing with the equation

$$x^p + y^p = z^p,$$

where p is an odd prime, it has been found convenient to consider two cases. The first case of Fermat's last theorem is that p divides none of the numbers x, y, and z. This is the easy case; the theorem has been proved in this case for all p less than 253 747 889. In the second case of Fermat's last theorem, p divides one of the numbers x, y, and z. The theorem has been proved in this case for all p less than 4001.

It is thought that Fermat was probably mistaken in thinking that he had a proof for the general theorem, but the more romantic still cling to the belief that Fermat was correct. In this regard, it may be worthwhile to mention the fate of a conjecture of Euler. On the basis of everything known at that time, Euler conjectured in 1778 that for $n \geq 3$, no perfect nth power can be expressed as the sum of fewer than n nth powers of positive integers. Fermat's last theorem is an immediate corollary of Euler's conjecture (if true). Although no real progress had been made on Euler's conjecture, it was widely believed until 1966 that it was correct. In 1966, a computer search by L. J. Lander and T. R. Parkin found the example

$$27^5 + 84^5 + 110^5 + 133^5 = 144^5,$$

which disproved Euler's conjecture. This is the second example we have seen of a conjecture of Euler which was widely believed to be true for almost 200 years but was ultimately disproved. This is why mathematicians insist on proofs of statements, no matter how reasonable they seem.

Since Fermat was the originator of the modern subject of Diophantine equations, it is not surprising that others of his time did not always realize what was involved in solving such equations. The English mathematician John Wallis (1616–1703)[1] provides an example. For example, when Fermat

---

[1] He was one of Newton's teachers at Cambridge. The reader may have seen Wallis' infinite product for π:

$$\frac{\pi}{2} = \frac{2}{1} \cdot \frac{2}{3} \cdot \frac{4}{3} \cdot \frac{4}{5} \cdot \frac{6}{5} \cdot \frac{6}{7} \cdots .$$

challenged the English mathematicians to solve the Fermat–Pell equation

$$x^2 - dy^2 = 1,$$

Wallis produced the solution $x = 1$, $y = 0$. It is customary today to call this the trivial solution to the Fermat–Pell equation; more generally, solutions to equations that are visible at a glance (particularly if they involve 0) are called trivial. In answer to Fermat's challenge to solve the equation (rational numbers allowed)

$$x^3 + y^3 = a^3 + b^3,$$

Frenicle (1605–1675) computed a few examples, such as

$$9^3 + 10^3 = 1^3 + 12^3, \quad 9^3 + 15^3 = 2^3 + 16^3.$$

Wallis thereupon gave the solutions

$$27^3 + 30^3 = 3^3 + 36^3, \qquad (4\tfrac{1}{2})^3 + (7\tfrac{1}{2})^3 = 1^3 + 8^3,$$

and others. In each case, Wallis had merely taken a solution of Frenicle and multiplied it by some rational number. Frenicle pointed out that Wallis had merely made a trivial alteration of known solutions, but Frenicle was not that much better off. What Fermat desired was a general solution to his problem and not just isolated computations.

On another instance, Fermat announced that the equation

$$x^2 + 2 = y^3$$

has only $x = 5$, $y = 3$ as a solution in positive integers and the equation

$$x^2 + 4 = y^3$$

has only $x = 11$, $y = 5$ as a solution in positive integers. Wallis replied that such negative theorems have no interest and that one could easily give many other such results. He then gave four examples, of which the equation

$$x^4 + 9 = y^2$$

is representative. Apart from his opinion on such theorems, the untrained eye may not notice anything particularly different between Fermat's equations and Wallis'. Nevertheless, there is an important difference between the equations: All Wallis' equations could be factored in such a way that they could be solved instantly. We illustrate by finding all solutions of Wallis' equation in positive integers. We see that Wallis' equation may be written in the form

$$y^2 - x^4 = 9$$

or

$$(y + x^2)(y - x^2) = 9.$$

Since we are interested in $y > 0$, $x > 0$, the factor $y + x^2$ is positive; thus the other factor $y - x^2$ is positive since the product of the two factors is positive. There are only two ways of factoring 9 into a product of positive integers, $9 = 1 \cdot 9 = 3 \cdot 3$. Thus we have only three possibilities:

(a) $y + x^2 = 1$, $y - x^2 = 9$,

(b) $y + x^2 = 3$, $y - x^2 = 3$,

(c) $y + x^2 = 9$, $y - x^2 = 1$.

In each of the cases, there are two equations and two unknowns; the problem is thus reduced to an algebra problem. In case (a), we find $x = \pm\sqrt{-4}$, which is not an integer. In case (b), we find $x = 0$, which is not a positive integer. In case (c), we find $y = 5$, $x = \pm 2$, and hence the only solution to Wallis' equation in positive integers is $x = 2$, $y = 5$.

It would be very interesting to know Fermat's proofs of his theorems (as with almost everything else he stated, he did not make his proofs public). The two equations $x^2 + 2 = y^3$ and $x^2 + 4 = y^3$ can be solved fairly easily using the ideas of quadratic fields developed two centuries after Fermat's announcement. The reader interested in more historical details is referred to Bell's very readable history, *The Last Theorem*.

In the remainder of this chapter, we present three of the elementary methods that have been most used in solving Diophantine equations. It should be understood that when we say, "solve a Diophantine equation," we mean that we should find *all* solutions to it in integers.

EXERCISES

1. Find all solutions in positive integers to $x^2 + 12 = y^4$ (Wallis).
2. Find all solutions in positive integers to $x^3 + y^3 = 20$ (Wallis).
3. Find all solutions in positive integers to $x^3 - y^3 = 19$ (Wallis).
4. Find all solutions to $x^2 - dy^2 = 1$ when d is a perfect square.
*5. Show that 2 is the only prime which is the sum of two positive cubes. The word positive is necessary and hence must play a role in your proof; consider the examples $7 = 2^3 + (-1)^3$, $61 = 5^3 + (-4)^3$.

5.2. The Use of Congruences in Solving Diophantine Equations

Congruences often provide an easy way of showing that certain Diophantine equations have no solutions. As a first example of this method, we consider one of the Fermat–Pell equations. The Diophantine equations

$$x^2 - dy^2 = 1, \qquad x^2 - dy^2 = -1$$

in the unknowns x and y are called the **Fermat–Pell equations.** These equations are often just called Pell's equations, but this seems to have been a historical error of Euler which has been handed down. In fact, Pell never did anything with these equations, and they should really have been called the Fermat equations. The Fermat–Pell equations are important in many areas of number theory; we will encounter them again in Section 5.4 as well as in Chapters 7 and 8.

Consider the equation $x^2 - 7y^2 = -1$. Suppose $x = x_0$, $y = y_0$ is a solution to this equation. Then

$$x_0^2 \equiv x_0^2 - 7y_0^2 \equiv -1 (\text{mod } 7),$$

which is impossible by Theorem 3.29 since 7 is a prime $\equiv 3 (\text{mod } 4)$. Thus the equation $x^2 - 7y^2 = -1$ has no solutions. The same thing can be done for a large number of d's.

Theorem 5.1. Suppose that d is divisible by a prime $p \equiv 3 (\text{mod } 4)$ or that d is divisible by 4. Then the equation $x^2 - dy^2 = -1$ has no solutions.

Proof. Let $x = x_0$, $y = y_0$ be a solution to the equation $x^2 - dy^2 = -1$. Suppose that $p \equiv 3 (\text{mod } 4)$ and $p|d$ so that $d \equiv 0 (\text{mod } p)$. Then

$$x_0^2 \equiv x_0^2 - dy_0^2 \equiv -1 (\text{mod } p),$$

and this is impossible by Theorem 3.29. In like manner, if $4|d$, then

$$x_0^2 \equiv x_0^2 - dy_0^2 \equiv -1 (\text{mod } 4),$$

which is also impossible since $x_0^2 \equiv 0$ or $1 (\text{mod } 4)$ [that is, $x_0 \equiv 0, 1, 2,$ or $3 (\text{mod } 4)$ and therefore, $x_0^2 \equiv 0^2, 1^2, 2^2,$ or $3^2 \equiv 0$ or $1 (\text{mod } 4)$]. Thus our assumption that the equation $x^2 - dy^2 = -1$ has a solution has led to a contradiction and therefore the equation has no solutions. ▲

A close look at the above reveals that it is not really necessary to replace x by x_0 and y by y_0. The main idea is that if a Diophantine equation has no solutions (mod n), then it certainly has no solutions. For example, the equation

(1) $$x^2 - 5y^2 = 2$$

leads to the congruence equation $x^2 \equiv 2 (\text{mod } 5)$; this latter equation is impossible since the squares (mod 5) are $0^2, 1^2, 2^2, 3^2, 4^2 \equiv 0, 1,$ or $4 (\text{mod } 5)$. Thus there are no integral solutions to equation (1).

Let us take a slightly more complicated example. We shall show that the only integral solution to the equation

(2) $x^2 - 5y^2 = 3z^2$

is $x = y = z = 0$. Suppose there is an integral solution with not all three of x, y, z equal zero. Let $d = (x,y,z)$ and then set

$$x = dx_0, \qquad y = dy_0, \qquad z = dz_0$$

so that not all three of x_0, y_0, z_0 are zero and $(x_0,y_0,z_0) = 1$. Also, by dividing equation (2) by d^2, we get

$$x_0^2 - 5y_0^2 = 3z_0^2.$$

Therefore,

$$x_0^2 + y_0^2 \equiv x_0^2 - 5y_0^2 \equiv 3z_0^2 \equiv 0(\text{mod } 3).$$

By Theorem 3.30, $x_0 \equiv y_0 \equiv 0(\text{mod } 3)$. Therefore, $3|x_0$, $3|y_0$, $9|x_0^2$, $9|y_0^2$ and thus

$$9|(x_0^2 - 5y_0^2); \quad \text{that is, } 9|3z_0^2.$$

Thus $3|z_0^2$ and $3|z_0$. In other words, $3|x_0$, $3|y_0$, $3|z_0$ even though $(x_0,y_0,z_0) = 1$. This is a contradiction. Thus equation (2) has only the trivial solution $x = y = z = 0$.

If Theorem 3.30 was not available, we could still show that if

$$x_0^2 + y_0^2 \equiv 0(\text{mod } 3),$$

then $x_0 \equiv y_0 \equiv 0(\text{mod } 3)$. One way would be to notice that the squares (mod 3) are $0^2 \equiv 0(\text{mod } 3)$ and $1^2 \equiv 2^2 \equiv 1(\text{mod } 3)$, so that

$$x_0^2 + y_0^2 \equiv \begin{cases} 0(\text{mod } 3) & \text{if } x_0 \equiv y_0 \equiv 0(\text{mod } 3), \\ 1(\text{mod } 3) & \text{if } x_0 \equiv 0, y_0 \not\equiv 0 \text{ or } x_0 \not\equiv 0, y_0 \equiv 0(\text{mod } 3), \\ 2(\text{mod } 3) & \text{if } x_0 \not\equiv 0, y_0 \not\equiv 0(\text{mod } 3). \end{cases}$$

If 3 was replaced by a much larger prime p, there would be too many squares (mod p) to use this method effectively. The amount of work could be considerably reduced by reducing the problem to one of the equations

$$t^2 \equiv -1(\text{mod } p) \qquad \text{or} \qquad y_0 \equiv 0(\text{mod } p)$$

in the same manner used in the proof of Theorem 3.30.

EXERCISES

1. In solving equation (2), we assumed that there is a solution with x, y, z not all zero and arrived at a contradiction. Why is this contradiction not achievable when $x = y = z = 0$ is used?

2. Show that equation (2) can also be solved by using congruences (mod 5).

3. Can the equation $x^2 - 11y^2 = 3$ be solved by the methods of this section using congruences (mod 3) and, if so, what is the solution? (mod 4)? (mod 11)?

4. Same as problem 3 with the equation $x^2 - 3y^2 = 2$. (mod 3)? (mod 4)? (mod 8)?

5. Same as problem 3 with the equation $x^2 + 3y^2 = 2z^2$. (mod 3)? (mod 4)?

6. Same as problem 3 with the equation $x^4 + y^4 = 13z^4$. (mod 4)? (mod 5)? (mod 8)? (mod 13)?

7. Solve: $x^2 + 7y^2 = 3$.

8. Solve: $x^2 - 7y^2 = 3z^2$.

9. Solve using congruences (mod 3): $11x^2 + 10x - y^2 + 2 = 0$.

10. Solve using congruences (mod 4): $11x^2 + 10x - y^2 + 2 = 0$.

11. Show that 7 is not the sum of two rational squares.

5.3. Pythagorean Triples

Many Diophantine equations can be solved by using Theorem 2.12. We illustrate the application of Theorem 2.12. by finding all **Pythagorean triples,** that is, all right triangles with sides and hypotenuse having integral lengths. Similar triangles are not very interesting. Thus from the 3,4,5 triangle come an infinite number of similar triangles $6, 8, 10$; $9, 12, 15$; $12, 16, 20$; .... For this reason, we seek only the **primitive triangles,** in other words, those triangles x, y, z with $(x,y,z) = 1$. We adopt the convention that in the triangle x, y, z, the hypotenuse is z. However, the triangle x, y, z is the same triangle as y, x, z. We show that if $(x,y,z) = 1$, then exactly one of x and y is even; then we may assume that y is even without losing any triangles. If x and y are both even, then $2|(x^2 + y^2)$ and thus $2|z^2$ so that z is even; thus $2|(x,y,z)$, which is false. If x and y are both odd, then

$$z^2 \equiv x^2 + y^2 \equiv 1 + 1 \equiv 2 (\mathrm{mod}\ 4),$$

which is also impossible. Therefore, exactly one of x and y is even, the other is odd. We assume that x is odd and y even.

Theorem 5.2. There are infinitely many solutions to the equation

(3) $$x^2 + y^2 = z^2$$

with $(x,y,z) = 1$, y even, $x > 0$, $y > 0$, $z > 0$. These are given by

(4) $x = u^2 - v^2, \quad y = 2uv, \quad z = u^2 + v^2,$

where u and v are any integers whatsoever satisfying the conditions that $u > v > 0$, $(u,v) = 1$, and one of u and v is even. (Example: $u = 2$, $v = 1$, then $x, y, z = 3, 4, 5$; $u = 3$, $v = 2$, then $x, y, z = 5, 12, 13$.)

Proof. Since x is odd and y even, it follows that $z^2 \equiv 1 + 0 \pmod 4$ and thus z is odd. Thus $z + x$ and $z - x$ are both even:

$$2|(z + x), \qquad 2|(z - x).$$

It follows from (3) that

$$(z + x)(z - x) = y^2$$

or

$$\left(\frac{z + x}{2}\right)\left(\frac{z - x}{2}\right) = \left(\frac{y}{2}\right)^2,$$

where $(z + x)/2$, $(z - x)/2$, and $y/2$ are all integers. We will now show that $(z + x)/2$ and $(z - x)/2$ are relatively prime positive integers and then we will use Theorem 2.12. Note that if x and z have a common prime factor, then y also has that factor and this contradicts the fact that $(x,y,z) = 1$. Therefore, $(x,z) = 1$. Let

$$d = \left(\frac{z + x}{2}, \frac{z - x}{2}\right).$$

Then

$$d \left| \left(\frac{z + x}{2} + \frac{z - x}{2}\right) \right. ; \quad \text{that is, } d|z$$

and

$$d \left| \left(\frac{z + x}{2} - \frac{z - x}{2}\right) \right. ; \quad \text{that is, } d|x.$$

Hence

$$d|(x, z); \quad \text{that is, } d|1.$$

Thus $d = 1$. Since $x^2 < x^2 + y^2 = z^2$, it follows that $x < z$ and therefore $(z + x)/2$ and $(z - x)/2$ are positive. Since they are also relatively prime, it follows from Theorem 2.12 that there are positive integers u and v such that

$$\frac{z + x}{2} = u^2, \qquad \frac{z - x}{2} = v^2.$$

Thus

$$x = \frac{z + x}{2} - \frac{z - x}{2} = u^2 - v^2,$$

$$y = 2\left(\frac{z + x}{2}\right)^{1/2}\left(\frac{z - x}{2}\right)^{1/2} = 2uv,$$

$$z = \frac{z + x}{2} + \frac{z - x}{2} = u^2 + v^2.$$

If $d|u$, $d|v$, then $d^2|x$, $d^2|y$, $d^2|z$ and thus $(u,v) = 1$. If u and v are both odd or both even, then x is even; since x is odd, it follows that one of the pair u,v is even and the other odd. Finally, since $x > 0$ and u and v are positive, it must be that $u > v > 0$.

We now show conversely that if $x = u^2 - v^2$, $y = 2uv$, $z = u^2 + v^2$ with $u > v > 0$, $(u,v) = 1$, and one of u and v is even and the other odd, then $x^2 + y^2 = z^2$ with $x > 0$, $y > 0$, $z > 0$, y even and $(x,y,z) = 1$. It is easily checked that

$$(u^2 - v^2)^2 + (2uv)^2 = (u^2 + v^2)^2.$$

Also $u^2 - v^2 > 0$, $2uv > 0$, $u^2 + v^2 > 0$ and $2uv$ is even. It remains only to show that $(u^2 - v^2, 2uv, u^2 + v^2) = 1$. Suppose that $(u^2 - v^2, 2uv, u^2 + v^2) = d > 1$. Then there is a prime p dividing d and therefore $p|(u^2 - v^2)$, $p|2uv$, $p|(u^2 + v^2)$. Therefore,

$$p|[(u^2 + v^2) + 2uv]; \quad \text{that is, } p|(u + v)^2$$

and

$$p|[(u^2 + v^2) - 2uv]; \quad \text{that is, } p|(u - v)^2.$$

By Theorem 2.8, $p|(u + v)$, $p|(u - v)$. Therefore,

$$p|[(u + v) + (u - v)] \quad \text{or } p|2u,$$

$$p|[(u + v) - (u - v)] \quad \text{or } p|2v.$$

As a result $p|(2u,2v) = 2(u,v) = 2$. Therefore, $p = 2$. But since one of u and v is even and the other odd, $u^2 + v^2$ is odd and $2 \nmid (u^2 + v^2)$. This is a contradiction, and therefore d is not greater than 1. Therefore, $d = 1$. ▲

Just as in Section 1.1, we were able to get solutions to $y^2 = (z + x)(z - x)$ by letting $z + x = 2u^2$, $z - x = 2v^2$. Mathematicians are not really happy with infinitely many solutions to an equation; the feeling is that while it is nice to know whether or not a given Diophantine equation has infinitely many solutions, it is nicer still to know all the solutions. When one is just trying to show that a given equation has solutions, any procedure or assumption, no matter how unjustified, that leads to solutions is legitimate. On the other hand, in trying to show that one has all the solutions to the equation, these procedures and assumptions must be justified. In this regard, the key to showing that (4) gives all the solutions to (3) is clearly the part of the proof that showed that $(z + x)/2$ and $(z - x)/2$ were relatively prime positive integers and then the use of Theorem 2.12 to prove that $(z + x)/2$ and $(z - x)/2$ must be squares.

To again illustrate the difference between infinitely many solutions to an equation and all solutions, consider the equation (geometrically it is the problem of finding a rectangular box with integral sides and diagonal)

$$(5) \qquad\qquad x^2 + y^2 + z^2 = w^2.$$

Again, we may assume that $(x,y,z,w) = 1$ and that x, y, z and w are positive. (If $w = 0$, then $x = y = z = 0$, if any of x, y, z are 0, then we are left with a Pythagorean triplet.) To find solutions, let us assume that

$$(6) \qquad\qquad w = x + y.$$

Then (5) becomes

$$z^2 = 2xy,$$

which is much simpler. Clearly, z is even, so we put $z = 2u$ and then we get

$$2u^2 = xy.$$

Thus at least one of x and y is even. Say x is even, $x = 2v$. Then

$$u^2 = vy.$$

We let

$$(7) \qquad\qquad v = a^2, \qquad y = b^2.$$

This gives as a solution to (5),

$$(8) \qquad x = 2a^2, \qquad y = b^2, \qquad z = 2ab, \qquad w = 2a^2 + b^2.$$

Thus there are infinitely many solutions to (5). Does (8) give all the solutions to (5)? In a trivial sense, certainly not. For instance, (8) gives the solution $(x,y,z,w) = (8,1,4,9)$ but not $(1,4,8,9)$. There are, in fact, six rearrangements of any given solution to (5) found simply by interchanging the positions of x, y and z. We agree to call two solutions of (5) the same if they differ only in the order of x, y, and z. With this meaning, does (8) give all solutions to (5)? If so, then our two assumptions, (6) and (7), must be justified. Assumption (6) says that the sum of two of the x, y, z equals w; since the order of x, y, and z does not really matter, we let $x + y = w$. This assumption was made because it converted equation (5) to one much simpler. But it is impossible to justify the assumption. For example,

$$x = 3, \qquad y = 4, \qquad z = 12, \qquad w = 13$$

gives a solution to (5) in which w is not the sum of two of the x, y, z. Thus we have shown only that (5) has infinitely many solutions; we have not found all of them [see problem 1 for another infinite family of solutions of (5) found by assuming that two of the x, y, z are equal].

EXERCISES

1. Find all solutions to the equation $x^2 + 2y^2 = w^2$ with $x > 0$, $y > 0$, $w > 0$.
2. Find all solutions to the equation $x^2 + 3y^2 = z^2$ with $x > 0$, $y > 0$. $z > 0$.
3. Find all solutions to the equation $x^2 + py^2 = z^2$ with $x > 0$, $y > 0$, $z > 0$ where p is a fixed odd prime.
4. Find all solutions to the equation $x^2 + y^2 = 4z^2$ with $x > 0$, $y > 0$, $z > 0$, $(x, y, z) = 1$.
5. We have given one geometrical interpretation of equation (5) in the text. Here is another. Let OA, OB, and OC be three mutually perpendicular line segments. Let x, y, z, and w be the areas of triangles OAB, OAC, OBC, and ABC, respectively. Show that the relation between x, y, z, and w is that of equation (5).
6. Show that all solutions to (5) with the restriction (6) and the assumptions that $x > 0$, $y > 0$, $z > 0$, $w > 0$, $(x,y,z,w) = 1$, $2|x$ are given in (8), where we restrict a and b so that $a > 0$, $b > 0$, $(a,b) = 1$, and b is odd.
7. Find all solutions to the equation $x^2 + y^3 = z^2$, $x > 0$, $y > 0$, $z > 0$, $(x,y,z) = 1$.

5.4. Fermat's Method of Descent

Another method often used in Diophantine equations is the method of descent employed originally by Fermat. The main idea is to show that one solution to a Diophantine equation leads to another "smaller" solution to the same equation, which then leads to a still "smaller" solution, and so on. Thus one gets an infinite sequence of ever "smaller" solutions to the equation and this is often impossible; for instance, there is no infinite sequence of *positive* integers each less than the preceding one. We illustrate Fermat's method by showing that there are no solutions to the equation

(9) $x^4 + y^4 = z^4$, $x \neq 0$, $y \neq 0$, $z \neq 0$.

Actually, we prove even more. Since any fourth power is also a square, the following theorem shows that equation (9) has no solutions.

Theorem 5.3.   There are no solutions to the equation

(10) $x^4 + y^4 = z^2$ $x \neq 0$, $y \neq 0$, $z \neq 0$.

Proof. If $x = x_0, y = y_0, z = z_0$ is a solution to (10), then so is $x = |x_0| > 0$, $y = |y_0| > 0$, $z = |z_0| > 0$. Thus it suffices to show that (10) has no solutions with x, y, and z all positive. Suppose there is such a solution. Again we wish to make the assumption that $(x,y) = 1$, but since the powers of x, y, and z in (10) are different, we cannot just divide out the fourth power of (x,y) without a word of justification. Let $d = (x,y)$. Then put $x = dx_1$, $y = dy_1$ so that

$$d^4(x_1^4 + y_1^4) = z^2,$$

and therefore $(z/d^2)^2 = x_1^4 + y_1^4$ is an integer. Since z/d^2 is a rational number whose square is an integer, Theorem 2.13 then tells us that z/d^2 is an integer, say $z/d^2 = z_1$. Thus

$$x_1^4 + y_1^4 = z_1^2$$

with $(x_1,y_1) = 1$ and x_1, y_1, and z_1 all positive integers.

Thus if (10) has any solutions at all; then it has a solution with x, y, z all positive and $(x,y) = 1$. Let $x = x_1$, $y = y_1$, $z = z_1$ be such a solution. We will now show how it leads to another solution x_2, y_2, z_2, all positive, with $(x_2,y_2) = 1$ and $0 < z_2 < z_1$. This is the key to the method of descent. We note that

$$(x_1^2)^2 + (y_1^2)^2 = z_1^2.$$

Also if a prime $p|x_1^2$ and $p|y_1^2$, then $p|x_1$ and $p|y_1$, which is false since $(x_1,y_1) = 1$. Therefore, $(x_1^2, y_1^2) = 1$, and thus x_1^2, y_1^2, z_1 is a primitive Pythagorean triplet. In such triplets, exactly one of x_1^2 and y_1^2 is even, since the order of x_1 and y_1 does not matter, we assume y_1^2 is even and x_1^2 is odd. By Theorem 5.2, there exist positive integers u and v with

(11) $x_1^2 = u^2 - v^2$, $y_1^2 = 2uv$, $z_1 = u^2 + v^2$,

where $(u,v) = 1$, u is odd and v even or u even and v odd. It follows from (11) that

$$x_1^2 + v^2 = u^2,$$

and since $(u,v) = 1$, $(x_1,v,u) = 1$ and so x_1,v,u is a primitive Pythagorean triplet also. Since x_1 is odd, v is even (and thus u is odd). Therefore, there exist positive integers a and b such that

(12) $\qquad a^2 + b^2 = u, \qquad 2ab = v, \qquad (a,b) = 1.$

Since u is odd, $(u,2v) = d$ is odd and thus since $d|2v$, $d|v$. Therefore, d divides both u and v and thus $d = 1$; that is, u and $2v$ are relatively prime. Since by equation (11),

$$y_1^2 = u(2v),$$

Theorem 2.12 says that u and $2v$ are both squares of positive integers, say

$$u = z_2^2, \qquad 2v = c^2.$$

Since c must be even, we set $c = 2d$ and then $v = 2d^2$. Thus

$$ab = \frac{v}{2} = d^2.$$

But $(a,b) = 1$ and therefore Theorem 2.12 says that there are positive integers x_2 and y_2 such that

$$a = x_2^2, \qquad b = y_2^2.$$

Since $(a,b) = 1$, $(x_2,y_2) = 1$. Replacing a, b, and u by x_2^2, y_2^2, and z_2^2, equation (12) becomes

$$x_2^4 + y_2^4 = z_2^2.$$

We have already shown that x_2, y_2, and z_2 are positive integers with $(x_2,y_2) = 1$. Finally, since z_2 and v are positive integers,

$$0 < z_2 \le z_2^4 = u^2 < u^2 + v^2 = z_1.$$

The key parts of this inequality are $0 < z_2 < z_1$. This completes the descent part of the proof. Now what?

We have shown that if x_1, y_1, z_1 provide a solution to (10) with x_1, y_1, z_1 all positive and $(x_1,y_1) = 1$, then there is another solution x_2, y_2, z_2 to (10) with x_2, y_2, z_2 all positive, $(x_2,y_2) = 1$, and $0 < z_2 <·z_1$. Repeating the descent argument, there is another solution x_3, y_3, z_3 with x_3, y_3, z_3 all positive, $(x_3,y_3) = 1$, and $0 < z_3 < z_2$. We repeat the descent again and get

x_4, y_4, z_4 and then x_5, y_5, z_5 and then x_6, y_6, z_6, and so on. Each solution leads to another solution and, in particular, we get an infinite sequence of positive integers

$$z_1 > z_2 > z_3 > z_4 > z_5 > z_6 > \cdots.$$

This is impossible, since there are exactly $z_1 - 1$ positive integers between z_1 and 0; the sequence must end after at most z_1 terms. Thus the assumption that (10) has a solution has led to a contradiction. Therefore, equation (10) has no solutions.   ▲

The method of descent is an ingenious one. It is clear that the major difficulty is in finding the descent; the rest is duck soup. Sometimes the descent works just so far and then the inequalities cease to be correct. In this case, one may have to check a certain range of numbers to see that there are no solutions in that range. It may also happen that a descent turns into an ascent. We illustrate this with equation (5) of Section 5.3. In order to find solutions to (5), we make the assumption (unjustifiably in view of $x = 2, y = 6, z = 9,$ $w = 11$) that two of the x, y, z form the first part of a Pythagorean triplet. Since the order of $x, y,$ and z does not matter, we assume that x, y, t is a Pythagorean triplet. Then

$$t^2 + z^2 = x^2 + y^2 + z^2 = w^2$$

and thus t, z, w is also a Pythagorean triplet. In order to be able to use Theorem 5.2, we further assume that x, y, t and t, z, w are primitive triplets. Thus we shall attempt to find solutions (or show that there are none) to the equations

(13)
$$x^2 + y^2 = t^2, \qquad t^2 + z^2 = w^2,$$
$$(x,y,t) = (t,z,w) = 1, \qquad t > 0, \quad x > 0, \quad y > 0, \quad z > 0, \quad w > 0.$$

We begin by attempting a descent. Since x, y, t is a primitive Pythagorean triple, t is odd; one of x and y is even and since the order does not matter, we let x be odd and y even. One of t and z is also even and since t is odd, z is even. Thus, by Theorem 5.2, there exist integers $r, s, u,$ and v such that $(r,s) = (u,v) = 1, r > s > 0, u > v > 0,$ exactly one of r and s is odd and exactly one of u and v is odd, and

$$x = r^2 - s^2, \qquad y = 2rs, \qquad t = r^2 + s^2,$$
$$t = u^2 - v^2, \qquad z = 2uv, \qquad w = u^2 + v^2.$$

The important part of these equations is the part involving t,

$$t = r^2 + s^2 = u^2 - v^2.$$

As a result,

$$r^2 + s^2 + v^2 = u^2,$$

with $r > 0$, $s > 0$, $v > 0$, $u > 0$. Also, $(r,s,v,u) = 1$ since $(r,s) = 1$. Finally, $u \leq u^2 < u^2 + v^2 = w$. Thus we have gone from one solution to equation (5) to another with a smaller last variable.

Further, it should be clear how to go in the other direction. For instance,

$$1^2 + 2^2 + 2^2 = 3^2.$$

Put $u = 3$. Since now v must be even, we let $v = 2$. Since $r > s$, we let $r = 2$, $s = 1$. The values r, s, u, v = 2, 1, 3, 2 give

$$x = 3, \qquad y = 4, \qquad t = 5, \qquad z = 12, \qquad w = 13$$

as a solution to equation (13) and thus incidentally,

$$3^2 + 4^2 + 12^2 = 13^2$$

as another solution to equation (5). Now let $u = 13$. Since v is now even, we put $v = 12$ [$v = 4$ leads to $(r,s) = (12,3) \neq 1$]. Since $r > s$, we let $r = 4$ and $s = 3$. We then get

$$x = 7, \qquad y = 24, \qquad t = 25, \qquad z = 312, \qquad w = 313$$

as another solution to (13), which then gives another solution to (5), and so on. Thus (13) has infinitely many solutions.

Geometrically, equation (13) (without the g.c.d. conditions) is describing a rectangular box with integral edges with a diagonal on one of the rectangular faces being an integer and the diagonal of the box being an integer. We have just shown that there are infinitely many such boxes. It is interesting to note that this brings us very near an unsolved problem. It is known that there are infinitely many boxes whose edges are integers and the diagonals of all the faces are integers (an example is the box with edges 44, 117, and 240), but it is not known if there are any such boxes such that in addition the diagonal of the box itself shall be an integer. In equations, we are asking for a solution in integers to the four equations

$$x^2 + y^2 = t^2,$$
$$x^2 + z^2 = s^2,$$
$$y^2 + z^2 = r^2,$$
$$x^2 + y^2 + z^2 = w^2,$$

with $x \neq 0$, $y \neq 0$, $z \neq 0$. When nothing is known about equations such as this, it is no wonder that higher-degree equations drive mathematicians to the same despair to which homework problems drive the student. There seems to be no way of distinguishing between the trivially solvable problem and the extremely difficult problem, between the problem solvable by advanced techniques and those solvable using elementary methods. Indeed, it has happened more than once that an extremely difficult solution to what was thought to be a hard problem has later been replaced by a trivial elementary solution (trivial, that is, after it has been read).

As another example of a proof by ascent we have

Theorem 5.4. Suppose d is a positive integer. If there is one solution to the Fermat–Pell equation

(14) $$x^2 - dy^2 = 1, \quad x > 0, \quad y > 0,$$

then there are infinitely many solutions. If there is one solution to the Fermat–Pell equation

(15) $$x^2 - dy^2 = -1$$

with x and y positive, then there are infinitely many solutions to both (14) and (15).

Proof. Suppose there are integers a and b,

(16) $$a^2 - db^2 = c, \quad a > 0, \quad b > 0,$$

where c will shortly be chosen to be ± 1. Suppose further that for some $n \geq 1$ there are integers x_n and y_n such that

(17) $$x_n^2 - dy_n^2 = c^n, \quad x_n > 0, \quad y_n > 0.$$

When $n = 1$ this is possible with $x_1 = a$ and $y_1 = b$. Set

$$x_{n+1} = ax_n + dby_n, \quad y_{n+1} = ay_n + bx_n.$$

These values are legal since

$$
\begin{aligned}
x_{n+1}^2 - dy_{n+1}^2 &= (a^2 x_n^2 + 2dabx_n y_n + d^2 b^2 y_n^2) - d(a^2 y_n^2 + 2abx_n y_n + b^2 x_n^2) \\
&= a^2 x_n^2 + d^2 b^2 y_n^2 - da^2 y_n^2 - db^2 x_n^2 \\
&= (a^2 - db^2)(x_n^2 - dy_n^2) \\
&= c \cdot c^n \\
&= c^{n+1}
\end{aligned}
$$

and

(18)
$$x_{n+1} = ax_n + dby_n > 1 \cdot x_n + d \cdot 0 \cdot y_n = x_n(> 0),$$
$$y_{n+1} = ay_n + bx_n > 1 \cdot y_n + 0 \cdot x_n = y_n(> 0).$$

Thus we have shown by induction that if (16) is possible, then (17) is possible for all $n \geq 1$ with the values of x_n and y_n increasing with n by (18) (so that there are infinitely many distinct pairs x_n and y_n involved).

When $c = 1$, we have just shown that if (14) has one solution in positive integers, then it has infinitely many solutions. When $c = -1$, we have just shown that if (15) has one solution in positive integers, then both (14) and (15) have infinitely many solutions. [The solutions to (14) having n even and the solutions to (15) having n odd.] ▲

We have already seen in Theorem 5.1 that there are instances when (15) has no solutions. On the other hand, we will show in Chapter 7 that (14) always has solutions when d is not a perfect square and give a method of finding them.

EXERCISES

1. In this section we descended from one solution x, y, z, w to equation (5) with x, y, z, w all positive and having no common factor to another solution r, s, v, u, again all being positive with no common factor, and in addition $w > u$. Why does this not lead us to an infinite descent and a resulting proof that equation (5) has no solutions with none of x, y, z, w being zero?

2. The attentive reader will have noticed that in our proof that (13) has infinitely many solutions, we failed to show that one solution to (13) leads to another with the restrictions that x,y,t and t,z,w are primitive Pythagorean triplets. Complete the proof by showing that given one solution to (13), one can always pick r, s, u, and v from this solution in such a way that the resulting new solution does have $(x,y,t) = (t,z,w) = 1$.

3. Show that the equation

$$x^4 + 4y^4 = z^2, \qquad x \neq 0, \quad y \neq 0, \quad z \neq 0$$

has no solutions. It may be helpful to reduce this to the case that $x > 0$, $y > 0$, $z > 0$, $(x,y) = 1$, and then by dividing by 4 (if necessary) to further reduce this to where x is odd.

4. Show that there is no right triangle with integral sides whose area is a perfect square by showing that it suffices to work with primitive triangles and with them, one is led to the equation of problem 3. (*Hint:* Do not use Theorem 5.2.)

5. Suppose $x = x_0 > 0$, $y = y_0 > 0$ is a solution to

$$x^2 - 3y^2 = 6.$$

Show that $x_1 = 2x_0 + 3y_0$, $y_1 = x_0 + 2y_0$ is also a solution to the equation with $x_1 > x_0 > 0$, $y_1 > y_0 > 0$. Find one solution to the equation experimentally and show as a result that the equation $x^2 - 3y^2 = 6$ has infinitely many solutions. Give the next three solutions in the ascent.

6. Suppose n is a fixed integer (positive or negative) $\neq 0$ and suppose

$$x_0^2 - 3y_0^2 = n$$

with $x_0 \geq 0$, $y_0 \geq 0$. Let $x_1 = 2x_0 + 3y_0$, $y_1 = x_0 + 2y_0$. Show that

$$x_1^2 - 3y_1^2 = n, \qquad x_1 > x_0 \geq 0, \quad y_1 > y_0 \geq 0.$$

Conclude that the equation $x^2 - 3y^2 = n$ either has no solutions or infinitely many solutions.

7. Suppose n is a fixed integer $\neq 0$ and d is a positive integer not a perfect square, and suppose

$$x_0^2 - dy_0^2 = n,$$

where $x_0 \geq 0$, $y_0 \geq 0$. Let $x_1 = ax_0 + bdy_0$, $y_1 = bx_0 + ay_0$, where $a > 0$, $b > 0$ provides a solution to the Fermat–Pell equation, $a^2 - db^2 = 1$. Show that

$$x_1^2 - dy_1^2 = n, \qquad x_1 > x_0 \geq 0, \quad y_1 > y_0 \geq 0.$$

Conclude that the equation $x^2 - dy^2 = n$ either has no solutions or infinitely many solutions.

8. Show that there are infinitely many solutions to the equation

$$x^2 + y^2 = z^4, \qquad x > 0, \quad y > 0, \quad z > 0, \quad (x,y,z) = 1.$$

MISCELLANEOUS EXERCISES

1. Show that $x = y = z = 0$ is the only solution of

$$3x^5 + 5y^5 = z^5.$$

2. There are infinitely many pairs of nonzero integers such that the sum of their squares is a square; there are also infinitely many pairs of nonzero integers such that the difference of their squares is a square. Show that these two sets do not overlap; that is, show that there is no pair of nonzero integers such that both the sum and difference of their squares are squares.

3. A right triangle ABC has integral sides with $AC > BC > AB$. The bisector of $\angle ABC$ meets AC at D. E is the projection of A upon BD (that is, $AE \perp BD$) and F is the midpoint of AC. Show that there are infinitely many such triangles with $EF = 49$ and give three explicitly. (It may be helpful to let G be the midpoint of AB and show that GEF is a straight-line segment.)

4. Find a connection between problems 3 and 9 of Section 5.2.

5. If $x^3 + y^3 + z^3 = 0$, show that

$$(x + y + z)^3 = 3(x + y)(x + z)(y + z).$$

Use this relation to show that 3 must divide one of the numbers x, y, z. This proves case 1 of Fermat's last theorem when $n = 3$.

6. If $x^3 + y^3 + z^3 = 0$ and $(x,y,z) = 1$, use the relation of problem 5 to show that

$$(x + y, x + z) = (x + y, y + z) = (x + z, y + z) = 1$$

and that one of the numbers $x + y$, $x + z$, $y + z$ is nine times the cube of an integer while the other two are cubes of integers.

7. The number $t_n = 1 + 2 + \cdots + n$ is called the nth triangular number (so called because t_n dots may be inserted in an orderly fashion in an equilateral triangle; the fourth triangular number is familiar to bowlers). Some triangular numbers are also squares (which means that the dots may be rearranged into a square), for example, $t_8 = 6^2$, $t_{49} = 35^2$. Show that there are infinitely many numbers that are simultaneously triangular and square.

8. Show that if a and b are integers, not both zero, and

$$\alpha^2 + 2(a + b)\alpha = a^2 + b^2,$$

then α is irrational. Fermat had trouble with this problem—this was before he became interested in Diophantine equations.

9. If x_0, y_0, and z_0 are integers which satisfy the equation

$$x^2 + y^2 + z^2 = 3xyz,$$

then $x_1 = x_0$, $y_1 = y_0$, $z_1 = 3x_0y_0 - z_0$ are also solutions. Show how this can be used to find infinitely many solutions in positive integers starting with the solution 1, 1, 1.

10. Show that if $x^5 + y^5 + z^5 = 0$, then

$$2(x + y + z)^5 = 5(x + y)(x + z)(y + z)[(x + y + z)^2 + x^2 + y^2 + z^2].$$

Use this to show that 5 divides one of the numbers x, y, z, thus proving the first case of Fermat's last theorem when $n = 5$.

Chapter 6

NUMBERS, RATIONAL AND IRRATIONAL

6.1. Rational Numbers[1]

Up to now, we have mainly considered properties of integers. In Chapters 6, 7, and 8 we are going to enlarge our horizons. Theorem 2.13 has shown us that not all real numbers are rational, an example of an irrational number being $\sqrt{2}$. The fact that there are irrational numbers came as a great shock to the Greeks, and it necessitated a complete revision in similarity theory to account for irrational ratios of lengths. In this section we will be primarily concerned with the decimal expansions of rational numbers. We will, however, learn more than enough here to be able to construct some irrational numbers through their decimal expansions.

Consider the decimal expansions of several sample rational numbers,

$$\tfrac{1}{3} = .333333\ldots,$$

$$\tfrac{1}{6} = .1666666\ldots,$$

$$\tfrac{1}{8} = .12500000\ldots,$$

$$\tfrac{2}{7} = .285714285714285714\ldots,$$

$$\tfrac{83}{74} = 1.1216216216216\ldots.$$

In each case, there is a digit or group of digits which repeat, seemingly forever. Assuming that this is true, we will write

$$\tfrac{1}{3} = .\overline{3}, \quad \tfrac{1}{6} = .1\overline{6}, \quad \tfrac{1}{8} = .125\overline{0}, \quad \tfrac{2}{7} = .\overline{285714}, \quad \tfrac{83}{74} = 1.1\overline{216}.$$

The digits under the bar are called the **period** of the expansion and the number of digits in the period is called the **length of the period.** Ordinarily, the

---

[1] This is a good point to note that mathematicians use the term "rational number" in place of "fraction." Technically there is a fine distinction; for example, $\tfrac{1}{2}$ and $\tfrac{2}{4}$ are different fractions that both denote the same rational number.

expansion of $\frac{1}{8}$ is written .125 and is said to terminate, but it will be convenient shortly to think of $\frac{1}{8}$ as an infinite decimal with all zeros from the fourth decimal place on.

It is natural to conjecture that all rational numbers have periodic decimal expansions. To prove this, we must examine the division process. For example, two key steps in the expansion of $\frac{83}{74}$ into a decimal are

$$
\begin{array}{r}
1.1 \\
74\overline{)83.000000,}
\end{array}
\qquad
\begin{array}{r}
1.1216 \\
74\overline{)83.000000000}
\end{array}
$$

$$
\begin{array}{r}
74 \\ \hline
90 \\
74 \\ \hline
16
\end{array}
\qquad
\begin{array}{r}
74 \\ \hline
90 \\
74 \\ \hline
160 \\
148 \\ \hline
120 \\
74 \\ \hline
460 \\
444 \\ \hline
16
\end{array}
$$

Now we see why the expansion of $\frac{83}{74}$ is periodic. We have reached the remainder 16 twice. We know what happens when a remainder of 16 is reached: We bring down a zero, divide by 74, and get a remainder of 12, bring down a zero, divide by 74, and get a remainder of 46, bring down a zero, divide by 74, and get a remainder of 16. The whole sequence then repeats and then repeats again, and so on. Thus the expansion of $\frac{83}{74}$ repeats because two remainders in the division of 83 by 74 are equal. Why should there be two equal remainders? Because there are only the 74 possible remainders $0, 1, 2, \ldots, 73$. After at most 75 divisions, we must have two equal remainders; it has happened much earlier here, but this was due to the choice of rational numbers.

Before we prove the general result, it will be convenient to convert the division algorithm into symbols. Let the rational number involved be m/n, where m and n are positive integers. As we are not really interested in the integer to the left of the decimal point, we simply indicate that much of the division all at once,

$$ m = qn + a_1, \qquad 0 \le a_1 \le n - 1, $$

where q is the integer to the left of the decimal point and a_1 the remainder (and hence the inequality on a_1). We now examine separately the divisions leading to each digit to the right of the decimal point. We will let the digits to the right of the decimal point be $q_1, q_2, q_3, q_4, \ldots$, where possibly all q_j's past a certain point are zero. To find q_1, we bring down the first zero to the right of the decimal point, place it at the end of a_1, and divide by n. In symbols, this means that we divide $10a_1$ by n,

$$10a_1 = q_1 n + a_2, \qquad 0 \leq a_2 \leq n - 1,$$

where a_2 is the remainder in the division. To find q_2, we bring down the second zero to the right of the decimal point, place it at the back of a_2, and divide by n. In symbols, this is,

$$10a_2 = q_2 n + a_3, \qquad 0 \leq a_3 \leq n - 1.$$

This process is continued indefinitely; we thus get the sequence of equations

$$m = qn + a_1, \qquad 0 \leq a_1 \leq n - 1,$$
$$10a_1 = q_1 n + a_2, \qquad 0 \leq a_2 \leq n - 1,$$
(1) $\qquad 10a_2 = q_2 n + a_3, \qquad 0 \leq a_3 \leq n - 1,$
$$10a_3 = q_3 n + a_4, \qquad 0 \leq a_4 \leq n - 1,$$
$$10a_4 = q_4 n + a_5, \qquad 0 \leq a_5 \leq n - 1, \ldots.$$

If $a_j = 0$, then

$$q_j = 0, q_{j+1} = 0, q_{j+2} = 0, \ldots.$$

This is a special case of the decimal expansion of a_j/n,

(2) $\qquad \dfrac{a_j}{n} = .q_j q_{j+1} q_{j+2} q_{j+3} \cdots.$

For example, in the expansion of $\frac{83}{74}$ above, we get

$$\tfrac{9}{74} = .1216216216\ldots (j = 1),$$
$$\tfrac{16}{74} = .216216216\ldots (j = 2),$$
$$\tfrac{12}{74} = .162162162\ldots (j = 3),$$
$$\tfrac{46}{74} = .621621621\ldots (j = 4),$$
$$\tfrac{16}{74} = .216216216\ldots (j = 5).$$

These are in agreement with (2). In the case of the second and fifth of these, we have

$$.\overline{q_2 q_3 q_4} = .\overline{q_5 q_6 q_7}.$$

Theorem 6.1. Let m and n be positive integers. Then m/n has a decimal expansion which either terminates or is periodic.

Proof. A terminating decimal will be considered as a periodic decimal with period, $\bar{0}$. Thus we need make no further mention of terminating decimals in the proof. After the above discussion, it is clear that we must show that $a_j = a_k$ for some j and k ($j \neq k$). [If $a_j = a_k$, then $10a_j = 10a_k$, the quotient of these (equal) numbers by n, will be $q_j = q_k$ with remainders $a_{j+1} = a_{k+1}$, and so on.] Each a_j is one of the n numbers $0, 1, 2, \ldots, n-1$. Thus two of the $n+1$ numbers $a_1, a_2, \ldots, a_{n+1}$ must be equal, as otherwise there would be $n+1$ different values of the a_j's. $\blacktriangle$

It follows from the proof of Theorem 6.1 that the length of the period of m/n is at most n. We will learn much more about the length of the period in the next theorem. If the period starts with the first digit to the right of the decimal point, the expansion is called **purely periodic.** For example, the expansions of $\frac{1}{3}$ and $\frac{2}{7}$ on page 164 are purely periodic, while the others are not.

Theorem 6.2. Suppose that $(m,n) = (10,n) = 1$, where m and n are positive integers. Then the rational number m/n has a purely periodic decimal expansion and the length of the period is $\text{ord}_n(10)$.

Proof. We will show that there is an integer s such that $a_1 = a_{1+s}$. If we can accomplish this, then clearly the block of digits $q_1, q_2, \ldots, q_s$ repeats and hence the expansion of m/n is purely periodic. We will then worry about whether the period length is actually s or if it is smaller.

In terms of congruences, equations (1) are

$$m \equiv a_1 (\text{mod } n),$$

(3) $$10a_1 \equiv a_2 (\text{mod } n),$$

$$10a_2 \equiv a_3 (\text{mod } n), \ldots.$$

In particular, for all $t > 0$,

(4) $$a_{1+t} \equiv 10a_t \equiv 10^2 a_{t-1} \equiv 10^3 a_{t-2} \equiv \cdots \equiv 10^t a_1 (\text{mod } n).$$

(Note that the subscript goes down one as the exponent goes up one; the subscript went down t altogether and thus the final exponent is t.) Let

$$s = \text{ord}_n(10).$$

Then by (4) and the definition of $\text{ord}_n(10)$,

$$a_{1+s} \equiv 10^s a_1 \equiv a_1 (\text{mod } n).$$

But by (1) a_{1+s} and a_1 are both members of the complete residue system (mod n): $0, 1, 2, \ldots, n - 1$; being congruent, they must therefore be equal,

$$a_{1+s} = a_1.$$

Thus the decimal expansion of m/n is purely periodic.

Suppose now that r is the length of the period, so that

(5) $r \leq s$

and the first and second complete periods are $q_1, q_2, \ldots, q_r$ and q_{r+1}, $q_{r+2}, \ldots, q_{2r}$, respectively. Since the second period is just a duplicate of the first period, we see from (2) that

$$\frac{a_1}{n} = \overline{.q_1 q_2 \cdots q_r} = \overline{.q_{r+1} q_{r+2} \cdots q_{2r}} = \frac{a_{r+1}}{n}$$

and hence

(6) $a_{r+1} = a_1.$

By (3), (4), and (6),

$$10^r m \equiv 10^r a_1 \equiv a_{1+r} \equiv a_1 \equiv m(\text{mod } n),$$

and Theorem 3.3 allows us to divide out the m,

$$10^r \equiv 1(\text{mod } n).$$

By Theorem 3.26 and the definition of s,

$$s \mid r$$

and thus

$$r \geq s.$$

Combining this with (5), we see that

$$r = s = \text{ord}_n(10). \qquad \blacktriangle$$

For example, since

$$\tfrac{1}{17} = .\overline{0588235294117647}$$

has a period of 16 digits,

$$\text{ord}_{17}(10) = 16.$$

Hence 10 is a primitive root of 17. This is, in fact, the quickest method of seeing whether or not 10 is a primitive root of n when n is relatively small (assuming the availability of a desk calculator that divides). One simply finds the first $\phi(n)$ digits of the decimal expansion of $1/n$. Either this block of digits is the period of the expansion (and 10 is a primitive root of n) or this block is broken up into several copies of a smaller block (and 10 is not a primitive root of n). As an example, $\phi(21) = 12$ and the first digits in the expansion of $\frac{1}{21}$ are

$$\tfrac{1}{21} = .047619047619\ldots.$$

We see that the block of 12 digits breaks up into the block of six digits, 047619, repeated twice. Hence

$$\tfrac{1}{21} = .\overline{047619}$$

and 10 is not a primitive root of 21 (which is in line with the result stated earlier that 21 has no primitive roots). The converse of Theorem 6.2 [if $(m,n) = 1$ and m/n has a purely periodic decimal expansion, then $(n,10) = 1$] is considerably easier and is left as an exercise for the reader (problem 9).

EXERCISES

1. In the notation of (1), prove that if a_k is the first of the remainders to be repeated among later remainders and if a_{k+r} is the first of the remainders after a_k to equal a_k, then the first full period in the decimal expansion of m/n is $\overline{q_k q_{k+1} \cdots q_{k+r-1}}$ and its length is r.
2. Verify the result of problem 1 for the rational numbers

$$\tfrac{1}{55}, \quad \tfrac{2}{13}, \quad \tfrac{1}{11111}.$$

3. Find the rational number (as a fraction) which has the decimal expansion $.\overline{1}$. (*Hint:* $.\overline{1} = 10^{-1} + 10^{-2} + 10^{-3} + 10^{-4} + \cdots$. Use the formula for the sum of a geometric progression.)
4. Find the rational number which has the decimal expansion $.\overline{027}$. [*Hint:* $.\overline{027} = 27(10^{-3} + 10^{-6} + 10^{-9} + 10^{-12} + \cdots).$]
5. Find the rational number which has the decimal expansion $.2\overline{713}$. [*Hint:* $.2\overline{713} = \tfrac{27}{100} + 13(10^{-4} + 10^{-6} + 10^{-8} + \cdots).$]
6. Let $\alpha = .q_1 q_2 \cdots q_{k-1} \overline{q_k \cdots q_{k+r-1}}$. Use the ideas in problems 3, 4, and 5 to show that α is rational.
7. What rational number is $.\overline{9}$? Does this suggest any difficulties about decimal expansions of numbers?
8. Since

$$\tfrac{7}{21} = .\overline{3}, \qquad \tfrac{1}{21} = .\overline{047619},$$

Theorem 6.2 does need the restriction that m and n are relatively prime (otherwise the period length of $\frac{7}{21}$ and $\frac{1}{21}$ would be the same). Where does the proof of Theorem 6.2 use this fact?

9. If $m/n = .\overline{q_1 q_2 \cdots q_r}$, show that

$$\frac{m}{n} = \frac{q_1 q_2 \cdots q_r}{10^r - 1}$$

(the numerator is a number of the r digits, $q_1, q_2, \ldots, q_r$ and not the product of $q_1, q_2, \ldots, q_r$), and use this to show that if $(m,n) = 1$, then

$$(10,n) = 1.$$

10. If $(m,n) = (10,n) = 1$, why do the first $\phi(n)$ digits in the decimal expansion of m/n constitute an integral number of copies of the period as opposed to, for example, $3\frac{1}{2}$ copies of the period?

6.2. Irrational Numbers

We may use Theorem 6.1 to show that there are irrational numbers. For example, the number

$$\alpha = .1011011101111011111011111110\ldots$$

(blocks of $1, 2, 3, 4, \ldots$ ones separated by single zeros)

is irrational. For if α were rational, its decimal expansion would be periodic and have a period of length r starting with the kth digit of the expansion. But by the very nature of α, there will be blocks of r digits, all 1, in this expansion after the kth digit and the periodicity would then guarantee that everything after such a block of r digits would also be all ones. This contradicts the fact that there will always be zeros occurring after any given point in the expansion of α. Hence α is irrational.

On the other hand, Theorem 6.1 is of no use in deciding whether or not $\sqrt{2}$ or π are irrational. Both numbers are irrational, but since we do not know the complete decimal expansion of $\sqrt{2}$ or π,[2] Theorem 6.1 cannot be used to say that they are irrational. In the number α above, we know that there are arbitrarily many ones with no intervening zeros; we know no such fact about $\sqrt{2}$ or π. Thus, while it is easy to produce irrational numbers, it is harder to prove that particular numbers are irrational. The following theorem shows that if we know one irrational, we may find infinitely many others.

---

[2] By now, somebody may well have programed a computer to find the first million digits of the expansion of π, but this would not show anything, as the period (if there were one) could have 1 trillion digits.

Theorem 6.3. If α is irrational and a and b are rational with $b \neq 0$, then

$$a + b\alpha, \qquad \frac{1}{\alpha}$$

are irrational also.

Proof. Let

$$x = a + b\alpha.$$

Then

$$b\alpha = x - a,$$

$$\alpha = \frac{x - a}{b}.$$

Hence, if x is rational, $x - a$ is rational, $(x - a)/b$ is rational (the division being allowed since $b \neq 0$). But α is irrational, and hence x must be irrational also. Let

$$y = \frac{1}{\alpha}$$

($1/\alpha$ is defined since 0 is rational, and hence $\alpha \neq 0$). Then $y \neq 0$ and since

$$\alpha = \frac{1}{y},$$

α would be rational if y were rational. Therefore, y is irrational.   ▲

Theorem 6.4. Suppose a and b are rational and α is irrational. If $a + b\alpha$ is rational, then $b = 0$. If $a + b\alpha = 0$, then $a = b = 0$.

Proof. If $b \neq 0$, then, by Theorem 6.3, $a + b\alpha$ would be irrational and therefore $b = 0$. Since 0 is rational, $a + b\alpha = 0$ implies that $b = 0$; therefore,

$$a = a + 0\alpha = a + b\alpha = 0$$

also.   ▲

EXERCISES
In problems 1–3, do not use Theorem 2.13.
1. Given that $\sqrt{2}$ is irrational, are $\frac{7}{2} + \frac{4}{5}\sqrt{2}$ and $[(1 + \sqrt{2})(2 + \sqrt{2})]/(4 + 3\sqrt{2})$ rational or irrational?

2. Given that $\sqrt{2}$ is irrational, show that $(1 - 2\sqrt{2})/(3 + 4\sqrt{2})$ and $\sqrt{128}$ are irrational.
3. Given that $\sqrt{6}$ is irrational, show that $\sqrt{2} + \sqrt{3}$ is irrational.
4. Find the rule of formation of the number

$$.1234567891011121314151617181920 21 \ldots$$

and show that it is irrational.

*6.3. Liouville's Theorem and Transcendental Numbers

We may split the set of irrational numbers into two sets: the (real) irrational algebraic numbers and the real transcendental numbers.

Definition. A number, real or complex, is said to be **algebraic** if it is the root of an equation of the form

$$a_0 x^n + a_1 x^{n-1} + \cdots + a_{n-1} x + a_n = 0,$$

where $a_0, a_1, \ldots, a_n$ are integers, $a_0 \neq 0$. If a number is not algebraic, it is said to be **transcendental**. (Clearly rational numbers are algebraic.)

Just as it was by no means obvious that irrational numbers exist, so it was by no means obvious that transcendental numbers exist. Liouville in 1844 was the first to prove the existence of transcendental numbers. He proved that real algebraic numbers cannot be too closely approximated by rational numbers. This enabled him to construct transcendental numbers.

Before we prove Liouville's theorems, we should note one theorem from outside that we will use. This is the

Fundamental Theorem of Algebra (first proved by Gauss in 1799). If

$$f(x) = a_0 x^n + a_1 x^{n-1} + \cdots + a_n$$

is a polynomial with complex numbers as coefficients, $a_0 \neq 0$, then there are complex numbers $\alpha_1, \ldots, \alpha_n$ (not necessarily distinct) such that

$$f(x) = a_0(x - \alpha_1)(x - \alpha_2) \cdots (x - \alpha_n).$$

Theorem 6.5 (Liouville). Suppose that α is a root of the equation

$$a_0 x^n + a_1 x^{n-1} + \cdots + a_n = 0,$$

where $a_0, \ldots, a_n$ are integers, $a_0 \neq 0$. There exists a number $\delta > 0$ (dependent upon a_0 and the roots of the equation but independent of everything else) such that for any rational number $p/q \neq \alpha, q > 0$, we have

$$|\alpha - p/q| > \delta/q^n.$$

(*Note:* If α is irrational, the hypothesis $p/q \neq \alpha$ is automatic.)

Proof. For convenience, set

$$f(x) = a_0 x^n + \cdots + a_n,$$

which factors by the fundamental theorem of algebra as

(7) $$f(x) = a_0(x - \alpha)^b (x - \alpha_1)^{b_1} \cdots (x - \alpha_r)^{b_r},$$

where $\alpha, \alpha_1, \ldots, \alpha_r$ are all distinct and $b, b_1, \ldots, b_r$ are their respective multiplicities. Let

(8) $$\beta = \tfrac{1}{2} \text{ minimum of } |\alpha - \alpha_1|, \ldots, |\alpha - \alpha_r|$$

and our first requirement on δ will be

(9) $$\delta \leq \beta.$$

Thus if

$$\left|\alpha - \frac{p}{q}\right| > \beta,$$

then automatically,

$$\left|\alpha - \frac{p}{q}\right| > \delta \geq \frac{\delta}{q^n}.$$

Hence we will now restrict our attention for the rest of this proof to those p/q for which

$$\left|\alpha - \frac{p}{q}\right| \leq \beta, \qquad \frac{p}{q} \neq \alpha.$$

Our choice of β in (8) is such that any such p/q is as close (or closer) to α as it is to any of the other α_i's and in particular it cannot equal any of the α_i. In terms of inequalities, for any i, $1 \leq i \leq r$,

$$\left|\frac{p}{q} - \alpha_i\right| = \left|(\alpha - \alpha_i) - \left(\alpha - \frac{p}{q}\right)\right| \geq \left|\alpha - \alpha_i\right| - \left|\alpha - \frac{p}{q}\right| \geq 2\beta - \beta = \beta > 0.$$

Since p/q is not a root of the equation $f(x) = 0$, we see that

$$q^n f(p/q) \neq 0,$$

but

$$q^n f(p/q) = a_0 p^n + a_1 p^{n-1} q + \cdots + a_n q^n$$

is an integer and hence

$$|q^n f(p/q)| \geq 1.$$

Therefore, by (7),

(10) $$\left| \alpha - \frac{p}{q} \right|^b = \left| \frac{p}{q} - \alpha \right|^b \geq \frac{1}{|a_0| \cdot |p/q - \alpha_1|^{b_1} \cdots |p/q - \alpha_r|^{b_r} q^n}.$$

We now wish to get an upper estimate on the denominator. Let

$$\gamma = \text{maximum of} \, |\alpha - \alpha_1|, \ldots, |\alpha - \alpha_r|.$$

Thus

$$\left| \frac{p}{q} - \alpha_i \right| = \left| \left(\frac{p}{q} - \alpha \right) + (\alpha - \alpha_i) \right| \leq \left| \frac{p}{q} - \alpha \right| + |\alpha - \alpha_i| \leq \gamma + \beta;$$

it follows from (10) that

(11) $$\left| \alpha - \frac{p}{q} \right|^b \geq \frac{1}{|a_0|(\gamma + \beta)^{b_1} \cdots (\gamma + \beta)^{b_r} \cdot q^n}.$$

If in addition to (9), we now also require that

(12) $$\delta < [|a_0|(\gamma + \beta)^{b_1 + \cdots + b_r}]^{-1/b},$$

then we see from (11) that

(13) $$\left| \alpha - \frac{p}{q} \right| > \frac{\delta}{q^{n/b}} \geq \frac{\delta}{q^n}. \qquad \blacktriangle$$

Theorem 6.5 not only allows us to prove that there are transcendental numbers but it allows us to exhibit some. As an example we have

Theorem 6.6. The number

$$\alpha = \sum_{k=1}^{\infty} 2^{-k!} (= 2^{-1} + 2^{-2} + 2^{-6} + 2^{-24} + 2^{-120} + \cdots)$$

is transcendental.

Proof. Suppose α is algebraic. Then there is a positive integer n and a number $\delta > 0$ such that for any rational number $p/q \neq \alpha, q > 0$,

(14) $$\left| \alpha - \frac{p}{q} \right| > \frac{\delta}{q^n}.$$

We will now choose a rational number ($\neq \alpha$) that violates this inequality. Let

(15)
$$q = 2^{K!}, p = \sum_{k=1}^{K} 2^{K!-k!},$$

so that

$$\frac{p}{q} = \sum_{k=1}^{K} 2^{-k!}.$$

The number K is still open to be chosen. We will ultimately choose it to be rather large. We now wish to estimate $\alpha - p/q$. We have

(16)
$$\alpha - \frac{p}{q} = \sum_{k=K+1}^{\infty} 2^{-k!} < \sum_{m=(K+1)!}^{\infty} 2^{-m}.$$

This last sum is a geometric progression and hence we can evaluate it,

$$\alpha - \frac{p}{q} < \frac{2^{-(K+1)!}}{1 - \frac{1}{2}} = 2 \cdot 2^{-(K+1)!} = 2 \cdot (2^{K!})^{-(K+1)} = 2q^{-(K+1)}.$$

Now by (16), $\alpha - p/q > 0$ (and hence $p/q \neq \alpha$), so that

$$\left| \alpha - \frac{p}{q} \right| < \frac{2q^{-1}}{q^K} = \frac{2 \cdot 2^{-K!}}{q^K}.$$

This contradicts (14) if $K \geq n$ and simultaneously K is chosen large enough that

$$2 \cdot 2^{-K!} < \delta$$

(this being clearly possible since $2^{-K!} \to 0$ as $K \to \infty$). Hence α is indeed transcendental.   ▲

As with irrational numbers, it is usually easier to construct a transcendental number than to prove that a given number is transcendental. For instance, e and π are both transcendental but the proofs are considerably more difficult than the proof of Theorem 6.6. The interested reader may find proofs in Chapter 11 of Hardy and Wright. The fact that $\sqrt{\pi}$ is transcendental is sufficient to show that the classical Greek problem of constructing (with straightedge and compass) a square with area equal to that of a given circle cannot be solved. This is because the problem is equivalent to constructing a line segment $\sqrt{\pi}$ times the radius of the circle in length. But through analytic geometry, we find that only algebraic multiples of the radius can be constructed (and not even all of them). A discussion of this and other classical problems can be found in the book by Courant and Robbins (see the Bibliography).

Theorem 6.5 is of interest in another way. Theorem 1.1 shows us that given a real algebraic number α, there are infinitely many p/q such that

$$\left| \alpha - \frac{p}{q} \right| \leq \frac{1}{2q}.$$

In Chapter 7 we will improve this: For real irrational algebraic α,

(17)
$$\left| \alpha - \frac{p}{q} \right| < \frac{1}{2q^2}$$

is true for infinitely many different p/q with q being arbitrarily large. But yet for each irrational algebraic α, there is an integer n and a positive δ such that

(18)
$$\left| \alpha - \frac{p}{q} \right| > \frac{\delta}{q^n}$$

is true for all rational p/q (here n is the degree of the equation satisfied by α). It is of some interest to close the gap between (17) and (18). In this vein, we may ask:

If α is an irrational algebraic number, for which θ does there exist $\delta > 0$ (depending on α and θ) such that for all rational numbers $p/q(q > 0)$

$$\left| \alpha - \frac{p}{q} \right| > \frac{\delta}{q^\theta}?$$

Many people have worked on this problem. The main historical results are:

$$\theta \geq n \qquad \text{(Liouville, 1844)},$$

$$\theta > \frac{n}{2} + 1 \qquad \text{(Thue, 1909)},$$

$$\theta > 2\sqrt{n} \qquad \text{(Siegel, 1921)},$$

$$\theta > 2 \qquad \text{(Roth, 1955)},$$

are all answers to this question. As we see from (17), if $\theta < 2$ then there cannot be any such δ. The problem of whether or not $\theta = 2$ is an answer to this question is still open (except for $n = 2$, when Liouville's result already answers the question in the affirmative). It is conjectured that $\theta = 2$ does not answer the question (except if α is the root of a quadratic equation).

An interesting fact about the later results of Thue, Siegel, and Roth is that while they say for certain values of θ there exist values of δ, their proofs do not allow one to determine values of δ. It is only recently that there has been any hope of giving explicit values of δ for the general irrational algebraic number (with $n \geq 3$) and values of $\theta < n$; however, this seems (as of now) still possible only for values of θ very close to n. These results have interesting implications in the theory of Diophantine equations (for a glimpse of these, see miscellaneous exercises 9 and 10).

EXERCISES
1. If $b > 1$, then the result of equation (13),

$$\left| \alpha - \frac{p}{q} \right| > \frac{\delta}{q^{n/b}},$$

   is a better inequality than the result of Theorem 6.5. Show that if n is the smallest number such that there is an equation of degree n (integral co-efficients) satisfied by α then $b = 1$. [*Hint:* Let $f(x) = 0$ be the equation and suppose $b > 1$. Show that α is a root of $f'(x) = 0$ also.] (Clearly what is desired is the minimum of n/b. It can be shown that this minimum occurs with the minimum n and $b = 1$).
2. In Theorem 6.5, if $f(x)$ factors as $f(x) = a_0(x - \alpha)^n$, then the proof must be modified. Show that the theorem holds with $\delta < |a_0|^{-1/n}$.
3. By examining the coefficient of x^{n-1}, show that the situation of problem 2 arises only if α is rational.
4. It can be shown that sums, differences, products, and quotients of algebraic numbers are algebraic. Use this result to show that if α is transcendental, β is algebraic and not 0, then $\alpha + \beta$, $\alpha\beta$, and α^{-1} are transcendental. Show further that for all positive integers n, $\alpha^{1/n}$ is transcendental.
5. Suppose α is transcendental. Show that if k is a positive integer, then α^k is transcendental. (*Hint:* Suppose α^k is algebraic and write an equation for α^k. Show how this gives an equation for α.)
6. Use the results of problems 4 and 5 to show that any nonzero rational power of a transcendental number is transcendental.

MISCELLANEOUS EXERCISES
1. Suppose that $(n,10) = 1$. Show that $\frac{1}{10}$ has a periodic expansion in the base n of period length 1, 2, or 4 according to whether $n \equiv 1 \pmod{10}$, $n \equiv 9 \pmod{10}$, or $n \equiv 3, 7 \pmod{10}$.
2. In each of the expansions

$$\tfrac{1}{7} = .\overline{142857}, \qquad \tfrac{1}{11} = .\overline{09}, \qquad \tfrac{1}{17} = .\overline{0588235294117647},$$

the first half of the period plus the second half of the period gives all nines. Show that if p is a prime and the length of the period of the decimal expansion of $1/p$ is even, then the sum of the first and second halves of the period has all nines.

3. Find all primes p such that the period of the decimal expansion of $1/p$ has exactly six digits.

4. If $(m,n) = 1$ and $n = 2^a 5^b n_0$ with $(n_0, 10) = 1$, show that the period in the decimal expansion of m/n begins with the jth digit after the decimal point and has length $\mathrm{ord}_{n_0}(10)$, where $j = 1 + \max(a,b)$. (*Hint:* Multiply m/n by appropriate powers of 10 and use problem 9, Section 6.1.)

5. Suppose the number $e = \sum_{n=0}^{\infty} 1/n!$ is rational, say $e = p/q$. Show that this implies that the number

$$\alpha = q!\left(e - \sum_{n=0}^{q} 1/n! \right)$$

is an integer. Show that, in fact, $0 < \alpha < 1$, so that e is irrational.

6. We have seen in problem 7, Section 6.1, that not every number has a unique decimal expansion; for example, $\frac{1}{8} = .125 = .124\overline{9}$. This fact temporarily negates the proof that $\alpha = .10110111011110\ldots$ is irrational, since α may have another decimal expansion which is periodic. This problem will show, among other things, that the decimal expansion of α is unique. Show that a number β has a unique decimal expansion unless β has a terminating decimal expansion. Show further that if β has a terminating decimal expansion, then there is exactly one other decimal representation of β and it is periodic with period, $\overline{9}$. (*Hints:* Let $\beta = b . b_1 b_2 b_3 \ldots = c . c_1 c_2 c_3 \ldots$, where $0 \leq b_j \leq 9, 0 \leq c_j \leq 9$ for all $j \geq 1$. This means that

$$\beta = b + \sum_{j=1}^{\infty} \frac{b_j}{10^j} = c + \sum_{j=1}^{\infty} \frac{c_j}{10^j}.$$

Show that if not all the b's and c's are the same, then the problem can be reduced to consideration of the above equation with $b \geq 1$, $c = 0$ by multiplying by an appropriate power of 10 and subtracting an appropriate integer. Show that if $b \geq 1$ and $c = 0$, then

$$b + \sum_{j=1}^{\infty} \frac{b_j}{10^j} \geq 1 \geq \sum_{j=1}^{\infty} \frac{c_j}{10^j}$$

with equality if and only if $b = 1$, $b_j = 0$, and $c_j = 9$ for all j.)

The remaining problems are related to Section 6.3.

7. Show that if α is transcendental, then so is $\sqrt{\alpha}$. [*Hint:* Suppose $f(x)$ is a polynomial with integral coefficients and $f(\sqrt{\alpha}) = 0$. Show that $g(x) = f(\sqrt{x})f(-\sqrt{x})$ is also a polynomial with integral coefficients and $g(\alpha) = 0.$]

8. (For those students familiar with complex numbers to complex powers.) The Gelfond–Schneider theorem states that if α and β are algebraic numbers, $\alpha \neq 0$, $\alpha \neq 1$, β irrational, then α^β is transcendental (no matter which value of the logarithm is used to define α^β). Correct to ten decimal places,

$$e^{\pi\sqrt{163}} = 262537412640768744\,.\,0000000000.$$

Show that, nevertheless, $e^{\pi\sqrt{163}}$ is not an integer.

9. It follows from the result of Thue that for all rational p/q, $q > 0$,

$$\left| \sqrt[3]{2} - \frac{p}{q} \right| > \frac{\delta}{q^{2.955}}$$

for some $\delta > 0$. Given an integer n, show that the Diophantine equation

$$x^3 - 2y^3 = n$$

has only finitely many solutions. [*Hints:* The other two cube roots of 2 are

$$\alpha_1 = \sqrt[3]{2}\left(\frac{-1 + i\sqrt{3}}{2}\right), \qquad \alpha_2 = \sqrt[3]{2}\left(\frac{-1 - i\sqrt{3}}{2}\right),$$

both of which have imaginary parts greater than 1 in absolute value. Show as a result that

$$\left| p^3 - 2q^3 \right| = q^3 \left| \frac{p}{q} - \sqrt[3]{2} \right| \cdot \left| \frac{p}{q} - \alpha_1 \right| \cdot \left| \frac{p}{q} - \alpha_2 \right| > \delta q^{.045}.]$$

10. (See problem 9.) We now know that there are only finitely many solutions to the equation

$$x^3 - 2y^3 = n.$$

How do we find them? Suppose we are prepared to check every value of y in absolute value less than Y and see if there is an integer x that corresponds to it. What value of Y is safe; in other words, for what value of Y can we guarantee there are no solutions with $|y| > Y$? We sadly cannot answer this question from problem 9 without knowing the value of δ

involved and unfortunately, Thue's proof of his result does not enable one to calculate a value of δ. His proof is an existence proof only. Baker showed in 1964 that $\delta = 10^{-6}$ will do in problem 9. Use this to show that if

$$|x^3 - 2y^3| \leq 10,$$

then

$$|y| < 10^{156}.$$

It is now feasible to use computers to check this range of y. To learn how this would be done, see miscellaneous exercise 25, Chapter 7.

11. Set $x_1 = .\overline{123456790}$, $x_2 = .\overline{101112\ldots969799000102\ldots0809}$,

    $x_3 = .\overline{100101102\ldots996997999000001002\ldots008009}$, etc. Show that

$$x_n = \frac{10^{n-1}(10^n - 1) + 1}{(10^n - 1)^2}.$$

[*Hint:* Multiply x_n by 10^n and subtract x_n from this; the result should be $(10^n - 1)x_n = 10^{n-1} + .00\ldots01$, where there are $n - 1$ zeros in the period.] Set

$$\alpha = .1234567891011121314151617181920 21\ldots.$$

We have seen in Problem 4 on page 172 that α is irrational. Here we shall learn more. Let N_n be the number of digits in α through $10^n - 1$ (for example, $N_1 = 9, N_2 = 9 + 2 \cdot 90 = 189$). Show that $N_n > 10N_{n-1}$. Show that $10^{N_{n-1}}\alpha$ and x_n agree after the decimal point for more than $8N_{n-1}$ digits. Let $q_n = 10^{N_{n-1}}(10^n - 1)^2$. Show that $q_n < 10^{2N_{n-1}}$. Show as a result that there is an integer p_n ($p_1 = 10, p_2 = 123\,456\,789 \cdot 99^2 + 991$) such that

$$\left| \alpha - \frac{p_n}{q_n} \right| < \frac{1}{10^{9N_{n-1}}} < \frac{1}{q_n^{4.5}}.$$

Show that Theorem 6.5 implies that α is not the root of a quadratic, cubic, or quartic equation with integral coefficients. Show that the result of Roth on page 176 implies that α is, in fact, transcendental.

Chapter 7

CONTINUED FRACTIONS FROM A
GEOMETRIC VIEWPOINT

7.1. Introduction

The goal of this chapter is to develop a method for finding "good" fractional approximations to a given real number. As we will proceed geometrically, we will first give precise definitions of some intuitive notions.

Definition. Let the point P have coordinates (b,a)[1] and suppose that (b,c) is the intersection of a line L with the vertical line $x = b$. If $a > c$, then we say that P is **over**, or **above**, L, while if $a < c$, we say that P is **under**, or **below**, L.

Definition. Let the point P have coordinates (b,a) with $b \neq 0$. We say that a/b is the **slope** of P. In other words, the slope of P is the slope of the line passing through the origin, $(0,0)$, and P.

Definition. The **first quadrant** is the set of points with positive x and y coordinates.

Theorem 7.1. Let L be a line passing through the origin having the slope $\alpha > 0$. If P is a point in the first quadrant, then P is above L if and only if slope $P > \alpha$, P is on L if and only if slope $P = \alpha$, P is below L if and only if slope $P < \alpha$.

Proof. Let the coordinates of P be (b,a). The equation of L is

$$y = \alpha x,$$

---

[1] We will not use the greatest common divisor notation in this chapter.

and thus the intersection of L with the line $x = b$ is the point $(b, \alpha b)$. Since $b > 0$ (P is in the first quadrant), we see that

$$a > b\alpha \qquad \text{if and only if } \frac{a}{b} > \alpha,$$

$$a = b\alpha \qquad \text{if and only if } \frac{a}{b} = \alpha,$$

$$a < b\alpha \qquad \text{if and only if } \frac{a}{b} < \alpha.$$

The theorem follows from the definitions of over, under, and the slope of P. ▲

Let α be a positive real number and L be the line $y = \alpha x$. Let the point $P = (q, p)$ be in the first quadrant and let L' be the line through the origin and P (see Figure 7.1).

It seems intuitively clear that the closer the point (q, p) is to the line L, the closer p/q (slope of L') will be to α (slope of L). We make this intuitive feeling precise in the following theorem.

Theorem 7.2. Let α be a positive real number and L be the line $y = \alpha x$. Suppose that the point (q, p) has integral coordinates, is in the first quadrant, and has the property that if (n, m) is also a point with integral coordinates and $0 < n \le q$, then the distance from (q, p) to L is less than or equal to

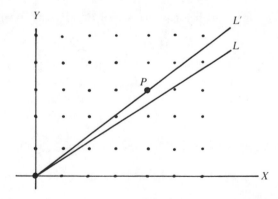

Figure 7.1.

the distance from (n,m) to L. Then

$$\left| \alpha - \frac{p}{q} \right| \leq \left| \alpha - \frac{m}{n} \right|$$

for all points (n,m) with $0 < n \leq q$ [in other words, $\alpha - (p/q)$ is numerically smaller than $\alpha - (m/n)$].

Proof. If the point (q,p) is on L (so that $p/q = \alpha$), then the result of the theorem is obvious. So we may as well assume that (q,p) is off the line L and then we let $d > 0$ be the distance from (q,p) to L [measured along the perpendicular line from (q,p) to L]. Let P be the point (q,p). P is either above or below L. Figure 7.2 shows the case where P is below L; it is this case that we consider now. Let L_1 be the line through P which is parallel to L; thus the distance between L_1 and L is d. Let L_2 be parallel to L and at the distance d from L but on the other side of L from L_1. Let Q and R be the intersections of the line $x = q$ with L_2 and L, respectively. Finally, let A and B be the points on L such that AP and BQ are perpendicular to L. Thus $AP \parallel BQ$, and therefore $\angle APR = \angle BQR$. Thus $\triangle APR \cong \triangle BQR$ (by the angle, side, angle theorem) and hence $QR = RP$. The x coordinate of R is q, and since R is on the line $y = \alpha x$, we see that $R = (q,q\alpha)$. Now we can find the coordinates of Q. The x coordinate of Q is q. Since $QR = RP$, the y coordinate of Q satisfies the equation

$$y - q\alpha = q\alpha - p,$$

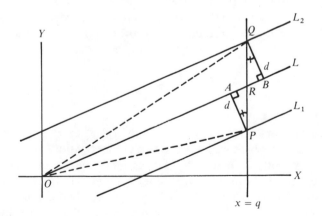

Figure 7.2.

and therefore

$$Q = (q, 2q\alpha - p).$$

The slopes of OP and OQ are p/q and $(2q\alpha - p)/q = 2\alpha - (p/q)$, respectively. Suppose now that

$$\left| \alpha - \frac{p}{q} \right| > \left| \alpha - \frac{m}{n} \right|,$$

where $C = (n,m)$ has integral coordinates, $0 < n \le q$. As shown in Figure 7.2, slope $L >$ slope OP, so that $\alpha > p/q$ and thus $|\alpha - (p/q)| = \alpha - (p/q)$. One of the numbers $\alpha - (m/n)$ and $(m/n) - \alpha$ is the positive number $|\alpha - (m/n)|$, the other is negative and therefore

$$\alpha - \frac{p}{q} > \alpha - \frac{m}{n}, \qquad \alpha - \frac{p}{q} > \frac{m}{n} - \alpha.$$

From the first of these inequalities, it follows that

$$\text{slope } C = \frac{m}{n} > \frac{p}{q} = \text{slope } OP,$$

and from the second,

$$\text{slope } OQ = 2\alpha - \frac{p}{q} > \frac{m}{n} = \text{slope } C.$$

Thus C is above OP and below OQ, so that C is between the lines OP and OQ (and not on either). Also since we are assuming that $0 < n \le q$, we see that C is actually inside the triangle OPQ (or on the edge PQ, but $C \ne P$ and $C \ne Q$, since C is not on OP or OQ). Therefore, the distance from C to L is less than d and this contradicts our hypothesis that this does not happen. Therefore,

$$\left| \alpha - \frac{p}{q} \right| \le \left| \alpha - \frac{m}{n} \right|$$

as asserted.   ▲

We have of course only proved Theorem 7.2 when P is under L. In this and later theorems where we have the two choices of putting a point over or under a line, we will take one of the choices and leave the other choice for the reader. As expected, the proofs are practically identical with the modifications usually restricted to the words "over" and "above" being interchanged with "under" and "below". There are also occasional changes of directions of inequalities and changes of sign.

We see from Theorem 7.2 that if we wish to find "good" fractional approximations to α, then it might well be useful if we could find the points (q,p) with the property in Theorem 7.2.

Definition. Let L be the line $y = \alpha x$. We define the **set of points closest to** L to be the set of all points (q,p) such that q and p are integers, $q \geq 1$, and having the property that if n and m are integers, $0 < n \leq q$, then the distance from (q,p) to L is less than or equal to the distance from (n,m) to L with these distances being unequal if $n < q$.

For example, when $\alpha = \sqrt{3}$, the first few members of the set of points closest to L are the points $(1,2), (3,5), (4,7), (11,19), (15,26)$ (see Figure 7.4). When $\alpha = \pi$, the first two closest points to L are $(1,3)$ and $(7,22)$ (we will verify this later). This to some degree shows why $\frac{22}{7}$ is used as an approximation to π; by Theorem 7.2, $\frac{22}{7}$ is closer to π numerically than any fraction with smaller denominator. In fact, anyone who wishes to waste time can discover that there is no point (n,m) with integral coordinates in the range $1 \leq n \leq 100$ that is closer to the line $y = \pi x$ than the point $(7,22)$. This brings up the question as to whether the set of points closest to L is finite or infinite. When α is rational, the set of points closest to L is finite (see problem 5), but when α is irrational, it is not clear what happens. We will see later that when α is irrational, the set of points closest to L is infinite.

EXERCISES
1. Complete the proof of Theorem 7.2 by considering the case that P is above L.
2. Is the proof to Theorem 7.1 valid when $\alpha < 0$ and P is in the first or fourth quadrants?
3. Find the set of closest points to the line $y = \frac{3}{2}x$.
4. Show that if there are two different integral points with x coordinate 1 equally close to the line $y = \alpha x$, then α is a rational number with denominator 2.
5. Show that if α is rational and L is the line $y = \alpha x$, then the set of points closest to L is a finite set.
6. Show that if (s,r) and (u,t) are distinct points with integral coordinates $(s > 0, u > 0)$ which are equally close to the line $L : y = \alpha x$, then either $(s + u, r + t)$ or $(s - u, r - t)$ is on L. Use this to show that α is rational.

7.2. The Continued Fraction Algorithm

After Theorem 7.2, it is desirable to find an algorithm for locating points close to the line $L : y = \alpha x$, in fact, the closer the better. In this section we shall describe an algorithm which by its very definition is producing points closer and closer to L. Thus it will be obvious from the start that we will be getting "good" fractional approximations to α. It will turn out that our algorithm actually produces the set of points closest to L and hence gives the "best" fractional approximations to α.

It is assumed that the reader is familiar with the material on vectors in Appendix A. Our notation for a vector is the same as our notation for a point, and we will use the two concepts interchangeably. It is for this reason that we will not use the customary boldface print or arrowheads in our drawings. In fact, we would not even introduce the word "vector" if the reader were familiar with the idea of adding points. When we seem to be adding points, the reader may interpret the points as vectors and add them by the usual method of vector addition. The final sum may then be reinterpreted as being a point. Thus we add points by adding their respective coordinates.

Our algorithm is based on the following theorem.

Theorem 7.3. Given a line L passing through the origin and points U_1 and U_2 on opposite sides of L, there is a unique integer a such that either $U_1 + aU_2$ is on L or $U_1 + aU_2$ and $U_1 + (a + 1)U_2$ are on opposite sides of L. The point $U_1 + aU_2$ is closer to L than U_2 and if it is not on L, then $U_1 + aU_2$ is on the same side of L as U_1. Further, if U_1 is closer to L than U_2, then $a = 0$ while if U_2 is as close or closer to L than U_1, then $a \geq 1$.

Proof: Figure 7.3 makes the theorem obvious, but we sketch the proof anyway. Let A be the point on L such that $AU_2 \perp L$. Let L_1 be the line through U_1 which is parallel to L and let B be the point on L_1 such that $B(U_1 + U_2) \perp L_1$. The parallelogram rule of adding vectors says that OU_2 is parallel to $U_1(U_1 + U_2)$ and has the same length. Thus since $L \| L_1$,

$$\angle U_2 OA = \angle (U_1 + U_2)U_1 B.$$

It follows that the right triangles $U_2 OA$ and $(U_1 + U_2)U_1 B$ are congruent. Thus

$$AU_2 = B(U_1 + U_2).$$

Let d_1 and d_2 be the distances of U_1 and U_2 from L. Then the length of $B(U_1 + U_2)$ is d_2 and thus if $d_1 < d_2$, the point $U_1 + U_2$ will be on the other side of L from U_1 (this is the situation in the theorem of $a = 0$). If $d_1 = d_2$, then the point $U_1 + U_2$ will be on L, and if $d_1 > d_2$, then $U_1 + U_2$ will be on the same side of L as U_1 but at a distance $d_1 - d_2$ from L. In the same way, we see that $U_1 - U_2, U_1 - 2U_2, \ldots$ are all on the same side of L as U_1 (and are actually at a distance $d_1 + d_2, d_1 + 2d_2, \ldots$ from L). Also, the points $U_1 + U_2, U_1 + 2U_2, \ldots$ get successively closer to L, the distance

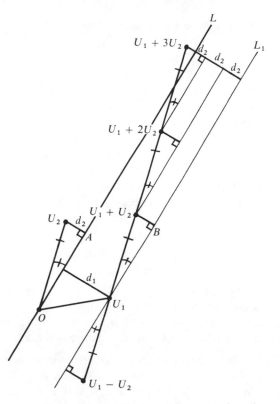

Figure 7.3. As illustrated here, $a = 2$.

from L being reduced by d_2 each time, until we reach a point $U_1 + aU_2$ which is either on L or on the same side of L as U_1 but within d_2 of L (and hence closer to L than U_2) while $U_1 + (a + 1)U_2$ is on the other side of L from U_1. The points $U_1 + (a + 2)U_2$, $U_1 + (a + 3)U_3$ are then all on the U_2 side of L (and are successively d_2 units further from L). ▲

Theorem 7.3 allows us to make the following definition.

Definition. Let α be a positive real number. The following process will be called **the continued fraction algorithm for approximating α.** Let L be the line $y = \alpha x$ and let $V_{-2} = (1,0)$, $V_{-1} = (0,1)$. We let

$$V_0 = V_{-2} + a_0 V_{-1} = (1, a_0),$$

where a_0 is the unique integer such that $V_{-2} + a_0 V_{-1}$ is either on L or on the same side of L as V_{-2}, but $V_{-2} + (a_0 + 1)V_{-1}$ is on the opposite side of L from V_{-2}. If V_0 is on L, the process terminates at V_0. If V_0 is not on L, the process is repeated. In general, if we have already defined V_{n-2}

and V_{n-1} (and V_{n-1} is not on L), then we let

(1) $V_n = V_{n-2} + a_n V_{n-1}$,

where a_n is the unique integer such that $V_{n-2} + a_n V_{n-1}$ is either on L or on the same side of L as V_{n-2}, but $V_{n-2} + (a_n + 1)V_{n-1}$ is on the opposite side of L from V_{n-2}. If V_n is on L, the process terminates at V_n. If V_n is not on L, the process is continued.

Notation. For the rest of this chapter, it will be assumed that α is a positive real number, L is the line $y = \alpha x$, and that the numbers $a_0, a_1, a_2, \ldots$ and the points $V_{-2} = (1,0)$, $V_{-1} = (0,1)$ $V_0, V_1, V_2, \ldots$ are given by the continued fraction algorithm for approximating α. We will soon wish to use the coordinates of V_n, and thus we let them be given by the letters q_n and p_n,

$$V_n = (q_n, p_n).$$

It is the slope of V_n, p_n/q_n, that is our (hopefully good) approximation to α.

Theorem 7.4. Although a_0 may be 0, $a_n \geq 1$ for $n \geq 1$. Further, $V_0, V_1, V_2, \ldots$ are successively closer to L and $V_0, V_2, V_4, \ldots$ are all below L while $V_1, V_3, V_5, \ldots$ all above L (unless, of course, the continued fraction process terminates with some V_n, in which case it is on L).

Proof. By Theorem 7.3 and the definition of V_0, it is on the same side of L as V_{-2} (namely, below L) and is closer to L than V_{-1}. Thus V_{-1} and V_0 are on opposite sides of L and, by Theorem 7.3, V_1 is on the same side of L as V_{-1} (namely, above L), V_1 is closer to L than V_0, and since V_0 is closer to L than V_{-1}, it follows that $a_1 \geq 1$. But now V_0 and V_1 are on opposite sides of L and, by Theorem 7.3, V_2 is on the same side of L as V_0 (namely, below L), V_2 is closer to L than V_1, and $a_2 \geq 1$. Figure 7.7 shows the general situation. The rest of the theorem follows in the same manner (that is, it follows by induction). ▲

Our algorithm may well go on indefinitely and if α is irrational, the algorithm certainly does not terminate, since the V_n's have integral coordinates and if some V_n were on L, then L would have a rational slope.

Let us present an example of the algorithm in action with $\alpha = \sqrt{3}$ (see Figure 7.4). We start with $V_{-2} = (1,0)$ below L and $V_{-1} = (0,1)$ above L. Since $V_{-2} + 1V_{-1} = (1,1)$ is below L while $V_{-2} + 2V_{-1} = (1,2)$ is above L, we see that $a_0 = 1$ and $V_0 = (1,1)$. Now $V_{-1} + 1V_0 = (1,2)$ is above L while $V_{-1} + 2V_0 = (2,3)$ is below L and thus $a_1 = 1$, $V_1 = (1,2)$. Continuing, $V_0 + 2V_1 = (3,5)$ is below L while $V_0 + 3V_1 = (4,7)$ is above L. Thus $a_2 = 2$ and $V_2 = (3,5)$. Repeating, $V_1 + 1V_2 = (4,7)$ is above L while $V_1 + 2V_2 = (7,12)$ is below L. Thus $a_3 = 1$ and $V_3 = (4,7)$. Now $V_2 + 2V_3 = (11,19)$ is below L while $V_2 + 3V_3 = (15,26)$ is above L. Therefore, $a_4 = 2$ and $V_4 = (11,19)$.

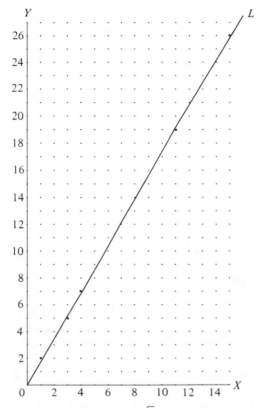

Figure 7.4. L is the line $y = x\sqrt{3}$. The bigger dots represent those members of the set of the closest points to L with x coordinate ≤ 15.

Last, $V_3 + 1V_4 = (15,26)$ is above L while $V_3 + 2V_4 = (26,45)$ (not in the figure) is below L (see problem 4 at the end of the section). Hence $a_5 = 1$ and $V_5 = (15,26)$. To summarize,

$$a_0 = 1, \quad a_1 = 1, \quad a_2 = 2, \quad a_3 = 1, \quad a_4 = 2, \quad a_5 = 1,$$

$$V_0 = (1,1), \quad V_1 = (1,2), \quad V_2 = (3,5), \quad V_3 = (4,7), \quad V_4 = (11,19), \quad V_5 = (15,26).$$

By Theorem 7.4 the points V_0, V_1, V_2, V_3, V_4, and V_5 are successively closer to L (and, of course, on alternating sides of L). In this case, when we throw away V_0, we come up with all the points with x coordinate ≤ 15 in the set of points closest to L. We will see soon that this is no accident. To three decimal places, $\sqrt{3} = 1.732$ and $\frac{26}{15} = 1.733$. The closeness of this approximation can best be appreciated by looking at some other denominator of roughly the same size, for example 10, where the best we can do is $\frac{17}{10} = 1.700$.

Since the points V_0, V_1, $V_2, \ldots$ are successively closer to L, we have every reason to believe after Theorem 7.2 that the fractions p_n/q_n are giving excellent approximations to α. Two questions arise. How good are these approximations and can we do better or are the V_n the set of closest points to L? We leave the second question to Section 7.4; the next two theorems will help to answer the first question.

Theorem 7.5. Let U_1 and U_2 be two points with nonnegative x and y coordinates, neither at the origin, and having different slopes. If $a > 0$, then the slope of

$$U_3 = U_1 + aU_2$$

is strictly between (not equal to either) the slopes of U_1 and U_2.

Proof. We will consider the case that slope $U_1 <$ slope U_2 (see Figure 7.5), the other case is virtually identical to this one (with the inequalities reversed, however) and is left as an exercise. Let L' be a line parallel to OU_2 and passing through U_1. The point U_3 is on this line and hence below the line OU_2 extended (or to the right of OU_2 extended if U_2 is on the positive Y axis). Since $a > 0$, aU_2 is on the same side of OU_1 extended as U_2, namely, above OU_1 extended. Thus U_3 is above OU_1 extended and hence U_3 is inside the angular sector U_1OU_2 extended. Therefore, by Theorem 7.1,

$$\text{slope } U_1 < \text{slope } U_3 < \text{slope } U_2. \qquad \blacktriangle$$

Theorem 7.6. Let $x_0, x_1, x_2, x_3, \ldots$ be a sequence of real numbers and let $U_{-2} = (1,0)$, $U_{-1} = (0,1)$,

$$U_0 = U_{-2} + x_0 U_{-1}, \qquad U_1 = U_{-1} + x_1 U_0, \ldots ;$$

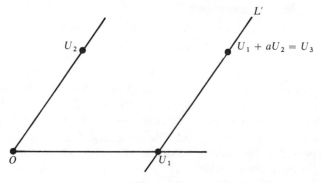

Figure 7.5.

In general, if U_{n-2} and U_{n-1} are already defined, let

$$U_n = U_{n-2} + x_n U_{n-1}.$$

If we let (s_n, r_n) be the coordinates of U_n, then for all $n \geq 0$,

$$r_{n-1} s_{n-2} - r_{n-2} s_{n-1} = (-1)^n.$$

Proof. The proof is by induction. Since $(s_{-2}, r_{-2}) = (1,0)$ and $(s_{-1}, r_{-1}) = (0,1)$, we see that

$$r_{-1} s_{-2} - r_{-2} s_{-1} = (1)(1) - (0)(0) = 1 = (-1)^0,$$

so the theorem is true for $n = 0$. Now suppose that the theorem is true for some particular value $n = k \geq 0$, so that we are assuming that

$$r_{k-1} s_{k-2} - r_{k-2} s_{k-1} = (-1)^k.$$

We then wish to show that

$$r_k s_{k-1} - r_{k-1} s_k = (-1)^{k+1}.$$

By the definition of U_k,

$$(s_k, r_k) = (s_{k-2}, r_{k-2}) + x_k(s_{k-1}, r_{k-1}),$$

so that

$$s_k = s_{k-2} + x_k s_{k-1}, \qquad r_k = r_{k-2} + x_k r_{k-1}.$$

Therefore,

$$
\begin{aligned}
r_k s_{k-1} - r_{k-1} s_k &= (r_{k-2} + x_k r_{k-1}) s_{k-1} - r_{k-1}(s_{k-2} + x_k s_{k-1}) \\
&= -(r_{k-1} s_{k-2} - r_{k-2} s_{k-1}) \\
&= -(-1)^k \\
&= (-1)^{k+1}.
\end{aligned}
$$

Thus if the theorem is true for $n = k$, then it is true for $n = k + 1$. Since we have verified the theorem for $n = 0$, it is true for all $n \geq 0$ by induction. ▲

Another proof (see problem 3 at the end of this section) of the inductive step in Theorem 7.6 can be given using the properties of determinants along with the fact that

$$r_{n-1} s_{n-2} - r_{n-2} s_{n-1} = \begin{vmatrix} s_{n-2} & r_{n-2} \\ s_{n-1} & r_{n-1} \end{vmatrix}.$$

Theorem 7.7. Suppose that $n \geq 0$ and V_n is defined. Then

(2) $q_n = q_{n-2} + a_n q_{n-1}, \qquad p_n = p_{n-2} + a_n p_{n-1}.$

The coordinates, p_n and q_n, of V_n are integers. Further,

$$p_{n-1}q_{n-2} - p_{n-2}q_{n-1} = (-1)^n.$$

It follows that p_n and q_n are relatively prime. Last,

$$1 = q_0 \leq q_1 < q_2 < q_3 < \cdots < q_{n-1} < q_n,$$

and if α is irrational (so that the continued fraction algorithm does not terminate), then

$$\lim_{n \to \infty} q_n = +\infty.$$

Proof. The first part of the theorem is simply the defining equation for V_n,

$$V_n = V_{n-2} + a_n V_{n-1},$$

written in terms of coordinates. Since $q_{-2}, q_{-1}, p_{-2}, p_{-1}$ are integers and since the a_n are also integers, we see from (2) that q_0 and p_0 are integers and then that q_1 and p_1 are integers, and continuing this way, we see inductively that q_n and p_n are integers for all n.

If in Theorem 7.6 we put $x_n = a_n$, then $U_n = V_n$ and as a result

$$p_{n-1}q_{n-2} - p_{n-2}q_{n-1} = (-1)^n.$$

When we replace n by $n+1$ in this last result, we get

$$p_n q_{n-1} - p_{n-1} q_n = (-1)^{n+1}.$$

Therefore, if $d|p_n, d|q_n$, then $d|(-1)^{n+1}$ and thus $d = \pm 1$. Therefore, p_n and q_n are relatively prime.

We have already noted that $q_0 = 1$. We recall from Theorem 7.4 that if $n \geq 1$, then $a_n \geq 1$. Thus

$$q_1 = q_{-1} + a_1 q_0 = a_1 q_0 \geq q_0 \geq 1,$$

and then

$$q_2 = q_0 + a_2 q_1 > a_2 q_1 \geq q_1,$$

and, in general, if for some $k \geq 2$, $q_k > q_{k-1} \geq 1$ (this is true for $k = 2$), then

$$q_{k+1} = q_{k-1} + a_{k+1} q_k > a_{k+1} q_k \geq q_k.$$

Therefore, by induction, if $n \geq 2$,

$$q_n > q_{n-1} \geq 1.$$

Using just the $q_n > q_{n-1}$ part of this, we see that

$$q_n > q_{n-1} > q_{n-2} > \cdots > q_2 > q_1 \geq q_0 = 1.$$

Last, if the continued fraction algorithm for α does not terminate, then we have an increasing sequence of integers, $q_0 \leq q_1 < q_2 < q_3 < \cdots$ and thus as n gets large, q_n increases without bound; that is, the sequence tends to infinity.   ▲

Theorem 7.8. Suppose that α is irrational. Then the continued fraction algorithm does not terminate and

$$\frac{p_0}{q_0} < \frac{p_2}{q_2} < \frac{p_4}{q_4} < \frac{p_6}{q_6} < \cdots < \alpha < \cdots < \frac{p_7}{q_7} < \frac{p_5}{q_5} < \frac{p_3}{q_3} < \frac{p_1}{q_1}.$$

In particular, if $n \geq 0$, then α is strictly between p_n/q_n and p_{n+1}/q_{n+1}.

Proof. By Theorem 7.7, the denominators of the above fractions are positive and the fractions are therefore defined. Recall that the slope of V_n is p_n/q_n. V_0, V_1, and V_2 meet the hypothesis of Theorem 7.5 and thus p_2/q_2 is strictly between p_0/q_0 and p_1/q_1. Since V_0 is below L while V_1 is above L, we see that $p_0/q_0 < p_1/q_1$ and therefore

$$(3) \qquad\qquad \frac{p_0}{q_0} < \frac{p_2}{q_2} < \frac{p_1}{q_1}.$$

Now we apply Theorem 7.5 to V_1, V_2, and V_3 and find that p_3/q_3 is strictly between p_1/q_1 and p_2/q_2. When we insert this into (3), we get

$$(4) \qquad\qquad \frac{p_0}{q_0} < \frac{p_2}{q_2} < \frac{p_3}{q_3} < \frac{p_1}{q_1}.$$

Again by Theorem 7.5, p_4/q_4 is strictly between p_2/q_2 and p_3/q_3, and it follows from (4) that

$$\frac{p_0}{q_0} < \frac{p_2}{q_2} < \frac{p_4}{q_4} < \frac{p_3}{q_3} < \frac{p_1}{q_1}.$$

The next step gives us

$$\frac{p_0}{q_0} < \frac{p_2}{q_2} < \frac{p_4}{q_4} < \frac{p_5}{q_5} < \frac{p_3}{q_3} < \frac{p_1}{q_1}.$$

If we continue in this manner, we get the sequence of the theorem except that α has not yet been placed in the sequence; we have only

$$\frac{p_0}{q_0} < \frac{p_2}{q_2} < \frac{p_4}{q_4} < \frac{p_6}{q_6} < \cdots < \frac{p_7}{q_7} < \frac{p_5}{q_5} < \frac{p_3}{q_3} < \frac{p_1}{q_1}.$$

But, by Theorem 7.4, V_{2n} is below L while V_{2n+1} is above L. Therefore, the slopes are ordered as follows:

$$\frac{p_{2n}}{q_{2n}} < \alpha < \frac{p_{2n+1}}{q_{2n+1}}.$$

Thus α occupies the spot claimed in our inequality.

Last, by what we have just proved, either $n + 1$ is odd and

$$\frac{p_n}{q_n} < \alpha < \frac{p_{n+1}}{q_{n+1}},$$

or $n + 1$ is even and

$$\frac{p_{n+1}}{q_{n+1}} < \alpha < \frac{p_n}{q_n}.$$

In either case, α is strictly between p_n/q_n and p_{n+1}/q_{n+1}. ▲

Theorem 7.8, while interesting, still does not give us any real information on how close p_n/q_n is to α. We now answer this question.

Theorem 7.9. Suppose that α is irrational and $n \geq 0$. Then

$$\left| \alpha - \frac{p_n}{q_n} \right| < \frac{1}{q_n q_{n+1}}.$$

It follows that

$$\left| \alpha - \frac{p_n}{q_n} \right| < \frac{1}{a_{n+1} q_n^2} \leq \frac{1}{q_n^2}.$$

Further,

$$\lim_{n \to \infty} \frac{p_n}{q_n} = \alpha.$$

Proof. By Theorem 7.8, α is strictly between p_n/q_n and p_{n+1}/q_{n+1}. Thus the numerical difference between α and p_n/q_n is less than the numerical difference between p_{n+1}/q_{n+1} and p_n/q_n; that is,

$$\left| \alpha - \frac{p_n}{q_n} \right| < \left| \frac{p_{n+1}}{q_{n+1}} - \frac{p_n}{q_n} \right| = \left| \frac{p_{n+1}q_n - p_n q_{n+1}}{q_n q_{n+1}} \right|$$

$$= \frac{|p_{n+1}q_n - p_n q_{n+1}|}{q_n q_{n+1}}$$

$$= \frac{|(-1)^{n+2}|}{q_n q_{n+1}}$$

$$= \frac{1}{q_n q_{n+1}}.$$

We have used Theorem 7.7 in the above inequality with the n of Theorem 7.7 replaced by $n + 2$.

By Theorem 7.7,

$$q_{n+1} = q_{n-1} + a_{n+1}q_n \geq a_{n+1}q_n,$$

and therefore

$$\left| \alpha - \frac{p_n}{q_n} \right| < \frac{1}{q_n q_{n+1}} \leq \frac{1}{q_n \cdot a_{n+1}q_n} = \frac{1}{a_{n+1}q_n^2} \leq \frac{1}{q_n^2},$$

the last inequality follows from the fact that $a_{n+1} \geq 1$ since $n + 1 \geq 1$ for $n \geq 0$. By Theorem 7.7,

$$\lim_{n \to \infty} \frac{1}{q_n^2} = 0.$$

It is now clear that $\lim_{n \to \infty} p_n/q_n = \alpha$, but we give the formal proof anyway. Given $\varepsilon > 0$, we must show that there is a number N such that if $n > N$, then

$$\left| \alpha - \frac{p_n}{q_n} \right| < \varepsilon.$$

Since $1/q_n^2 \to 0$ as $n \to \infty$, there is a number N such that if $n > N$, then

$$\frac{1}{q_n^2} = \left| 0 - \frac{1}{q_n^2} \right| < \varepsilon.$$

This number (N) will do fine, since if $n > N$, then

$$\left| \alpha - \frac{p_n}{q_n} \right| < \frac{1}{q_n^2} < \varepsilon.$$

Thus, by the definition of limit, $\lim_{n \to \infty} p_n/q_n = \alpha$. ▲

The last part of Theorem 7.9 is the reason that p_n/q_n is often called the nth **convergent** to α. Theorem 7.9 demonstrates that p_n/q_n gives a considerably better approximation to α than that guaranteed by Theorem 1.1. For example, in approximating $\sqrt{3}$ by a fraction with denominator 15, Theorem 1.1 guarantees only that with the proper choice of numerator, we will come within $\frac{1}{30}$ of $\sqrt{3}$. On the other hand, we have seen earlier that when $\alpha = \sqrt{3}$, $V_5 = (15,26)$ and thus Theorem 7.9 shows us that $\frac{26}{15}$ is within $\frac{1}{225}$ of $\sqrt{3}$. It is a fact (and one that the reader will soon be able to check for himself) that $V_6 = (41,71)$, and hence we can even guarantee by the first part of Theorem 7.9, without having to calculate the difference, that $\frac{26}{15}$ is within $\frac{1}{615}$ of $\sqrt{3}$.

EXERCISES

1. Show that the statement, "If s_1 and s_2 are positive and $r_1/s_1 < r_2/s_2$, then for any positive a,

$$\frac{r_1}{s_1} < \frac{r_1 + ar_2}{s_1 + as_2} < \frac{r_2}{s_2},"$$

   is nothing more than Theorem 7.5.

2. Prove Theorem 7.5 when slope $U_1 >$ slope U_2.

3. Suppose that $s_k = s_{k-2} + x_k s_{k-1}$, $r_k = r_{k-2} + x_k r_{k-1}$. Using the properties of determinants, show that

$$\begin{vmatrix} s_{k-1} & r_{k-1} \\ s_k & r_k \end{vmatrix} = - \begin{vmatrix} s_{k-2} & r_{k-2} \\ s_{k-1} & r_{k-1} \end{vmatrix}.$$

   Do not actually evaluate any of the determinants.

4. Verify that $3 \cdot 26^2 > 45^2$. Use this to show that the point $(26, 45)$ is below the line $L : y = x\sqrt{3}$.

5. Use Figure 7.6 to find all V_n, with $q_n \le 12$, in the continued fraction algorithm for $\alpha = (1 + \sqrt{5})/2$. What are the corresponding a_n?

6. Same problem as 5 with $\alpha = \sqrt{2}$.

7. Same problem as 5 with $\alpha = \pi - 2$ (the 2 is here just to keep things in the picture).

8. Same problem as 5 with $\alpha = \frac{39}{51}$, except find all V_n with $q_n \le 17$.

9. Show that

$$\frac{1}{(a_{n+1} + 2)q_n^2} < \left| \alpha - \frac{p_n}{q_n} \right| < \frac{1}{a_{n+1}q_n^2}$$

   (when everything is defined). Loosely speaking, this says that the larger a_{n+1}, the better the approximation of p_n/q_n to α.

10. Give the induction argument referred to at the end of the proof of Theorem 7.4.

7.3. Computation of a_n

If we are lucky enough to be given a (correctly drawn!) graph of the line $y = \alpha x$, then we can fairly easily find a few good fractional approximations to α from our geometric algorithm. But what happens if we desire more good approximations or we do not even have a graph provided? The definition of V_n involves knowing a_n, and a_n depends on the relative distances of V_{n-2} and V_{n-1} from L. Let d_n be the distance of V_n from L. Then (see

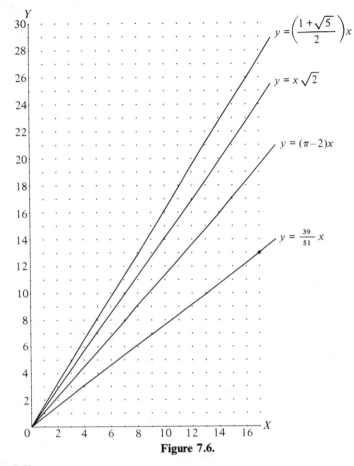

Figure 7.6.

Figure 7.7)

(5) $$d_n = d_{n-2} - a_n d_{n-1},$$

and since V_n is closer to L than V_{n-1},

(6) $$d_{n-1} > d_{n-2} - a_n d_{n-1} \geq 0.$$

It follows from (6) that

$$1 > \frac{d_{n-2}}{d_{n-1}} - a_n \geq 0;$$

that is,

$$a_n \leq \frac{d_{n-2}}{d_{n-1}}$$

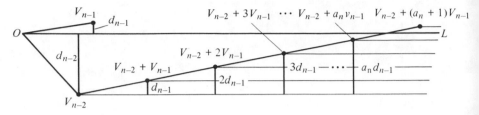

Figure 7.7.

but

$$a_n + 1 > \frac{d_{n-2}}{d_{n-1}}.$$

Thus, by the definition of the greatest integer function (page 120),

$$a_n = \left[\frac{d_{n-2}}{d_{n-1}} \right].$$

To simplify this result, we define for $n \geq 0$,

(7) $$\alpha_n = \frac{d_{n-2}}{d_{n-1}}$$

so that

(8) $$a_n = [\alpha_n].$$

Further, it follows by dividing both sides of (5) by d_{n-1} that

$$\frac{d_{n-2}}{d_{n-1}} = a_n + \frac{d_n}{d_{n-1}},$$

or, if V_n is not on L (so that $d_n \neq 0$ and α_{n+1} is defined),

(9) $$\alpha_n = a_n + \frac{1}{\alpha_{n+1}}.$$

When V_n is on L, $d_n = 0$ and $\alpha_n = a_n$. Thus the process terminates with V_n if α_n is an integer and continues otherwise.

If we can find a simple expression for α_0, then it will be a simple matter to find the a_n's from (8) and (9). The key to everything lies in the following theorem:

Theorem 7.10. $\alpha_0 = \alpha$. Thus the numbers a_n may be calculated from the following formulas: For $n \geq 0$, if α_n is defined, then

$$a_n = [\alpha_n],$$

and if α_n is not an integer, then α_{n+1} is defined and can be found from the equation

$$\alpha_n = a_n + \frac{1}{\alpha_{n+1}}.$$

Proof. After the previous discussion, it is clear that we need only show that $\alpha_0 = \alpha$. Recall that

$$\alpha_0 = \frac{d_{-2}}{d_{-1}}.$$

We could use the general formula for the distance from a point to a line, but it is not really necessary here. Let θ be the angle from the x axis to L (see Figure 7.8) so that

$$\alpha = \text{slope } L = \tan \theta.$$

Then $\angle OV_{-1}A = \theta$ also. Hence

$$\sin \theta = \frac{BV_{-2}}{OV_{-2}} = \frac{d_{-2}}{1} = d_{-2}, \qquad \cos \theta = \frac{AV_{-1}}{OV_{-1}} = \frac{d_{-1}}{1} = d_{-1}$$

and, therefore,

$$\alpha_0 = \frac{d_{-2}}{d_{-1}} = \frac{\sin \theta}{\cos \theta} = \tan \theta = \alpha. \qquad \blacktriangle$$

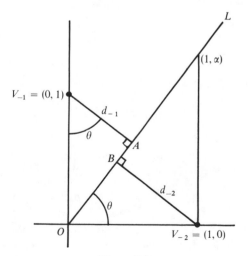

Figure 7.8.

The next theorem will be used to help justify naming our algorithm "the continued fraction algorithm."

Theorem 7.11. Let L' be the line $y = \beta x$. Suppose we are given a sequence of real numbers $b_0, b_1, b_2, \ldots$. We define a sequence of numbers $\beta_0, \beta_1, \beta_2, \ldots$ by letting $\beta_0 = \beta$, and if β_n is already known and $\beta_n \neq b_n$, then we let β_{n+1} be defined by

(10)
$$\beta_n = b_n + \frac{1}{\beta_{n+1}}.$$

(If $\beta_n = b_n$, then the sequence $\beta_0, \beta_1, \ldots$ is terminated at β_n). Let

$$U_{-2} = (1,0), \qquad U_{-1} = (0,1)$$

and if U_{n-1} and U_{n-2} have already been defined and $n \geq 0$, then we let

$$U_n = U_{n-2} + b_n U_{n-1}.$$

If $n \geq 0$ and β_n is defined, then $U_{n-2} + \beta_n U_{n-1}$ is on L'.

Proof. We give a proof by induction. When $n = 0$,

$$U_{-2} + \beta_0 U_{-1} = (1,0) + \beta(0,1) = (1,\beta)$$

is on L'. Suppose that for some $k \geq 0$, $U_{k-2} + \beta_k U_{k-1}$ is on L' and β_{k+1} is defined. Then

$$U_{k-1} + \beta_{k+1} U_k = U_{k-1} + \beta_{k+1}(U_{k-2} + b_k U_{k-1})$$

$$= \beta_{k+1} U_{k-2} + (\beta_{k+1} b_k + 1) U_{k-1}$$

$$= \beta_{k+1} \left[U_{k-2} + \left(b_k + \frac{1}{\beta_{k+1}} \right) U_{k-1} \right]$$

$$= \beta_{k+1}(U_{k-2} + \beta_k U_{k-1})$$

is also on L' (and is, in fact; β_{k+1} times further from the origin than $U_{k-2} + \beta_k U_{k-1}$). Since the result is true for $n = 0$, it is now true for all n. ▲

Theorem 7.12. For each $n \geq 0$ such that V_n is defined,

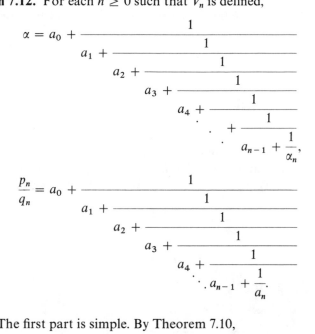

$$\alpha = a_0 + \cfrac{1}{a_1 + \cfrac{1}{a_2 + \cfrac{1}{a_3 + \cfrac{1}{a_4 + \cfrac{1}{\ddots + \cfrac{1}{a_{n-1} + \cfrac{1}{\alpha_n}}}}}}},$$

$$\frac{p_n}{q_n} = a_0 + \cfrac{1}{a_1 + \cfrac{1}{a_2 + \cfrac{1}{a_3 + \cfrac{1}{a_4 + \cfrac{1}{\ddots a_{n-1} + \cfrac{1}{a_n}}}}}}.$$

Proof. The first part is simple. By Theorem 7.10,

$$\alpha = a_0 + \cfrac{1}{\alpha_1} = a_0 + \cfrac{1}{a_1 + \cfrac{1}{\alpha_2}} = a_0 + \cfrac{1}{a_1 + \cfrac{1}{a_2 + \cfrac{1}{\alpha_3}}} = \cdots$$

$$\cdots = a_0 + \cfrac{1}{a_1 + \cfrac{1}{\ddots a_{n-1} + \cfrac{1}{\alpha_n}}}.$$

For the second part, we use Theorem 7.11 with L' being the line $y = (p_N/q_N)x$, the sequence of b_n's being given by $b_n = a_n$ and the β_n's defined by (10) with $\beta_0 = p_N/q_N$. Then

$$\frac{p_N}{q_N} = a_0 + \cfrac{1}{\beta_1} = a_0 + \cfrac{1}{a_1 + \cfrac{1}{\beta_2}} = \cdots = a_0 + \cfrac{1}{a_1 + \cfrac{1}{a_2 + \cfrac{1}{\ddots a_{N-1} + \cfrac{1}{\beta_N}}}},$$

and thus our task is to show that $\beta_N = a_N$. The U_n of Theorem 7.11 are clearly our V_n [see the defining equation (1) for V_n] and thus by Theorem 7.11, the point

$$V_{N-2} + \beta_N V_{N-1} = U_{N-2} + \beta_N U_{N-1}$$

is on L'. But

$$(q_N, p_N) = V_N = V_{N-2} + a_N V_{N-1}$$

is certainly also on L'. Let L'' be the line through V_{N-2} with the slope of V_{N-1} (see Figure 7.9). The slope of L' is the slope of V_N, which is different from the slope of L'', and thus L' and L'' have exactly one point of intersection and this point is $V_N = V_{N-2} + a_N V_{N-1}$ or, alternatively, $V_{N-2} + \beta_N V_{N-1}$. Hence

$$V_{N-2} + a_N V_{N-1} = V_{N-2} + \beta_N V_{N-1},$$

$$a_N V_{N-1} = \beta_N V_{N-1}$$

and since $V_{N-1} \neq (0,0)$,

$$a_N = \beta_N. \quad \blacktriangle$$

Theorem 7.12 gives the explanation for naming our algorithm the continued fraction algorithm. The name continued fraction seems very appropriate for the two expressions in Theorem 7.12. In fact, the subject of continued fractions is usually introduced by defining $\alpha_0 = \alpha$ and then defining the later α_n's and the a_n's by equations (8) and (9). The number p_n/q_n is then defined by the expression given in Theorem 7.12. One can even find p_n and q_n

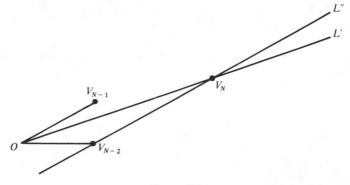

Figure 7.9.

by reducing the fraction to lowest terms with the denominator being positive; the numerator is then p_n and the denominator is q_n.

The expressions in Theorem 7.12 are rather unwieldy, and thus it is customary to compactify them. We let

$$\langle a_0, a_1, a_2, \ldots, a_n \rangle$$

stand for

$$a_0 + \cfrac{1}{a_1 + \cfrac{1}{a_2 + \cfrac{1}{\ddots\ a_{n+1} + \cfrac{1}{a_n}}}}$$

and

$$\langle a_0, a_1, a_2, \ldots \rangle$$

stand for

$$a_0 + \cfrac{1}{a_1 + \cfrac{1}{a_2 + \cfrac{1}{\ddots}}}$$

We will write

(11) $$\alpha = \langle a_0, a_1, a_2, \ldots \rangle$$

This is deserving of some interpretation, as it is not even immediately clear how to evaluate the infinite continued fraction on the right-hand side. By

$$\pi = 3.14159265\ldots$$

we mean that π is the limit of the sequence

$$3, 3.1, 3.14, 3.141, 3.1415, 3.14159, \ldots.$$

Thanks to Theorem 7.9, we may interpret (11) in the same way. Since

$$\frac{p_n}{q_n} = \langle a_0, a_1, \ldots, a_n \rangle$$

and

$$\lim_{n \to \infty} \frac{p_n}{q_n} = \alpha,$$

we may write

$$\alpha = \lim_{n \to \infty} \langle a_0, a_1, \ldots, a_n \rangle,$$

and this is the meaning to be attached to the equality of (11).

After Theorem 7.10, we can now find the V_n's from α without resorting to an illustration. It may seem tempting to find p_n/q_n from the expression in Theorem 7.12, but it is just as fast to find it from V_n, which is calculated from our definition (1). The latter method has the additional advantage that one gets all the earlier V_n's at the same time. We give three examples to illustrate the process. For our first example, we let $\alpha = \frac{42}{57}$. Then

$$\alpha_0 = \tfrac{42}{57} = 0 + \tfrac{42}{57}, \qquad a_0 = 0,$$

$$\alpha_1 = \tfrac{57}{42} = 1 + \tfrac{15}{42}, \qquad a_1 = 1,$$

$$\alpha_2 = \tfrac{42}{15} = 2 + \tfrac{12}{15}, \qquad a_2 = 2,$$

$$\alpha_3 = \tfrac{15}{12} = 1 + \tfrac{3}{12}, \qquad a_3 = 1,$$

$$\alpha_4 = \tfrac{12}{3} = 4, \qquad\qquad a_4 = 4,$$

and the algorithm terminates.

The arrangement shown in Figure 7.10 makes it very simple to calculate V_n from V_{n-2} and V_{n-1}; V_n is simply a_n times the row above it plus the row

	$\alpha = \frac{42}{57}$			$\alpha = \sqrt{3}$			$\alpha = \pi$			$\alpha \lessapprox \pi$		
n	a_n	q_n	p_n	a_n	q_n	p_n	a_n	q_n	p_n	a_n	q_n	p_n
-2		1	0		1	0		1	0		1	0
-1		0	1		0	1		0	1		0	1
0	0	1	0	1	1	1	3	1	3	3	1	3
1	1	1	1	1	1	2	7	7	22	7	7	22
2	2	3	2	2	3	5	15	106	333	15	106	333
3	1	4	3	1	4	7	1	113	355	1	113	355
4	4	19	14	2	11	19	292	33 102	103 993	288	32 650	102 573
5				1	15	26						
6				2	41	71						
7				1	56	97						
8				2	153	265						

Figure 7.10. The complete continued fraction expansion of $\frac{42}{57}$ and the beginnings of the expansions of $\sqrt{3}$ and π. In the last column, the result of each calculation was rounded off to eight decimal places before the next calculation was made.

above that. Since $\alpha_4 = a_4$, $V_4 = (19,14)$ is on the line $y = (42/57)x$. This shows incidentally that $\frac{42}{57}$ is not reduced and can be simplified to $\frac{14}{19}$, which is reduced. We have found that the continued fraction expansion of $\frac{42}{57}$ is

$$\tfrac{42}{57} = \langle 0,1,2,1,4 \rangle.$$

For our second example, we let $\alpha = \sqrt{3} \approx 1.7$ (so that $a_0 = 1$),

$$\alpha_0 = \sqrt{3} = 1 + (\sqrt{3} - 1),$$

$$\alpha_1 = \frac{1}{\sqrt{3} - 1} = \frac{\sqrt{3} + 1}{2} \approx \frac{2.7}{2}, \qquad a_1 = 1,$$

$$\alpha_1 = \frac{\sqrt{3} + 1}{2} = 1 + \frac{\sqrt{3} - 1}{2},$$

$$\alpha_2 = \frac{2}{\sqrt{3} - 1} = \sqrt{3} + 1 \approx 2.7, \qquad a_2 = 2,$$

$$\alpha_2 = 2 + (\sqrt{3} - 1),$$

$$\alpha_3 = \frac{1}{\sqrt{3} - 1} = \frac{\sqrt{3} + 1}{2}.$$

A very curious thing has just occurred, namely $\alpha_3 = \alpha_1$. Therefore, $a_3 = [\alpha_3] = [\alpha_1] = a_1$. Thus

$$\frac{1}{\alpha_4} = \alpha_3 - a_3 = \alpha_1 - a_1 = \frac{1}{\alpha_2},$$

so that $\alpha_4 = \alpha_2$. Then $a_4 = a_2$ and then $\alpha_5 = \alpha_3 = \alpha_1$. Then we get $a_5 = a_1$ and $\alpha_6 = \alpha_2$. This can be continued indefinitely. We have thus found the continued fraction expansion of $\sqrt{3}$,

$$\sqrt{3} = \langle 1,1,2,1,2,1,2,1,2,1,2, \ldots \rangle.$$

Several of the V_n are given in Figure 7.10; needless to say, the results here agree with the results obtained from Figure 7.4.

For our third example, we let $\alpha = \pi = 3.141592653589793\ldots$. Here we have

$$\alpha_0 = 3.14159265\ldots, \qquad\qquad\qquad\qquad a_0 = 3,$$

$$\alpha_1 = \frac{1}{.14159265\ldots} = \quad 7.06251330\ldots, \qquad a_1 = 7,$$

$$\alpha_2 = \frac{1}{.06251330\ldots} = \quad 15.99659440\ldots, \qquad a_2 = 15,$$

$$\alpha_3 = \frac{1}{.99659440\ldots} = \quad 1.00341723\ldots, \qquad a_3 = 1,$$

$$\alpha_4 = \frac{1}{.00341723\ldots} = 292.63459101\ldots, \qquad a_4 = 292,\ \text{etc.}$$

The corresponding V_n are given in Figure 7.10.

Each of the three previous examples is instructive. In the first, we unintentionally found that $(42,57) = 3$. In fact, the whole process of our first example is very reminiscent of the Euclidean algorithm. This is because the process is the Euclidean algorithm in a slightly different form (see problem 14, page 26). In this form, the Euclidean algorithm can be used as a test for irrationality: If the algorithm for α terminates, then α is rational and otherwise α is irrational.

The second example is very interesting because of the periodicity of the a_n. Once we found that $\alpha_3 = \alpha_1$, we were enabled to calculate as many of the V_n as we desired. This in turn enables us to find extremely good rational approximations to $\sqrt{3}$ relatively easily. Thus for example, by Theorem 7.9,

$$\left| \sqrt{3} - \frac{97}{56} \right| < \frac{1}{56 \cdot 153} = .000116\ldots.$$

Thus if we did not already know the decimal expansion of $\sqrt{3}$ (which we actually do not; no one knows what the one billionth digit of the expansion is), we could now say that with an error of about 1 ten-thousandth or less, the square root of three is $\frac{97}{56}$. This happy circumstance arose because of the periodicity of the a_n. To indicate periodicity, we shall put a line over the block of a_n's that repeat. Thus, for example,

$$\sqrt{3} = \langle 1,\overline{1,2} \rangle$$

and

$$\langle 2,3,1,4,2,5,7,4,2,5,7,4,2,5,7,4,2,5,7,\ldots \rangle,$$

where the block 4, 2, 5, 7 is repeated indefinitely, is abbreviated as

$$\langle 2,3,1,\overline{4,2,5,7} \rangle.$$

Needless to say, if the sequence of a_n becomes periodic, then we know far more about α than if we merely know the first hundred a_n's. We will show in Section 7.6 that the continued fraction expansion of α becomes periodic if and only if α is an irrational root of a quadratic equation with integral coefficients.

In the third example, the situation is worse. The complete sequence of a_n's is not known. In order to calculate later a_n's, it is necessary to know the decimal expansion of α ever more accurately. In our example, we have given the correct values of the α_n's (or at least eight places past the decimal point of correct values). Suppose we only knew that $\pi = 3.14159265$ was rounded off in the last place. We present here the same calculations with the stipulation that after each calculation, we round off at the eighth decimal place. The results are then (the equal signs are thus actually only approximations)

$$\alpha_0 = 3.14159265, \qquad\qquad a_0 = 3,$$

$$\alpha_1 = \frac{1}{.14159265} = \quad 7.06251348, \qquad a_1 = 7,$$

$$\alpha_2 = \frac{1}{.06251348} = \quad 15.99654986, \qquad a_2 = 15,$$

$$\alpha_3 = \frac{1}{.99654986} = \quad 1.00346208, \qquad a_3 = 1,$$

$$\alpha_4 = \frac{1}{.00346208} = 288.84370090, \qquad a_4 = 288, \text{ etc.}$$

Again the corresponding V_n's are given in Figure 7.10. When we compare these numbers with our earlier numbers, we find that the roundoff error has crept forward so fast that a_4 is already incorrect. This is explainable since if a_4 were correct we would be claiming that

$$\left| \pi - \frac{102\ 573}{32\ 650} \right| < \frac{1}{32\ 650^2} < 10^{-9},$$

even though we were starting only with the knowledge that π is contained in a given interval of length 10^{-8}. Since our calculations made no use of any properties of π other than it being $3.14159265 \pm 5 \cdot 10^{-9}$, we cannot possibly guarantee the accuracy of $a_4 = 288$ (nor can we say that $a_4 = 288$ is incorrect without further information). All this does not really matter if all one wants is a fraction that is a very good approximation to the first eight decimal places of π. Then we simply crank out the continued fraction expansion of 3.14159265 and take whichever p_n/q_n we wish. In this case there is no use taking any $n > 4$, since all $n \geq 4$ will give 3.14159265 when rounded off and the correct next place is unknown anyway.

Our example also shows why $\frac{22}{7}$ is used so often in place of π. In fact,

$$\left| \pi - \frac{22}{7} \right| < \frac{1}{7 \cdot 106} = \frac{1}{742}.$$

This is a far better approximation to π than can be obtained by any other fraction with denominator near 7. We have also been helped along by having a_2 being large. Again, since $a_4 = 292$, p_3/q_3 gives an excellent approximation to π:

$$\left| \pi - \frac{355}{113} \right| < \frac{1}{113 \cdot 33\,102} < 3 \cdot 10^{-7}.$$

The continued fraction expansion of π is

$$\pi = \langle 3,7,15,1,292,\ldots \rangle;$$

the next several a_n are known, but no formula is known that gives all the a_n. The situation is thus different from what happens with $\sqrt{3}$, where all the a_n are known.

EXERCISES

1. Find the continued fraction expansion of $\frac{97}{40}$. Give all the V_n also.
2. Find the continued fraction expansion of $(1 + \sqrt{5})/2$.
3. Find the continued fraction expansion of $\sqrt{2}$.
4. Find the continued fraction expansion of $\sqrt{7}$.
5. Give a fraction with denominator < 100 which is within 10^{-4} of $e = 2.7182818\ldots$.
6. Show that if a and b are integers with $b > 0$ and α is real, then

$$\left[\frac{a + \alpha}{b} \right] = \left[\frac{a + [\alpha]}{b} \right].$$

This is useful in calculating continued fraction expansions of roots of quadratic equations.

7. Let the decimal expansion of α be

$$\alpha = b + \frac{b_1}{10} + \frac{b_2}{10^2} + \frac{b_3}{10^3} + \frac{b_4}{10^4} + \cdots,$$

where $0 \le b_n \le 9$ for all n. Suppose that for some convergent p_k/q_k we have $q_k = 100$. Prove that either $b_3 = b_4 = 0$ or $b_3 = b_4 = 9$.

8. Find p_4/q_4 if $\alpha = \sqrt[12]{2} = 1.05946309\ldots$. This number is important in music. (There is an interesting ad related to this problem on page 88 of the March 1967 issue of *Scientific American*.)

9. Suppose the continued fraction expansion of α is

$$\alpha = \langle a_0, a_1, a_2, a_3, \ldots \rangle.$$

What is the continued fraction expansion of α_n?

10. (a) If a and b are rational numbers greater than 1 and $c \ge 0$, show that $1/\log_a b = \log_b a$ and that $(\log_b a) - c = \log_b(a/b^c)$.

   (b) Use the relations of problem 10(a) and the following inequalities to find the first five convergents (that is, find p_n/q_n for $0 \le n \le 4$) of the continued fraction expansion of $\log_{10} 2$;

$$10^0 < 2 < 10^1,$$

$$8 = 2^3 < 10 < 2^4,$$

$$\frac{125}{64} = \left(\frac{10}{8}\right)^3 < 2 < \left(\frac{10}{8}\right)^4,$$

$$\left(\frac{128}{125}\right)^9 < \frac{5}{4} < \left(\frac{128}{125}\right)^{10},$$

$$\left(\frac{5^{28}}{2^{65}}\right)^2 < \frac{128}{125} < \left(\frac{5^{28}}{2^{65}}\right)^3.$$

11. Using the method of problem 10, find p_2/q_2 for $\alpha = \log_{10} 6$. Given that

$$\left(\frac{2 \cdot 3^4}{5^3}\right)^1 < \frac{5}{3} < \left(\frac{2 \cdot 3^4}{5^3}\right)^2,$$

$$\left(\frac{5^4}{2 \cdot 3^5}\right)^1 < \frac{2 \cdot 3^4}{5^3} < \left(\frac{5^4}{2 \cdot 3^5}\right)^2,$$

$$\left(\frac{2^2 \cdot 3^9}{5^7}\right)^{32} < \frac{5^4}{2 \cdot 3^5} < \left(\frac{2^2 \cdot 3^9}{5^7}\right)^{33},$$

find p_5/q_5 for $\alpha = \log_{10} 6$.

7.4. The Best Approximations

In Section 7.1 we saw that the set of closest points to the line $L: y = \alpha x$ gives the best fractional approximations to α. In the next two sections we

developed a simple algorithm which gives excellent fractional approximations to α. We may well wonder whether we are in fact finding the set of
closest points to L or are there even better fractional approximations to α
available than the ones we get from our algorithm? In this section we will
show that our algorithm gives the best approximations to α. We first give
two preliminary results which will be very useful later on.

Definition. Let U_1 and U_2 be nonzero vectors with different slopes. Let
$L_n (n = 0, \pm 1, \pm 2, \ldots)$ be the line through nU_1 which is parallel to U_2, and
let $l_m (m = 0, \pm 1, \pm 2, \ldots)$ be the line through mU_2 which is parallel to U_1
(see Figure 7.11). The set of points of intersection of the L_n with the l_m is
called the **lattice** generated by U_1 and U_2; this set is also the set of vertices
of the parallelograms formed by the L_n and L_m.

For example, the lattice generated by $(1,0)$ and $(0,1)$ is the set of all points
with integral coordinates.

Theorem 7.13. Let $U_1 = (s_1, r_1)$ and $U_2 = (s_2, r_2)$ have integral coordinates
and $r_2 s_1 - r_1 s_2 = \pm 1$. Then the lattice generated by U_1 and U_2 is
merely the set of all points with integral coordinates.

Proof. The lattice generated by U_1 and U_2 is the set of points of the form
$nU_1 + mU_2$, where n and m are integers. Since U_1 and U_2 have integral

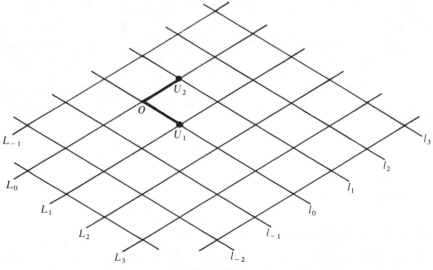

Figure 7.11.

coordinates, any point of the lattice has integral coordinates. Thus our problem is to show that any point with integral coordinates is a point of the lattice. Let (b,a) be a point with integral coordinates. If we set

$$xU_1 + yU_2 = (b,a),$$

then in terms of coordinates, we get the two equations

$$s_1x + s_2y = b,$$

$$r_1x + r_2y = a,$$

and thus

$$x = \frac{br_2 - as_2}{r_2s_1 - r_1s_2}, \qquad y = \frac{as_1 - br_1}{r_2s_1 - r_1s_2}.$$

Since $br_2 - as_2$ and $as_1 - br_1$ are integers and $r_2s_1 - r_1s_2 = \pm 1$ by hypothesis, x and y are integers, and thus (b,a) is a lattice point.   ▲

Theorem 7.14. Suppose that $A = (s_1,r_1)$ and $B = (s_2,r_2)$ are in the first quadrant. Then

$$\text{area } \triangle OAB = \tfrac{1}{2}|r_2s_1 - r_1s_2|.$$

Proof. Since

$$|r_2s_1 - r_1s_2| = |-(r_2s_1 - r_1s_2)| = |r_1s_2 - r_2s_1|,$$

the result is the same when the subscripts, 1 and 2, are interchanged. Therefore, we may assume (after interchanging subscripts if necessary) that $s_2 \le s_1$. Let C be the intersection of OA and the line $x = s_2$ (see Figure 7.12). B is

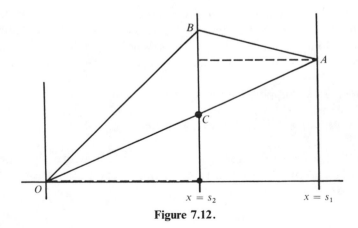

Figure 7.12.

either above or below C, and we examine the case with B above C. Since BC is a vertical line, the altitudes of triangles OBC and ABC to BC are horizontal and hence have lengths s_2 and $s_1 - s_2$, respectively. Further, the equation of OA extended is $y = (r_1/s_1)x$ and since the x coordinate of C is s_2,

$$C = \left(s_2, \frac{r_1 s_2}{s_1}\right).$$

Thus

$$BC = r_2 - \frac{r_1 s_2}{s_1}.$$

Therefore,

$$\text{area } \triangle OAB = \text{area } \triangle OBC + \text{area } \triangle ABC = \tfrac{1}{2}(BC)s_2 + \tfrac{1}{2}(BC)(s_1 - s_2)$$

$$= \frac{1}{2}(BC)(s_1) = \frac{1}{2}\left(r_2 - \frac{r_1 s_2}{s_1}\right)s_1 = \frac{1}{2}(r_2 s_1 - r_1 s_2).$$

When B is under C, the proof is the same except that we come up with

$$BC = (y \text{ coord of } C) - (y \text{ coord of } B) = \frac{r_1 s_2}{s_1} - r_2,$$

which gives

$$\text{area } \triangle OAB = \tfrac{1}{2}(r_1 s_2 - r_2 s_1).$$

Whether B is above C or below C, we get

$$\text{area } \triangle OAB = \tfrac{1}{2}|r_2 s_1 - r_1 s_2|. \qquad \blacktriangle$$

Theorem 7.14 is actually true without any restrictions on the locations of A and B; the reader familiar with vector cross products has probably seen a simpler proof covering all cases. The essence of our claim that the V_n's essentially give the set of closest points to L is the following theorem.

Theorem 7.15. Suppose that α is irrational and $0 < q_{n-1} < q_n$ (in other words, either $n \geq 2$ or $n \geq 1$ and $a_1 > 1$). Let $U = (s,r)$, where $U \neq V_{n-1}$, $U \neq V_n$, and $0 < s \leq q_n$. Then U is further from L than V_{n-1}.

Proof. We consider the case that the slope of V_n is greater than the slope of V_{n-1}. The proof is the same for the other case except that the roles of the words above and below are reversed. By Theorem 7.7,

$$p_n q_{n-1} - p_{n-1} q_n = \pm 1,$$

and hence the lattice generated by V_{n-1} and V_n is the set of points in the XY plane with integral coordinates. The lattice generated by V_{n-1} and V_n is shown in Figure 7.13. We have subscripted the lines L_j and l_k so that the intersection of L_j and l_k is the point $jV_{n-1} + kV_n$.

Since U has integral coordinates, there are integers j and k such that

$$U = jV_{n-1} + kV_n.$$

We now investigate the various possibilities for j and k, always remembering that U is to the right of the line $x = 0$ and either on or to the left of the line $x = q_n$. If $j < 0$, then U is on one of the lines $L_{-1}, L_{-2}, L_{-3}, \ldots$. In the range $0 < x \le q_n$, these lines are all above L. In addition, since the slope of L_j is greater than the slope of L, as we move from the Y axis along L_j ($j < 0$) in the direction of increasing x, we get further from the line L. Thus U is further from L than the intersection of L_j and the Y axis, this point being farther ($j \le -2$) or at the same distance from ($j = -1$) L as the intersection of L_{-1} and the Y axis, and this point is further from L than $-V_{n-1}$, which is as far from L as V_{n-1}. In other words, if $j < 0$, then U is further from L than V_{n-1}.

Now suppose that $j \ge 2$. If $k \ge 1$, then s, the x coordinate of U, satisfies the inequality

$$s = jq_{n-1} + kq_n > q_{n-1} + q_n > q_n,$$

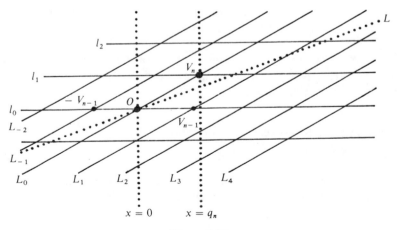

$x = 0$ $x = q_n$

Figure 7.13.

and thus U would be to the right of the line $x = q_n$. Since this is outside the desired region, it follows that $k \leq 0$. But again, U is on L_j and the slope of L_j is greater than the slope of L, so that as we move from U in the direction of increasing x (which is the same thing as increasing k) along L_j, we get nearer to L. Thus jV_{n-1} is as close to L as U (and, in fact, closer unless $k = 0$) and this is j times farther from L than V_{n-1}. Since $j \geq 2$, U is further from L than V_{n-1}.

This leaves us with $j = 0$ and $j = 1$. On the line L_0, the points of the lattice are either at 0 and to the left of $x = 0$ or at V_n and to the right of $x = q_n$. Since $U \neq V_n$, U is not on the line L_0 so $j \neq 0$. On the line L_1, we note that $V_{n-1} - V_n$ has a negative x coordinate and $V_{n-1} + V_n$ has an x coordinate greater than q_n. Thus the only lattice point on L_1 with an x coordinate in the range $0 < x \leq q_n$ is V_{n-1}. Since we are supposing that $U \neq V_{n-1}$, we see that U is not on L_1. Thus wherever U is in the region $0 < s \leq q_n$, U is farther from L than V_{n-1}.   ▲

Theorem 7.16. If α is irrational, then the set of closest points to L is an infinite set and in fact is either the set of points V_n ($n \geq 0$) with $a_1 > 1$ or the set of points V_n ($n \geq 1$) with $a_1 = 1$.

Proof. When $a_1 = 1$, $q_1 = q_0 = 1$ while $0 < q_1 < q_2$. Thus, by Theorem 7.15, V_1 is closer to L than V_0 and hence the first member of the set of closest points to L is V_1. When $a_1 > 1$, $0 < q_0 < q_1$, and thus V_0 is the first member of the set of closest points to L. Theorem 7.15 shows that no point is closer to L until we come to the next V_n, and then no point is closer until we come to the one after that, and so on. Thus the set of closest points to L and the set of V_n ($n \geq 0$) are the same except that we throw out V_0 when $a_1 = 1$.   ▲

Theorem 7.15 can naturally be put in terms of symbols; since the distance from a point to L is a somewhat messy expression, it is customary to use the vertical distance from a point to L.

Theorem 7.17. Suppose that α is irrational and

$$0 < q_{n-1} < q_n$$

(that is, $n \geq 1$ and $a_1 > 1$ or $n \geq 2$ and $a_1 = 1$). If $V = (q,p)$, $V \neq V_{n-1}$, $V \neq V_n$, and $0 < q \leq q_n$, then the vertical distances of V_{n-1} and V from L satisfy

$$|q_{n-1}\alpha - p_{n-1}| < |q\alpha - p|.$$

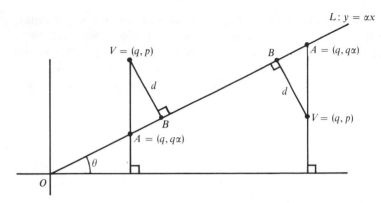

Figure 7.14. Shown are the two cases of V above and below L.

Proof. The number $|q\alpha - p|$ is actually the vertical distance from (q,p) to L (see Figure 7.14). We let A be the intersection of the lines $x = q$ and $y = \alpha x$, so that $A = (q, q\alpha)$. Thus

$$VA = |q\alpha - p|.$$

Let B be the point on L such that $VB \perp L$ and let d be the distance of V from L so that

$$d = VB.$$

Finally, we let θ be the angle between the x axis and L. Since $VA \perp x$ axis and $VB \perp L$, it follows that

$$\angle AVB = \theta.$$

(The geometry theorem actually says: If the initial sides of two angles are perpendicular and the terminal sides are perpendicular, then the angles are equal.) Therefore,

$$\frac{d}{|q\alpha - p|} = \frac{VB}{VA} = \cos\theta$$

or

$$|q\alpha - p| = d \sec\theta.$$

In words, the vertical distance of a point from L is a positive constant (sec θ) times the distance of the point from L. The theorem now follows immediately from Theorem 7.15. ▲

We shall devote the rest of the section to the following question: How close to L does a point have to be before we can guarantee that it is a member of the set of closest points to L? Another way of looking at this question is:

"The set of closest points to L give excellent fractional approximations to α but these points are sometimes widely separated (for example, $\alpha = \pi$). Perhaps there are other fractions which also give very good approximations to α." We will examine this problem from the second viewpoint.

After Theorem 7.9, it seems reasonable to start by asking if the inequality

$$\left| \alpha - \frac{p}{q} \right| < \frac{1}{q^2}$$

implies that $p/q = p_n/q_n$ for some n. The answer is no; there are sometimes fractions other than p_n/q_n which can give this good an approximation to α. Since we have already considered the example of $\alpha = \sqrt{3}$ in some detail, we furnish the example

$$\left| \sqrt{3} - \frac{12}{7} \right| < \frac{1}{7^2},$$

even though $\frac{12}{7}$ is not one of the p_n/q_n. This inequality can be easily checked by means of Figure 7.4, where we see that $(7,12)$ is below L while $(15,26)$ is above L. Thus

$$\tfrac{12}{7} < \sqrt{3} < \tfrac{26}{15},$$

and therefore

$$\left| \sqrt{3} - \frac{12}{7} \right| < \frac{26}{15} - \frac{12}{7} = \frac{2}{15 \cdot 7} < \frac{2}{14 \cdot 7} = \frac{1}{7^2}.$$

On the other hand, we have better approximations from our algorithm when a_{n+1} is large. It follows from Theorem 7.9 that if $a_{n+1} \geq 2$, then

$$\left| \alpha - \frac{p_n}{q_n} \right| \leq \frac{1}{2q_n^2}.$$

This is a better approximation to α than $1/q_n^2$, and it is actually this level of approximation that guarantees that $p/q = p_n/q_n$. In the case of $\alpha = (1 + \sqrt{5})/2$, every $a_n = 1$ and it is possible that there is no fraction p/q within $\frac{1}{2}q^2$ of α. The theorem, "If p/q is within $\frac{1}{2}q^2$ of α, then $p/q = p_n/q_n$ for some n," would then be meaningless. We therefore prove two results. The first will state that there are V_n which give

$$\left| \alpha - \frac{p_n}{q_n} \right| < \frac{1}{2q_n^2},$$

and, in fact, if α is irrational, there are infinitely many such V_n. The second result will show that if

$$\left| \alpha - \frac{p}{q} \right| < \frac{1}{2q^2},$$

then $p/q = p_n/q_n$ for some n.

Theorem 7.18. Suppose that $n \geq 2$ and α is irrational. Then either

$$\left| \alpha - \frac{p_{n-1}}{q_{n-1}} \right| < \frac{1}{2q_{n-1}^2} \qquad \text{or} \qquad \left| \alpha - \frac{p_n}{q_n} \right| < \frac{1}{2q_n^2}$$

(or, of course, both).

Proof. We suppose that both

$$\left| \alpha - \frac{p_{n-1}}{q_{n-1}} \right| \geq \frac{1}{2q_{n-1}^2}, \qquad \left| \alpha - \frac{p_n}{q_n} \right| \geq \frac{1}{2q_n^2}$$

and derive a contradiction. After multiplication by q_{n-1} and q_n, our supposition can be written

(12) $$|q_{n-1}\alpha - p_{n-1}| \geq \frac{1}{2q_{n-1}}, \qquad |q_n\alpha - p_n| \geq \frac{1}{2q_n}.$$

We will take the case that V_{n-1} is under L and V_n is over; the proof in the other case is identical, the figure being the only thing that changes. Let A be the intersection of L and $x = q_{n-1}$, B be the intersection of L and $x = q_n$, and C be the intersection of L and $V_{n-1}V_n$ (see Figure 7.15). Since $n \geq 2$, Theorem 7.7 guarantees that

$$0 < q_{n-1} < q_n.$$

This means that $V_{n-1}V_n$ is not a vertical line and hence C is strictly between

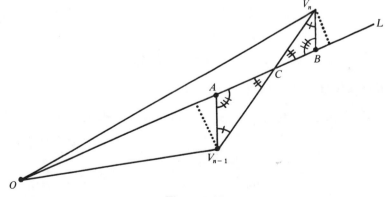

Figure 7.15.

A and *B*. Since

$$p_n q_{n-1} - p_{n-1} q_n = \pm 1$$

(Theorem 7.7), we see from Theorem 7.14 that

(13)   area $\triangle OV_{n-1}V_n = \frac{1}{2}$.

Our contradiction will be that the inequalities (12) require a larger area for $\triangle OV_{n-1}V_n$. The lengths of AV_{n-1} and BV_n are the vertical distances from V_{n-1} and V_n to L. Thus

$$AV_{n-1} = |q_{n-1}\alpha - p_{n-1}| \ge \frac{1}{2q_{n-1}},$$

$$BV_n = |q_n\alpha - p_n| \ge \frac{1}{2q_n}.$$

Since AV_{n-1} and BV_n are vertical, the altitudes of triangles OAV_{n-1} and OBV_n from O are q_{n-1} and q_n, respectively. Therefore,

(14)   area $\triangle OAV_{n-1} = \frac{1}{2}q_{n-1}(AV_{n-1}) \ge \frac{1}{4}$,

(15)   area $\triangle OBV_n = \frac{1}{2}q_n(BV_n) \ge \frac{1}{4}$.

Since AV_{n-1} and BV_n are parallel, triangles $AV_{n-1}C$ and BV_nC are similar; the triangle with the greater altitude (from corresponding vertices) will therefore have the greater area. The continued fraction algorithm gives V_n closer to L than V_{n-1}, and thus the altitude of $\triangle AV_{n-1}C$ from V_{n-1} is greater than the altitude of $\triangle BV_nC$ from V_n. Therefore,

area $\triangle AV_{n-1}C >$ area $\triangle BV_nC$

or

(16)   area $\triangle AV_{n-1}C -$ area $\triangle BV_nC > 0$.

When we add (14), (15), and (16), we get

$$\begin{aligned}
\text{area } \triangle OV_{n-1}V_n &= \text{area } \triangle OAV_{n-1} + \text{area } \triangle OBV_n \\
&\quad + (\text{area } \triangle AV_{n-1}C - \text{area } \triangle BV_nC) \\
&> \tfrac{1}{4} + \tfrac{1}{4} + 0,
\end{aligned}$$

and this contradicts (13). Thus the assumption (12) is in error and this proves the theorem.   ▲

Theorem 7.19. Suppose that p and q are integers and $q > 0$ so that p/q is defined. If α is irrational and

$$\left| \alpha - \frac{p}{q} \right| < \frac{1}{2q^2},$$

then, for some n,

$$\frac{p}{q} = \frac{p_n}{q_n}.$$

Proof. Since $q_0 = 1$, and

$$\lim_{n \to \infty} q_n = +\infty,$$

we may define n by the condition that

(17) $$q_n \leq q < q_{n+1}.$$

We will prove that for this n, $p/q = p_n/q_n$. Suppose $p/q \neq p_n/q_n$. By multiplying through by qq_n and transposing, we may put our assumption in the form

(18) $$pq_n - qp_n \neq 0.$$

Let $V = (q,p)$. Since $pq_n - qp_n$ is an integer,

$$|pq_n - qp_n| \geq 1$$

and thus, by Theorem 7.14,

(19) $$\text{area } \triangle OVV_n \geq \tfrac{1}{2}.$$

The idea of the proof is that the hypothesis

$$\left| \alpha - \frac{p}{q} \right| < \frac{1}{2q^2}$$

says that V is very close to L; by (17) and Theorem 7.15, V_n is even closer to L, thus triangle OVV_n will be too thin to enclose an area as large as $\tfrac{1}{2}$.

Let A be the intersection of L with the line $x = q$, B be the intersection of L with the line $x = q_n$, C be the intersection of L with VV_n, and D be the intersection of the lines OV and $x = q_n$ (see Figure 7.16). The proof is the same whether V_n is above or below L, but the pictures differ markedly according as to which of the numbers, slope L, slope V_n, slope V, is between the other two. Before considering each of the three possibilities, we make two general remarks. First, the hypothesis

$$\left| \alpha - \frac{p}{q} \right| < \frac{1}{2q^2}$$

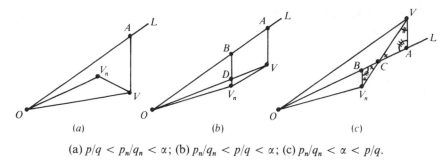

(a) $p/q < p_n/q_n < \alpha$; (b) $p_n/q_n < p/q < \alpha$; (c) $p_n/q_n < \alpha < p/q$.

Figure 7.16.

may be rewritten

(20)
$$AV = |\alpha q - p| < \frac{1}{2q}.$$

Second, since AV and BV_n are vertical, the altitudes of triangles OAV and OBV_n from O are horizontal and thus have the lengths q and q_n, respectively.

The first case is that of the slope of V_n being between the slope of V and the slope of L [see Figure 7.16(a)]. Since $q_n \le q$, V_n is either inside $\triangle OAV$ or on the edge AV and thus

$$\text{area } OVV_n < \text{area } OAV = \tfrac{1}{2}q(AV) < \tfrac{1}{4},$$

by (20). This contradicts (19).

The second case is that of the slope of V being between the slope of V_n and the slope of L [Figure 7.16(b)]. Since DV_n is vertical, the altitudes of triangles ODV_n and VDV_n from O and V are q_n and $q - q_n$, respectively. Thus

$$\begin{aligned}
\text{area } \triangle OVV_n &= \text{area } \triangle ODV_n + \text{area } \triangle VDV_n \\
&= \tfrac{1}{2}q_n(DV_n) + \tfrac{1}{2}(q - q_n)(DV_n) \\
&= \tfrac{1}{2}q(DV_n).
\end{aligned}$$

But clearly,

$$DV_n < BV_n$$

and, by (17) and Theorem 7.17,

$$BV_n \le AV.$$

Therefore, by (20),

$$\text{area } \triangle OVV_n = \tfrac{1}{2}q(DV_n) < \tfrac{1}{2}q(AV) < \tfrac{1}{4},$$

which contradicts (19).

The third case has the slope of L between the slopes of V and V_n. This is the only case that makes full use of the inequality in the hypothesis; the other cases can be proved with just

$$\left| \alpha - \frac{p}{q} \right| < \frac{1}{q^2}.$$

We see from (20) that

(21) $$\text{area } \triangle OAV = \tfrac{1}{2}q(AV) < \tfrac{1}{4}.$$

It follows from (17) and Theorem 7.17 that

$$BV_n \le AV \, ;$$

(17) also says that

$$q_n \le q$$

and therefore, by these inequalities and (20),

(22) $$\text{area } \triangle OBV_n = \tfrac{1}{2}q_n(BV_n) \le \tfrac{1}{2}q(AV) < \tfrac{1}{4}.$$

Triangles ACV and BCV_n are similar, and since it follows from (17) and Theorem 7.15 that the altitude of $\triangle ACV$ from V is greater than or equal to the altitude of $\triangle BCV_n$ from V_n, we see that

$$\text{area } \triangle ACV \ge \text{area } \triangle BCV_n$$

or

(23) $$\text{area } \triangle BCV_n - \text{area } \triangle ACV \le 0.$$

When we add (21), (22), and (23) we see that

$$\text{area } \triangle OVV_n = \text{area } \triangle OAV + \text{area } \triangle OBV_n + (\text{area } \triangle BCV_n - \text{area } \triangle ACV)$$
$$< \tfrac{1}{4} + \tfrac{1}{4} + 0,$$

which contradicts (19). ▲

EXERCISES
1. When $a_1 > 1$, $0 < q_0 < q_1$ and the proof of Theorem 7.18 holds as given for $n = 1$. Prove Theorem 7.18 when $n = 1$ and $a_1 = 1$.
2. Prove Theorem 7.15 for the case that V_n is under L and V_{n-1} is over L.
3. Use Theorems 7.16 and 7.17 to prove Theorem 7.2 when α is irrational.
4. Find a point, other than $(0,0)$, with integral coordinates whose distance from the line $21y = x\sqrt{7}$ is less than .001.

5. If V_n is on L and $q \geq q_n$, show that

$$\left| \alpha - \frac{p}{q} \right| < \frac{1}{q^2}$$

   implies (q,p) is on L also.
*6. Show that if

$$\frac{p}{q} \neq \frac{22}{7}$$

   and $0 < q \leq 50$, then

$$\left| \pi - \frac{22}{7} \right| < \left| \pi - \frac{p}{q} \right|.$$

   (In other words, $\frac{22}{7}$ is closer to π than p/q.) Does your method allow the
   bound of 50 to be increased and if so, to what?
7. Find the first four points in the set of closest points to the line $y = \alpha x$
   when $\alpha = \sqrt{231}$.
8. Repeat problem 7 with $\alpha = 11 + \sqrt{130}$.
*9. Suppose that α is rational, $0 < q_{n-1} < q_n$ and V_n is off L. Verify that
   Theorem 7.15 and its proof hold as given. Show, however, that if V_n is
   on L, then there is one exception to Theorem 7.15; it is with
   $U = V_n - V_{n-1}$, and in this case the distance from U to L is equal to
   the distance from V_{n-1} to L and $q_{n-1} \leq s < q_n$.
10. Prove that if $p > 0, q > 0, d > 4$ and

$$p^2 - dq^2 = \pm 1,$$

   then there is a V_n in the continued fraction expansion of $\sqrt{d}$ such that
   $V_n = (q,p)$.

$$\left[\text{Hint: Use the relation} \right.$$

$$\left. \left| \sqrt{d} - \frac{p}{q} \right| = \frac{1}{q(p + q\sqrt{d})}. \right]$$

11. Same as Problem 10 except $d = 2$ or 3.
12. It can be shown (miscellaneous exercise 17) that if d is a positive integer
   which is not a perfect square, then there are infinitely many n such that
   $a_{n+1} = 2[\sqrt{d}]$. Show that if $d \geq 7$ and $a_{n+1} = 2[\sqrt{d}]$, then
$$p_n^2 - dq_n^2 = (-1)^{n+1}.$$

$\Bigg[$ *Hint :* First show that $|p_n^2 - dq_n^2| < 2$ by means of the equality

$$p_n^2 - dq_n^2 = q_n^2 \left(\frac{p_n}{q_n} - \sqrt{d} \right) \left[\left(\frac{p_n}{q_n} - \sqrt{d} \right) + 2\sqrt{d} \right],$$

and the inequality

$$\frac{2\sqrt{d} + 1}{2[\sqrt{d}]} < \frac{2\sqrt{d} + 1}{2\sqrt{d} - 2} . \Bigg]$$

13. Same as problem 12 except with $d = 2, 3, 5, 6$. The inequalities must be refined.

7.5. A Commentary on Proof by Picture

There are many pitfalls in proving things from pictures. This is the reason that we used so much space in Section 7.4 justifying the way the figures were drawn. In this section we give two illustrations of what can happen when one argues from an incorrect picture. We begin by proving that every triangle is equilateral! Actually, if we can prove that sides AB and AC of triangle ABC are equal without any extra conditions, then we can also show $AB = BC$ by the same method ; we therefore content ourselves with showing that in any triangle ABC, $AB = AC$. The proof follows.

Let M be the midpoint of BC and let D be the intersection of the perpendicular bisector of BC and the angle bisector of $\angle BAC$ (see Figure 7.17). Let E be the point on AB such that $AB \perp DE$ and let F be the point on AC such that $AC \perp DF$. Now $BM = CM$ by construction, $\angle BMD = \angle CMD$ $(= 90°)$ by construction and $DM = DM$. Therefore, triangles BMD and CMD are congruent (angle, side, angle theorem). Thus $BD = CD$ and

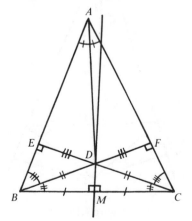

Figure 7.17.

$\angle MBD = \angle MCD$ (corresponding parts of congruent triangles are equal). Next, $DE = DF$ (a point on an angle bisector is equidistant from the sides) and $\angle BED = \angle CFD$ by construction. Therefore, triangles BED and CFD are congruent (two right triangles with two equal sides and equal hypotenuses are congruent—or, by the Pythagorean Theorem, the other sides are also equal and then we can use the side, angle, side, or the side, side, side theorems). Therefore, $\angle DBE = \angle DCF$ (corresponding sides of congruent triangles are equal). Thus

$$\angle MBE = \angle MCF$$

(sums of equals are equal). This says that $\triangle ABC$ is isosceles with equal angles at B and C and thus the sides opposite these angles, AB and AC, are equal.   Q.E.D.[2]

The error in the previous argument is not in any of the steps or reasons given but rather in Figure 7.17! Figure 7.17 is incorrectly drawn and the result has been an absurdity. It is commonly stated that the error consists of the fact that D is drawn above BC, whereas D is actually below BC. Thus, the reader may enjoy proving that $AB = AC$ from Figure 7.18 (the letters have the same meaning as before). The proof is practically identical to the one given using Figure 7.17.

I leave to the reader the joy of discovering the correct picture. Also left to the reader is the problem of finding the correct picture for the following "proof" that all angles are zero.

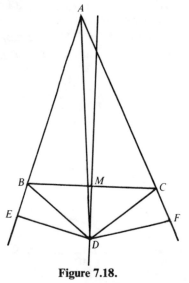

Figure 7.18.

[2] This comes from the English, Quite Easily Done.

Given $\angle A$, we pick B and C on the sides of $\angle A$ so that $AB = AC$. We will show that $\angle CAB = 0$ (see Figure 7.19). Let $ABED$ be a rectangle so that $AB = DE$ and $\angle ADE = \angle DAB = 90°$. Let the perpendicular bisectors of AD and CE meet at F. Then $AF = DF$ and $CF = EF$ (a point on the perpendicular bisector of a line segment is equidistant from the end points of the line segment). Therefore, triangles ACF and DEF are congruent (side, side, side theorem). Thus

$$\angle FDE = \angle FAC$$

(corresponding parts of congruent triangles are equal). But since $AF = DF$, triangle AFD is isosceles and therefore $\angle FDA = \angle FAD$. Subtracting equals from equals gives

$$\angle ADE = \angle DAC$$

or

$$90° = 90° + \angle CAB$$

and thus

$$\angle CAB = 0,$$

as claimed.   Q.E.D.

It is perhaps because of difficulties such as these that most mathematicians avoid arguing from pictures. One picture is worth a thousand words, provided one uses another thousand words to justify the picture. This is the spirit in which we presented Section 7.4. The usual development of continued

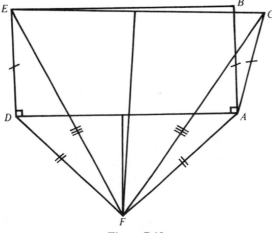

Figure 7.19.

fractions has no need of pictures; however, it takes some time by the usual method before one sees that the algorithm for α gives decent approximations to α. Further, the ideas involved seem simpler when pictures are used as aids. The reader wishing to compare the various methods of development can find the usual presentation in almost all the number theory texts listed in the bibliography.

*7.6. Periodic Continued Fractions

In this and the next section, it will be assumed that the reader is familiar with the material in Appendix B on matrices. We begin by putting the continued fraction algorithm in matrix notation (the earlier notation is given on page 188 and in Theorem 7.10). For $n \geq 0$, let

$$A_n = \begin{pmatrix} 0 & 1 \\ 1 & a_n \end{pmatrix}, \qquad M_n = \begin{pmatrix} q_{n-2} & p_{n-2} \\ q_{n-1} & p_{n-1} \end{pmatrix}, \qquad \gamma_n = q_{n-2} + \alpha_n q_{n-1}.$$

In this and the following section, the letters A_n, M_n, and γ_n will always have the above meaning.

Theorem 7.20. Suppose that α is irrational. Then for all $n \geq 0$,

$$M_{n+1} = A_n M_n,$$

$$\gamma_n(1, \alpha) = (1, \alpha_n) M_n,$$

$$\gamma_n \neq 0,$$

$$\det M_n = (-1)^n,$$

$$M_{n+1} = A_n A_{n-1} \cdots A_1 A_0.$$

Further, M_n^{-1} is a matrix of integers and, last, if $M_n = M_m$, then $n = m$.

Proof. It follows from the definition of matrix multiplication and Theorem 7.7 that

$$A_n M_n = \begin{pmatrix} 0 & 1 \\ 1 & a_n \end{pmatrix} \begin{pmatrix} q_{n-2} & p_{n-2} \\ q_{n-1} & p_{n-1} \end{pmatrix} = \begin{pmatrix} q_{n-1} & p_{n-1} \\ q_{n-2} + a_n q_{n-1} & p_{n-2} + a_n p_{n-1} \end{pmatrix}$$

$$= \begin{pmatrix} q_{n-1} & p_{n-1} \\ q_n & p_n \end{pmatrix} = M_{n+1}.$$

If, in Theorem 7.11, we put $\beta = \alpha$, $b_n = a_n$, then the conclusion of Theorem 7.11 says that the point $V_{n-2} + \alpha_n V_{n-1}$ is on the line $y = \alpha x$. Thus the coordinates of $V_{n-2} + \alpha_n V_{n-1}$ satisfy the equation

$$p_{n-2} + \alpha_n p_{n-1} = \alpha(q_{n-2} + \alpha_n q_{n-1}).$$

Therefore,

$$(1,\alpha_n)M_n = (1,\alpha_n)\begin{pmatrix} q_{n-2} & p_{n-2} \\ q_{n-1} & p_{n-1} \end{pmatrix}$$

$$= (q_{n-2} + \alpha_n q_{n-1}, p_{n-2} + \alpha_n p_{n-1})$$

$$= (q_{n-2} + \alpha_n q_{n-1})(1,\alpha)$$

$$= \gamma_n(1,\alpha).$$

Since α is irrational, α_n is also irrational. Thus

$$q_{n-2} + \alpha_n q_{n-1} = 0$$

implies by Theorem 6.4 that $q_{n-2} = q_{n-1} = 0$. But $q_{-2} = 1$, $q_{-1} = 0$, and $q_n \geq 1$ for $n \geq 0$. Thus it is impossible for two consecutive q_n's to be zero and hence

$$\gamma_n = q_{n-2} + \alpha_n q_{n-1} \neq 0.$$

Next, by Theorem 7.7 and the definition of M_n,

$$\det M_n = p_{n-1}q_{n-2} - p_{n-2}q_{n-1} = (-1)^n.$$

Since

$$M_0 = \begin{pmatrix} q_{-2} & p_{-2} \\ q_{-1} & p_{-1} \end{pmatrix} = \begin{pmatrix} 1 & 0 \\ 0 & 1 \end{pmatrix} = I,$$

we see from the first part of the theorem that

$$M_{n+1} = A_n M_n$$

$$= A_n A_{n-1} M_{n-1}$$

$$\vdots$$

$$= A_n A_{n-1} \cdots A_1 M_1$$

$$= A_n A_{n-1} \cdots A_1 A_0 M_0$$

$$= A_n A_{n-1} \cdots A_1 A_0.$$

Since $\det M_n = (-1)^n$,

$$M_n^{-1} = \frac{1}{\det M_n}\begin{pmatrix} p_{n-1} & -p_{n-2} \\ -q_{n-1} & q_{n-2} \end{pmatrix}$$

$$= (-1)^n\begin{pmatrix} p_{n-1} & -p_{n-2} \\ -q_{n-1} & q_{n-2} \end{pmatrix},$$

which is a matrix of integers. Finally, suppose that $n \neq m$ but $M_n = M_m$; that is,

$$\begin{pmatrix} q_{n-2} & p_{n-2} \\ q_{n-1} & p_{n-1} \end{pmatrix} = \begin{pmatrix} q_{m-2} & p_{m-2} \\ q_{m-1} & p_{m-1} \end{pmatrix}.$$

Therefore,

$$q_{n-2} = q_{m-2}, \qquad q_{n-1} = q_{m-1}.$$

Since

$$q_{-1} = 0 < 1 = q_{-2} = q_0 \leq q_1 < q_2 < \cdots,$$

we see that $n - 2$ equals either -2, 0, or 1, so that n equals either 0, 2, or 3. Likewise $n - 1$ equals either -2, 0, or 1, so that n equals either -1, 1, or 2. Hence $n = 2$. In like manner, $m = 2$ and thus $n = m$. ▲

Let us now give an example of the method of finding α from its periodic continued fraction expansion. Suppose we are given

$$\alpha = \langle 4,\overline{1,3,1,8} \rangle = \langle 4,1,3,1,8,1,3,1,8,\ldots \rangle$$

Then

$$\alpha_1 = \langle 1,3,1,8,1,3,1,8,\ldots \rangle = \langle \overline{1,3,1,8} \rangle,$$

$$\alpha_2 = \langle 3,1,8,1,3,1,8,1,\ldots \rangle = \langle \overline{3,1,8,1} \rangle,$$

$$\alpha_3 = \langle 1,8,1,3,1,8,1,3,\ldots \rangle = \langle \overline{1,8,1,3} \rangle,$$

$$\alpha_4 = \langle 8,1,3,1,8,1,3,1,\ldots \rangle = \langle \overline{8,1,3,1} \rangle,$$

$$\alpha_5 = \langle 1,3,1,8,1,3,1,8,\ldots \rangle = \langle \overline{1,3,1,8} \rangle.$$

In other words,

$$\alpha_5 = \alpha_1.$$

We now calculate M_1 and M_5 by first finding V_0 through V_4 by our usual process:

n	a_n	q_n	p_n
-2		1	0
-1		0	1
0	4	1	4
1	1	1	5
2	3	4	19
3	1	5	24
4	8	44	211.

Therefore,

$$M_1 = \begin{pmatrix} q_{-1} & p_{-1} \\ q_0 & p_0 \end{pmatrix} = \begin{pmatrix} 0 & 1 \\ 1 & 4 \end{pmatrix},$$

$$M_5 = \begin{pmatrix} q_3 & p_3 \\ q_4 & p_4 \end{pmatrix} = \begin{pmatrix} 5 & 24 \\ 44 & 211 \end{pmatrix},$$

and also,

$$M_1^{-1} = \begin{pmatrix} -4 & 1 \\ 1 & 0 \end{pmatrix}.$$

By Theorem 7.20,

$$\gamma_1(1,\alpha) = (1,\alpha_1)M_1, \qquad \gamma_5(1,\alpha) = (1,\alpha_5)M_5.$$

From the first of these equations,

$$(1,\alpha_1) = [(1,\alpha_1)M_1]M_1^{-1} = \gamma_1(1,\alpha)M_1^{-1},$$

and thus the second equation gives (since $\alpha_5 = \alpha_1$),

$$\gamma_5(1,\alpha) = (1,\alpha_5)M_5 = (1,\alpha_1)M_5 = \gamma_1(1,\alpha)M_1^{-1}M_5$$

or, letting $\delta = \gamma_5/\gamma_1$,

$$\delta(1,\alpha) = (1,\alpha)M_1^{-1}M_5 = (1,\alpha)\begin{pmatrix} -4 & 1 \\ 1 & 0 \end{pmatrix}\begin{pmatrix} 5 & 24 \\ 44 & 211 \end{pmatrix} = (1,\alpha)\begin{pmatrix} 24 & 115 \\ 5 & 24 \end{pmatrix}.$$

This says that

$$\delta = 24 + 5\alpha, \qquad \delta\alpha = 115 + 24\alpha,$$

and thus

$$(24 + 5\alpha)\alpha = \delta\alpha = 115 + 24\alpha.$$

Therefore,

$$24\alpha + 5\alpha^2 = 115 + 24\alpha,$$

or, in other words,

$$\alpha^2 = 23.$$

Thus α is a root of a quadratic equation. In this case,

$$\alpha = \pm\sqrt{23}$$

and since

$$[\alpha] = a_0 = 4,$$

we see that $\alpha > 0$ and thus $\alpha = \sqrt{23}$.

The previous example illustrates the major idea in the proof of the fact that if the continued fraction expansion of α becomes ultimately periodic, then α is the root of a quadratic equation with integral coefficients. There is one detail that creeps into the general case that never shows up in any numerical examples. Suppose we end up showing that

$$a\alpha^2 + b\alpha + c = 0,$$

where a, b, and c are integers. Does this mean that α is the root of a quadratic equation? Of course, but the equation may end up being the identity

$$0\alpha^2 + 0\alpha + 0 = 0$$

satisfied by all real numbers α. Thus we will have to show that the coefficients in the equation are not all zero or we will have proved nothing. We will actually show that the coefficient of α^2 is not zero. The next two theorems will be used for that purpose.

Theorem 7.21. Suppose $n \geq 0$ and $j > 0$. Let

$$M_n^{-1} M_{n+j} = \begin{pmatrix} r & s \\ t & u \end{pmatrix}.$$

Then

$$\begin{pmatrix} r & s \\ t & u \end{pmatrix} \neq rI.$$

Proof. Suppose

$$M_n^{-1} M_{n+j} = rI.$$

Then

$$M_{n+j} = M_n[M_n^{-1} M_{n+j}] = M_n(rI) = rM_n.$$

When we equate the first rows of M_{n+j} and rM_n, we get

$$q_{n+j-2} = rq_{n-2}, \qquad p_{n+j-2} = rp_{n-2}.$$

Therefore, r divides the relatively prime numbers q_{n+j-2} and p_{n+j-2} and hence $r = \pm 1$. Since M_{n+j} has no negative entries, $r = 1$. Thus $M_n = M_{n+j}$

and therefore, by Theorem 7.20, $n = n + j$. But since $j > 0$, this is impossible. Therefore, $M_n^{-1}M_{n+j} \neq rI$, as advertised.   ▲

Theorem 7.22. Suppose α is irrational, r, s, t, and u are integers and δ is a real number such that

$$\delta(1, \alpha) = (1, \alpha)\begin{pmatrix} r & s \\ t & u \end{pmatrix}.$$

If $t = 0$, then

$$\begin{pmatrix} r & s \\ t & u \end{pmatrix} = rI.$$

Proof. When $t = 0$, we have

$$(\delta, \delta\alpha) = (1, \alpha)\begin{pmatrix} r & s \\ 0 & u \end{pmatrix} = (r, s + u\alpha).$$

Thus

$$\delta = r, \qquad \delta\alpha = s + u\alpha.$$

The last equation therefore becomes

$$r\alpha = s + u\alpha$$

or

$$s + (u - r)\alpha = 0.$$

Therefore, by Theorem 6.4,

$$u = r, \qquad s = 0,$$

and hence

$$\begin{pmatrix} r & s \\ t & u \end{pmatrix} = \begin{pmatrix} r & 0 \\ 0 & r \end{pmatrix} = rI. \qquad ▲$$

Theorem 7.23. Suppose that the continued fraction expansion of α is ultimately periodic. More exactly, suppose that there are numbers $N \geq 0$, $j > 0$ such that for all $n \geq N$,

$$a_{n+j} = a_n.$$

Then α is an irrational root of a quadratic equation with integral coefficients and the coefficient of α^2 is not 0.

Proof. By hypothesis

$$\alpha_{N+j} = \overline{\langle a_{N+j}, a_{N+J+1}, \ldots, a_{N+2j-1} \rangle}$$

$$= \overline{\langle a_N, a_{N+1}, \ldots, a_{N+j-1} \rangle} = \alpha_N.$$

Hence α is irrational since its continued fraction expansion does not terminate. By Theorem 7.20,

$$\gamma_N(1,\alpha) = (1,\alpha_N)M_N, \qquad \gamma_{N+j}(1,\alpha) = (1,\alpha_{N+j})M_{N+j}.$$

From the first of these equations,

$$(1,\alpha_N) = [(1,\alpha_N)M_N]M_N^{-1} = \gamma_N(1,\alpha)M_N^{-1},$$

and thus the second equation gives (since $\alpha_{N+j} = \alpha_N$),

$$\gamma_{N+j}(1,\alpha) = (1,\alpha_{N+j})M_{N+j} = (1,\alpha_N)M_{N+j} = \gamma_N(1,\alpha)M_N^{-1}M_{N+j}$$

or, letting $\delta = \gamma_{N+j}/\gamma_N$ (the division is legal since $\gamma_N \neq 0$),

(24) $$\delta(1,\alpha) = (1,\alpha)M_N^{-1}M_{N+j}.$$

Let

(25) $$M_N^{-1}M_{N+j} = \begin{pmatrix} r & s \\ t & u \end{pmatrix},$$

where $r, s, t,$ and u are integers since M_N^{-1} is a matrix of integers. Since α is irrational, Theorems 7.21 and 7.22 tell us that

(26) $$t \neq 0.$$

Equation (24) may be rewritten

$$(\delta,\delta\alpha) = (1,\alpha)\begin{pmatrix} r & s \\ t & u \end{pmatrix} = (r + t\alpha, s + u\alpha),$$

and therefore

$$\delta = r + t\alpha,$$

$$\delta\alpha = s + u\alpha.$$

Thus

$$(r + t\alpha)\alpha = \delta\alpha = s + u\alpha,$$

so that

(27) $$t\alpha^2 + (r - u)\alpha - s = 0.$$

Thus α satisfies a quadratic equation with integral coefficients and the coefficient of α^2 is not zero.   ▲

Now we shall tackle the converse of Theorem 7.23. This is the more interesting and more difficult result, first proved around 1766 by Lagrange, that if α is an irrational root of a quadratic equation with integral coefficients (leading coefficient $\neq 0$), then the continued fraction algorithm for α repeats. There are several proofs of this theorem, but they all are based on the same ideas. The goal is to prove that for some $n \geq 0$ and $j > 0$, $\alpha_{n+j} = \alpha_n$. The periodicity follows immediately from this, since then $a_{n+j} = a_n$, $\alpha_{n+j+1} = \alpha_{n+1}$, and so on. The method of showing that for some n and j, $\alpha_{n+j} = \alpha_n$, is to show that the set of numbers $\alpha_0, \alpha_1, \alpha_2, \ldots$ is actually a finite set. This means that there are repetitions in the sequence $\alpha_0, \alpha_1, \ldots$, and this is what we desire to prove. It will be shown that each α_n satisfies a quadratic equation with integral coefficients, the coefficient of α_n^2 not equaling zero. Thus if we can show that there are only finitely many such equations, then since each equation has only two roots (we use here the fact that the coefficient of α_n^2 is not zero so as to avoid the equation $0\alpha_n^2 + 0\alpha_n + 0 = 0$), there will be only finitely many α_n. Since the coefficients of the equations for the α_n are integers, if we can show that the absolute values of the coefficients are bounded (for example, if they are all less than 1 billion), then there will be only a finite number of possibilities for each of the three coefficients and hence only a finite number of equations for the α_n. This is what the proof does.

Since we will be dealing with matrices, we introduce a symbol for measuring the size of the entries.

Definition. Let

$$A = \begin{pmatrix} a & b \\ c & d \end{pmatrix}.$$

We define the **norm** of A by

$$\|A\| = \text{the maximum of the four numbers } |a|, |b|, |c|, |d|.$$

It follows that $|a| \leq \|A\|, |b| \leq \|A\|$, and so on. As an example, if

$$A = \begin{pmatrix} 1 & -5 \\ 4 & 2 \end{pmatrix},$$

then $\|A\| = 5$. We use the notation $\|A\|$ instead of $|A|$ because many authors reserve $|A|$ to mean the determinant of A.

Theorem 7.24. If A, B, and C are 2×2 matrices, then

$$\|A + B\| \le \|A\| + \|B\|, \|AB\| \le 2\|A\| \cdot \|B\|,$$
$$\|ABC\| \le 4\|A\| \cdot \|B\| \cdot \|C\|.$$

Proof. Let

$$A = \begin{pmatrix} a_{11} & a_{12} \\ a_{21} & a_{22} \end{pmatrix}, \qquad B = \begin{pmatrix} b_{11} & b_{12} \\ b_{21} & b_{22} \end{pmatrix},$$

$$A + B = \begin{pmatrix} d_{11} & d_{12} \\ d_{21} & d_{22} \end{pmatrix}, \qquad AB = \begin{pmatrix} e_{11} & e_{12} \\ e_{21} & e_{22} \end{pmatrix}.$$

Then for $i = 1$ or $2, j = 1$ or 2,

$$|d_{ij}| = |a_{ij} + b_{ij}| \le |a_{ij}| + |b_{ij}| \le \|A\| + \|B\|,$$

and thus since all four numbers $|d_{11}|, |d_{12}|, |d_{21}|, |d_{22}|$ satisfy the above inequality,

$$\|A + B\| \le \|A\| + \|B\|.$$

Also for $i = 1$ or $2, j = 1$ or 2,

$$e_{ij} = a_{i1}b_{1j} + a_{i2}b_{2j},$$

and thus

$$|e_{ij}| = |a_{i1}b_{1j} + a_{i2}b_{2j}| \le |a_{i1}| \cdot |b_{1j}| + |a_{i2}| \cdot |b_{2j}|$$
$$\le \|A\| \cdot \|B\| + \|A\| \cdot \|B\| \le 2\|A\| \cdot \|B\|.$$

We may use the result for the product of two matrices to get the result for the product of three matrices:

$$\|ABC\| = \|(AB)C\| \le 2\|AB\| \cdot \|C\| \le 2 \cdot 2\|A\| \cdot \|B\| \cdot \|C\|. \qquad \blacktriangle$$

Theorem 7.25. Suppose that a, b, and c are integers, $a \ne 0$, and α is a positive irrational number satisfying the equation

$$a\alpha^2 + b\alpha + c = 0.$$

Then the continued fraction expansion of α is ultimately periodic. (*Remark:* This theorem is true for negative α also, but we have not considered the continued fraction expansions of negative numbers.)

Proof. Since $a\alpha^2 = -b\alpha - c$,

$$a\alpha(1,\alpha) = (a\alpha, a\alpha^2) = (a\alpha, -c - b\alpha) = (1,\alpha)\begin{pmatrix} 0 & -c \\ a & -b \end{pmatrix}.$$

Let

$$B = \begin{pmatrix} 0 & -c \\ a & -b \end{pmatrix},$$

so that

(28) $$a\alpha(1,\alpha) = (1,\alpha)B,$$

We now find the quadratic equation satisfied by α_n. By Theorem 7.20,

$$\gamma_n(1,\alpha) = (1,\alpha_n)M_n,$$

it follows from this and (28) that

(29) $$\begin{aligned} a\alpha(1,\alpha_n) &= a\alpha[(1,\alpha_n)M_n]M_n^{-1} \\ &= a\alpha[\gamma_n(1,\alpha)]M_n^{-1} \\ &= \gamma_n[a\alpha(1,\alpha)]M_n^{-1} \\ &= \gamma_n[(1,\alpha)B]M_n^{-1} \\ &= [\gamma_n(1,\alpha)]BM_n^{-1} \\ &= (1,\alpha_n)M_nBM_n^{-1}. \end{aligned}$$

Let

$$M_nBM_n^{-1} = \begin{pmatrix} r_n & s_n \\ t_n & u_n \end{pmatrix};$$

this is a matrix of integers since M_n, B, and M_n^{-1} are matrices of integers. Then

$$(a\alpha, a\alpha\alpha_n) = (1,\alpha_n)\begin{pmatrix} r_n & s_n \\ t_n & u_n \end{pmatrix} = (r_n + t_n\alpha_n, s_n + u_n\alpha_n),$$

so that

(30) $$a\alpha = r_n + t_n\alpha_n, \qquad a\alpha\alpha_n = s_n + u_n\alpha_n.$$

It follows from (30) that

$$(r_n + t_n\alpha_n)\alpha_n = a\alpha\alpha_n = s_n + u_n\alpha_n$$

and hence

$$t_n\alpha_n^2 + (r_n - u_n)\alpha_n - s_n = 0.$$

Therefore, α_n satisfies a quadratic equation with integral coefficients and further t_n, the coefficient of α_n^2, is not zero since $t_n = 0$ implies [see (30)] that $\alpha = r_n/a$ is rational, which is false.

Thus, given the matrix $M_n B M_n^{-1}$, there are only two possible choices for the number α_n and hence if the set of matrices $M_n B M_n^{-1}$ $(n = 0, 1, 2, \ldots)$ is finite, then so is the set of α_n $(n = 0, 1, 2, \ldots)$ and then for two different subscripts $n \neq m$, we would have $\alpha_n = \alpha_m$ and the continued fraction expansion for α would be ultimately periodic. Therefore, it suffices to prove that the set of matrices $M_n B M_n^{-1}$ is finite. Let

(31) $$\theta = |a\alpha| + 4(1 + |\alpha|) \cdot \|B - a\alpha I\|$$

and set

$$k = [\theta].$$

Note that θ is a number independent of the variable n. Suppose we can show that for all $n \geq 2$,

$$\|M_n B M_n^{-1}\| \leq \theta.$$

Then by definition of the norm of a matrix, the numbers r_n, s_n, t_n, u_n are between $-\theta$ and θ. Thus there are $2k + 1$ possibilities for each of the numbers r_n, s_n, t_n, u_n and hence $(2k + 1)^4$ possibilities for the matrix $M_n B M_n^{-1}$ $(n \geq 2)$. Thus the set of matrices $M_n B M_n^{-1}$ would be finite as desired (the reason for using $n \geq 2$ instead of $n \geq 0$ is that we will be using $1/q_{n-2}$ and $1/q_{n-1}$ in the course of the proof and $q_{-1} = 0$). Thus our theorem will be proved if we can show that for $n \geq 2$,

$$\|M_n B M_n^{-1}\| \leq \theta.$$

Let

$$E_n = \begin{pmatrix} 0 & p_{n-2} - q_{n-2}\alpha \\ 0 & p_{n-1} - q_{n-1}\alpha \end{pmatrix}.$$

By Theorem 7.9, if $j \geq 0$, then

$$\left| \alpha - \frac{p_j}{q_j} \right| \leq \frac{1}{q_j q_{j+1}}$$

or

$$|q_j \alpha - p_j| \leq \frac{1}{q_{j+1}}$$

so that, for $n \geq 2$,

$$|q_{n-2}\alpha - p_{n-2}| \leq \frac{1}{q_{n-1}}$$

and

$$|q_{n-1}\alpha - p_{n-1}| \le \frac{1}{q_n} < \frac{1}{q_{n-1}}.$$

Therefore,

(32)
$$\|E_n\| \le \frac{1}{q_{n-1}}.$$

The matrix E_n enters into our calculations as follows:

(33)
$$M_n = \begin{pmatrix} q_{n-2} & p_{n-2} \\ q_{n-1} & p_{n-1} \end{pmatrix}$$

$$= \begin{pmatrix} q_{n-2} & q_{n-2}\alpha \\ q_{n-1} & q_{n-1}\alpha \end{pmatrix} + \begin{pmatrix} 0 & p_{n-2} - q_{n-2}\alpha \\ 0 & p_{n-1} - q_{n-1}\alpha \end{pmatrix}$$

$$= \begin{pmatrix} q_{n-2} & 0 \\ 0 & q_{n-1} \end{pmatrix}\begin{pmatrix} 1 & \alpha \\ 1 & \alpha \end{pmatrix} + E_n.$$

It follows from (33), (28), and (33), again, that

$$M_n B = \begin{pmatrix} q_{n-2} & 0 \\ 0 & q_{n-1} \end{pmatrix}\begin{pmatrix} 1 & \alpha \\ 1 & \alpha \end{pmatrix} B + E_n B$$

$$= \begin{pmatrix} q_{n-2} & 0 \\ 0 & q_{n-1} \end{pmatrix} \cdot a\alpha \begin{pmatrix} 1 & \alpha \\ 1 & \alpha \end{pmatrix} + E_n B$$

$$= a\alpha \begin{pmatrix} q_{n-2} & 0 \\ 0 & q_{n-1} \end{pmatrix}\begin{pmatrix} 1 & \alpha \\ 1 & \alpha \end{pmatrix} + E_n B$$

$$= a\alpha(M_n - E_n) + E_n B$$

$$= a\alpha M_n + E_n(B - a\alpha I).$$

Therefore,

$$M_n B M_n^{-1} = a\alpha M_n M_n^{-1} + E_n(B - a\alpha I)M_n^{-1}$$

$$= a\alpha I + E_n(B - a\alpha I)M_n^{-1}.$$

As a result, it follows from Theorem 7.24 that

(34)
$$\|M_n B M_n^{-1}\| \le \|a\alpha I\| + \|E_n(B - a\alpha I)M_n^{-1}\|$$

$$\le |a\alpha| + 4\|E_n\| \cdot \|B - a\alpha I\| \cdot \|M_n^{-1}\|.$$

We have already found an estimate on $\|E_n\|$. Since

$$M_n^{-1} = (-1)^n \begin{pmatrix} p_{n-1} & -p_{n-2} \\ -q_{n-1} & q_{n-2} \end{pmatrix}$$

and since for $j \geq 0$,

$$|p_j| = |\alpha q_j + (p_j - \alpha q_j)| \leq |\alpha q_j| + |(p_j - \alpha q_j)|$$

$$\leq |\alpha| q_j + \frac{1}{q_{j+1}} \leq |\alpha| q_j + q_j = q_j(1 + |\alpha|),$$

we see that, for $n \geq 2$,

$$|p_{n-2}| \leq q_{n-2}(1 + |\alpha|) \leq q_{n-1}(1 + |\alpha|),$$

$$|p_{n-1}| \leq q_{n-1}(1 + |\alpha|).$$

Therefore,

$$\|M_n^{-1}\| \leq q_{n-1}(1 + |\alpha|).$$

If we put this and (32) into (34), we get our desired result:

$$\|M_n B M_n^{-1}\| \leq |a\alpha| + 4 \cdot \frac{1}{q_{n-1}} \cdot \|B - a\alpha I\| \cdot q_{n-1}(1 + |\alpha|)$$

$$= |a\alpha| + 4(1 + |\alpha|) \cdot \|B - a\alpha I\|$$

$$= \theta. \qquad \blacktriangle$$

EXERCISES

1. The main difficulty in Theorem 7.25 is to prove that $\|M_n B M_n^{-1}\|$ is bounded by something independent of n. What is wrong with the following argument?

   "By Theorem 7.24,

$$\|M_n B M_n^{-1}\| \leq 4\|M_n\| \cdot \|B\| \cdot \|M_n^{-1}\|$$

$$= 4\|M_n\| \cdot \|B\| \cdot \frac{1}{\|M_n\|} = 4\|B\|,$$

   which is independent of n."

In problems 2–10, find α from its continued fraction expansion.

2. $\alpha = \langle \overline{4,1,1,1} \rangle$.

3. $\alpha = \langle 10,\overline{1,2,1,1,1,1,2,1,20} \rangle$.

4. $\alpha = \langle 1,2,3,\overline{1,2} \rangle$.

5. $\alpha = \langle 0,1,1,\overline{2} \rangle$.

6. $\alpha = \langle 1,1,1,\overline{1,8,1,18} \rangle$.

7. $\alpha = \langle 1,2,1,2,\overline{1,2,3,4} \rangle$.

8. $\alpha = \langle 0,6,\overline{1,4} \rangle$.

9. $\alpha = \langle 0,18,\overline{1,12} \rangle$.

10. $\alpha = \langle 2,2,4,3,\overline{22,3,5,3} \rangle$.

In problems 11–15, find the continued fraction expansion of α.

11. $\alpha = \dfrac{1 + \sqrt{17}}{4}$.

12. $\alpha = \dfrac{9 + \sqrt{5}}{5}$.

13. $\alpha = \dfrac{176 + \sqrt{2}}{2}$.

14. $\alpha = \dfrac{47 - 2\sqrt{5}}{6}$.

15. $\alpha = \dfrac{8 + 7\sqrt{2}}{34}$.

16. Let k and n be positive integers such that $k|n$. Find the continued fraction expansion of all numbers of the form

$$\sqrt{n^2 + k}.$$

In your answer, can the restriction that $k|n$ be relaxed (or even eliminated)?

*7.7. The Fermat–Pell Equation and the Continued Fraction Expansion of $\sqrt{d}$

In this section d will stand for a positive integer which is not a perfect square.[3] Thus $\sqrt{d}$ is irrational. The Fermat–Pell equation

(35) $$x^2 - dy^2 = 1$$

has the trivial solutions $x = \pm 1$, $y = 0$ while the Fermat–Pell equation

(36) $$x^2 - dy^2 = -1$$

has no obvious solutions.

[3] When d is a perfect square, the methods of Section 5.1 are sufficient to solve the Fermat–Pell equations. See problem 4, page 148.

In particular instances (for example, $d = 2$), it is easy to find solutions to each equation by inspection. In other instances (for example, $d = 3$), it is easy to find solutions to (35) by inspection but not (36). In still other instances (for example, $d = 67$), it is safe to say that the reader will not find solutions to either equation by inspection [other than the trivial solution to (35)]. Yet (35) always has an infinite number of solutions. Sometimes (36) has no solutions (as we have already learned in Theorem 5.1); other times it also has infinitely many solutions. The smallest solutions to these equations can be amazingly large. For example, the smallest nontrivial solution to

$$x^2 - 67y^2 = 1$$

is given by $|x| = 48\,842, |y| = 5967$. This happens even though when $d = 66$, we have a solution with $y = 8$ and when $d = 68$, we have a solution with $y = 4$. The solutions to (35) and (36) are intimately bound up with the continued fraction expansion of $\sqrt{d}$.

Theorem 7.26. The Fermat–Pell equation

$$x^2 - dy^2 = 1$$

has infinitely many integral solutions. If the length of the period in the continued fraction expansion of $\sqrt{d}$ is odd, then the Fermat–Pell equation

$$x^2 - dy^2 = -1$$

has infinitely many integral solutions.

Proof. Let the continued fraction expansion of $\sqrt{d}$ be given by

$$\sqrt{d} = \langle a_0, a_1, a_2, \ldots \rangle.$$

By Theorem 7.25, this expansion is ultimately periodic; thus there are integers $N \geq 0$ and $j \geq 1$ such that for $n \geq N$,

(37) $$a_{n+j} = a_n.$$

Here j represents the length of the period. But it follows from (37) that for any positive integer k,

(38) $$a_{n+kj} = a_n.$$

If we did not already know that

$$\langle a_0, \ldots, a_{N-1}, \overline{a_N, \ldots, a_{N+kj-1}} \rangle = \sqrt{d},$$

we could find this out by using the method of Theorem 7.23. We shall do this anyway. As shown in equations (24) of Theorem 7.23, there is a real number δ_k such that

$$\delta_k(1,\sqrt{d}) = (1,\sqrt{d})M_N^{-1}M_{N+kj}.$$

Further, if we put

$$M_N^{-1}M_{N+kj} = \begin{pmatrix} r_k & s_k \\ t_k & u_k \end{pmatrix},$$

as in (25), we get the equivalents of (26) and (27),

$$t_k(\sqrt{d})^2 + (r_k - u_k)\sqrt{d} - s_k = 0$$

with

$$t_k \neq 0.$$

In this case, our equation simplifies to

$$(dt_k - s_k) + (r_k - u_k)\sqrt{d} = 0.$$

Since $\sqrt{d}$ is irrational, it follows from Theorem 6.4 that

$$u_k = r_k, \qquad s_k = dt_k.$$

Thus, in the expansion of $\sqrt{d}$,

$$(39) \qquad M_N^{-1}M_{N+kj} = \begin{pmatrix} r_k & dt_k \\ t_k & r_k \end{pmatrix}.$$

Therefore,

$$(40) \qquad \begin{aligned} r_k^2 - dt_k^2 &= \det(M_N^{-1}M_{N+kj}) = (\det M_N)^{-1}(\det M_{N+kj}) \\ &= (-1)^{-N}(-1)^{N+kj} = (-1)^{kj}. \end{aligned}$$

There are infinitely many values of k involved in (40). When j is even, $(-1)^{jk} = 1$ for all k; if j is odd, then $(-1)^{jk} = 1$ for all even k and $(-1)^{jk} = -1$ for all odd k. Thus our theorem is proved if we can show that different values of k give different pairs of numbers r_k and t_k. Suppose

$$r_k = r_m, \qquad t_k = t_m.$$

Then, by (39),

$$M_N^{-1}M_{N+kj} = M_N^{-1}M_{N+mj}$$

and, therefore,

$$M_{N+kj} = M_N(M_N^{-1}M_{N+kj}) = M_N(M_N^{-1}M_{N+mj}) = M_{N+mj}.$$

Thus

$$N + kj = N + mj, \qquad k = m. \qquad \blacktriangle$$

As an example, let us consider the case of $d = 3$. The expansion of $\sqrt{3}$ is

$$\sqrt{3} = \langle 1,\overline{1,2} \rangle = \langle 1,\overline{1,2,1,2} \rangle$$

$$= \langle 1,\overline{1,2,1,2,1,2} \rangle$$

$$= \langle 1,\overline{1,2,1,2,1,2,1,2} \rangle$$

$$= \cdots.$$

Here we may take $N = 1, j = 2$. A short calculation gives

$$M_N = M_1 = \begin{pmatrix} 0 & 1 \\ 1 & 1 \end{pmatrix}, \qquad M_N^{-1} = M_1^{-1} = \begin{pmatrix} -1 & 1 \\ 1 & 0 \end{pmatrix},$$

$$M_{N+j} = M_3 = \begin{pmatrix} 1 & 2 \\ 3 & 5 \end{pmatrix},$$

$$M_{N+2j} = M_5 = \begin{pmatrix} 4 & 7 \\ 11 & 19 \end{pmatrix},$$

$$M_{N+3j} = M_7 = \begin{pmatrix} 15 & 26 \\ 41 & 71 \end{pmatrix},$$

$$M_{N+4j} = M_9 = \begin{pmatrix} 56 & 97 \\ 153 & 265 \end{pmatrix},\ldots.$$

Thus

$$M_N^{-1}M_{N+j} = \begin{pmatrix} -1 & 1 \\ 1 & 0 \end{pmatrix}\begin{pmatrix} 1 & 2 \\ 3 & 5 \end{pmatrix} = \begin{pmatrix} 2 & 3 \\ 1 & 2 \end{pmatrix},$$

$$M_N^{-1}M_{N+2j} = \begin{pmatrix} -1 & 1 \\ 1 & 0 \end{pmatrix}\begin{pmatrix} 4 & 7 \\ 11 & 19 \end{pmatrix} = \begin{pmatrix} 7 & 12 \\ 4 & 7 \end{pmatrix},$$

$$M_N^{-1}M_{N+3j} = \begin{pmatrix} -1 & 1 \\ 1 & 0 \end{pmatrix}\begin{pmatrix} 15 & 26 \\ 41 & 71 \end{pmatrix} = \begin{pmatrix} 26 & 45 \\ 15 & 26 \end{pmatrix},$$

$$M_N^{-1}M_{N+4j} = \begin{pmatrix} -1 & 1 \\ 1 & 0 \end{pmatrix}\begin{pmatrix} 56 & 97 \\ 153 & 265 \end{pmatrix} = \begin{pmatrix} 97 & 168 \\ 56 & 97 \end{pmatrix},\ldots.$$

Since j is even, we are led to solutions of

$$x^2 - 3y^2 = 1,$$

the first four being $x = 2$, $y = 1$; $x = 7$, $y = 4$; $x = 26$, $y = 15$; $x = 97$, $y = 56$.

As another example, we let $d = 13$. Here

$$\sqrt{13} = \langle 3, \overline{1,1,1,1,6} \rangle = \langle 3, \overline{1,1,1,1,6,1,1,1,1,6} \rangle = \cdots.$$

We may take $N = 1$ and $j = 5$. We quickly find that

$$M_N = M_1 = \begin{pmatrix} 0 & 1 \\ 1 & 3 \end{pmatrix}, \qquad M_N^{-1} = M_1^{-1} = \begin{pmatrix} -3 & 1 \\ 1 & 0 \end{pmatrix},$$

$$M_{N+j} = M_6 = \begin{pmatrix} 5 & 18 \\ 33 & 119 \end{pmatrix},$$

$$M_{N+2j} = M_{11} = \begin{pmatrix} 180 & 649 \\ 1189 & 4287 \end{pmatrix}, \cdots.$$

Therefore,

$$M_N^{-1} M_{N+j} = \begin{pmatrix} -3 & 1 \\ 1 & 0 \end{pmatrix} \begin{pmatrix} 5 & 18 \\ 33 & 119 \end{pmatrix} = \begin{pmatrix} 18 & 65 \\ 5 & 18 \end{pmatrix},$$

$$M_N^{-1} M_{N+2j} = \begin{pmatrix} -3 & 1 \\ 1 & 0 \end{pmatrix} \begin{pmatrix} 180 & 649 \\ 1189 & 4287 \end{pmatrix} = \begin{pmatrix} 649 & 2340 \\ 180 & 649 \end{pmatrix}, \cdots.$$

This time j is odd and there are infinitely many solutions to both Fermat–Pell equations. We see from the above that

$$x = 18, y = 5 \qquad \text{is a solution to } x^2 - 13y^2 = -1$$

and that

$$x = 649, y = 180 \qquad \text{is a solution to } x^2 - 13y^2 = 1.$$

A computationally more difficult example is $d = 1141$. This is because the period in the expansion of $\sqrt{1141}$ is rather long (but the expansion *is* periodic):

$$\sqrt{1141} = \langle 33, \overline{1,3,1,1,12,1,21,1,1,2,5,4,3,7,5,16,1,2,3,}$$

$$\overline{1,1,1,2,1,2,1,4,1,8,1,4,1,2,1,2,1,1,1,}$$

$$\overline{3,2,1,16,5,7,3,4,5,2,1,1,21,1,12,1,1,3,1,66} \rangle.$$

Here we may take $N = 1$ and $j = 58$. The matrix $M_N^{-1}M_{N+j}$ gives a solution to the equation

$$x^2 - 1141y^2 = 1,$$

which is rather large:

$$x = 1\,036\,782\,394\,157\,223\,963\,237\,125\,215$$

and

$$y = 30\,693\,385\,322\,765\,657\,197\,397\,208.$$

If you think this is large, when $d = 1\,000\,099$, the smallest value of y that comes out of this method has 1115 digits!

The examples of $d = 1141$ and $1\,000\,099$ make it questionable that our method is giving all solutions to the Fermat–Pell equations. Indeed, we may well wonder if we are even always getting the smallest solutions. Amazingly enough, our method does give all the solutions to the Fermat–Pell equations. (More exactly, it gives all positive solutions. All other solutions come by merely changing signs.) Thus the equation $x^2 - dy^2 = -1$ has no solutions unless the length of the period in the continued fraction expansion of $\sqrt{d}$ is an odd integer.

In Table 3 at the end of the book we have given the continued fraction expansions of all $\sqrt{d}$ with d in the range $2 \le d < 100$. In all the expansions listed there and in all examples given thus far in the text, the continued fraction expansion of $\sqrt{d}$ has always been of the form

(41) $\sqrt{d} = \langle a_0, \overline{a_1, a_2, \ldots, a_j} \rangle.$

This is, in fact, what always happens; the periodic part of the continued fraction expansion of $\sqrt{d}$ begins with a_1. The reader may have already noticed two other things which always occur. In the expansion (41) of $\sqrt{d}$, it is always true that

$$a_j = 2a_0.$$

Further, if $j > 1$ and we delete a_j from the period,

$$a_1, a_2, \ldots, a_j,$$

then what is left reads the same forward as backward. The proofs of all these facts are left to miscellaneous problems 16–18.

The fact that the periodic part of the continued expansion of $\sqrt{d}$ begins with a_1 (so that in our previous discussion $N = 1$) has an important consequence. Let $e = [\sqrt{d}] = a_0$, so that

$$M_N = M_1 = \begin{pmatrix} 0 & 1 \\ 1 & e \end{pmatrix}.$$

If $M_N^{-1}M_{N+kj}$ gives the solution x_k, y_k to

$$x^2 - dy^2 = (-1)^{jk},$$

then we have seen that

$$M_N^{-1}M_{N+kj} = \begin{pmatrix} x_k & dy_k \\ y_k & x_k \end{pmatrix}.$$

Therefore,

$$M_{kj+1} = M_N(M_N^{-1}M_{N+kj})$$

$$= \begin{pmatrix} 0 & 1 \\ 1 & e \end{pmatrix}\begin{pmatrix} x_k & dy_k \\ y_k & x_k \end{pmatrix}$$

(42)
$$= \begin{pmatrix} y_k & x_k \\ x_k + ey_k & dy_k + ex_k \end{pmatrix}.$$

Thus

$$V_{kj-1} = (y_k, x_k).$$

(Compare this with the examples of $d = \sqrt{3}$ and $d = \sqrt{13}$ presented earlier in this section.) As a result, our solution to the Fermat–Pell equation can be found directly from the process of finding the various V_n. If j is large, then we have to wait awhile before we reach V_{j-1}, the components of which are naturally large.

EXERCISES

In problems 1–5, find two positive solutions to the equation $x^2 - dy^2 = 1$.

1. $d = 6, \sqrt{6} = \langle 2, \overline{2, 4} \rangle$.

2. $d = 11, \sqrt{11} = \langle 3, \overline{3, 6} \rangle$.

3. $d = 23, \sqrt{23} = \langle 4, \overline{1, 3, 1, 8} \rangle$.

4. $d = 14, \sqrt{14} = \langle 3, \overline{1, 2, 1, 6} \rangle$.

5. $d = 5, \sqrt{5} = \langle 2, \overline{4} \rangle$.

In problems 6–10, find a solution to $x^2 - dy^2 = -1$.

6. $d = 29, \sqrt{29} = \langle 5,\overline{2,1,1,2,10} \rangle$.

7. $d = 58, \sqrt{58} = \langle 7,\overline{1,1,1,1,1,1,14} \rangle$.

8. $d = 73, \sqrt{73} = \langle 8,\overline{1,1,5,5,1,1,16} \rangle$.

9. $d = 106, \sqrt{106} = \langle 10,\overline{3,2,1,1,1,1,2,3,20} \rangle$.

10. $d = 97, \sqrt{97} = \langle 9,\overline{1,5,1,1,1,1,1,1,5,1,18} \rangle$.

In problems 11–15, find a solution (with x and y both positive) to one of the equations $x^2 - dy^2 = \pm 1$ and state to which equation your solution applies.

11. $d = 10, \sqrt{10} = \langle 3,\overline{6} \rangle$.

12. $d = 53, \sqrt{53} = \langle 7,\overline{3,1,1,3,14} \rangle$.

13. $d = 46, \sqrt{46} = \langle 6,\overline{1,3,1,1,2,6,2,1,1,3,1,12} \rangle$.
14. $d = 130$.
15. $d = 125$.

MISCELLANEOUS EXERCISES
(Problems marked with an asterisk are in some manner related to the last two sections.)

    1. If p and q are positive integers, show that the inequality

$$\pi < \frac{p}{q} < \frac{22}{7}$$

       implies that $q \geq 113$.

    2. Suppose that α is a positive real number and r_1, r_2, s_1, and s_2 are positive integers such that

$$\frac{r_1}{s_1} < \alpha < \frac{r_2}{s_2}, \qquad r_2 s_1 - r_1 s_2 = 1.$$

       Prove that either r_1/s_1 or r_2/s_2 is a convergent in the continued fraction expansion of α. (The example, $\frac{12}{7} < \sqrt{3} < \frac{7}{4}$, shows that there are times when not both r_1/s_1 and r_2/s_2 are convergents.)

    3. Let S be the set of points (x,y) such that x and y are integers, $2|x$, $3|y$, and $0 < x \leq 1000$. Find the closest point in the set S to the line, $y = x\sqrt{18}$. How is it related to the set of closest points to the line $y = x\sqrt{8}$?

*4. Find polynomials $p(n)$ and $q(n)$ such that

$$p(n)^2 - (169n^2 + 198n + 58)q(n)^2 = -1.$$

It may be useful to use the relation

$$169n^2 + 198n + 58 = (13n + 7)^2 + (16n + 9).$$

It may also be helpful to assume that n is large until the final answer is reached.

*5. Find polynomials $p(n)$ and $q(n)$ [with $q(n)$ not the zero polynomial] such that

$$p(n)^2 - (9n^2 + 8n + 2)q(n)^2 = 1.$$

It may be useful to use the relation

$$9n^2 + 8n + 2 = (3n + 1)^2 + (2n + 1).$$

*6. Suppose that a, b, and c are integers, $(a, b, c) = 1$ and the roots of

$$ax^2 + bx + c = 0$$

are real irrational numbers. Use the method of Section 7.7 to show that the equation

$$x^2 - bxy + acy^2 = 1$$

has infinitely many integral solutions and give a condition that ensures that the equation

$$x^2 - bxy + acy^2 = -1$$

has infinitely many integral solutions. Make a change of variable to show that there are infinitely many pairs of integers x and y such that

$$ax^2 + bxy + cy^2 = a.$$

7. Let the continued fraction expansions of α and β be given by

$$\alpha = \langle a, 1, b, c, d, \ldots \rangle,$$

$$\beta = \langle a, 1 + b, c, d, \ldots \rangle,$$

where it is presumed that both continue in exactly the same manner. There is an extremely simple equation relating α and β. Find such a relation and show that it is correct.

8. We have actually assigned two different uses to the notation $\langle a_0, \ldots, a_n \rangle$. For example, the third convergent in the expansion of π is

$$355/113 = \langle 3, 7, 15, 1 \rangle,$$

but the continued fraction expansion of 355/113 is

$$355/113 = \langle 3,7,16 \rangle.$$

In fact by Theorem 7.3, no finite continued fraction expansion ends with a one. Show that if $a_0, a_1, \ldots, a_n, b_0, b_1, \ldots, b_m$ are integers with $a_j \geq 1$ and $b_j \geq 1$ if $j \geq 1$ and $n \geq m$ and

$$\langle a_0, a_1, \ldots, a_n \rangle = \langle b_0, b_1, \ldots, b_m \rangle$$

then either

$$n = m \text{ and } a_j = b_j \qquad \text{for all } j \text{ in the range } 0 \leq j \leq m$$

or

$$n = m + 1, a_j = b_j \qquad \text{for } 0 \leq j \leq m - 1, \quad a_m = b_m - 1, \quad a_{m+1} = 1.$$

9. Problem 8 brings to mind the question as to whether there is an infinite sequence of a_n's with $a_n \geq 1$ for $n \geq 1$ such that

$$\langle a_0, a_1, \ldots \rangle$$

is not the continued fraction expansion of any real number. This and the next problem answer this question. Suppose $x_0, x_1, x_2, x_3, \ldots$ are real numbers such that $x_n \geq 1$ if $n \geq 1$. Let

$$U_{-2} = (1,0), \quad U_{-1} = (0,1), \quad U_n = U_{n-2} + x_n U_{n-1}.$$

Set $U_n = (s_n, r_n)$. Show that

$$\lim_{n \to \infty} s_n = +\infty.$$

Show that

$$\frac{r_0}{s_0} < \frac{r_2}{s_2} < \frac{r_4}{s_4} < \frac{r_6}{s_6} < \cdots < \frac{r_7}{s_7} < \frac{r_5}{s_5} < \frac{r_3}{s_3} < \frac{r_1}{s_1}.$$

Thus the sequence $r_0/s_0, r_2/s_2, \ldots$ is an increasing sequence bounded above by r_1/s_1. Therefore,

$$\lim_{n \to \infty} \frac{r_{2n}}{s_{2n}}$$

exists. Likewise,

$$\lim_{n \to \infty} \frac{r_{2n+1}}{s_{2n+1}}$$

exists. Let

$$\alpha = \lim_{n \to \infty} \frac{r_{2n}}{s_{2n}}, \qquad \beta = \lim_{n \to \infty} \frac{r_{2n+1}}{s_{2n+1}}.$$

Show that

$$\frac{r_{2n+1}}{s_{2n+1}} - \frac{r_{2n}}{s_{2n}} = \frac{1}{s_{2n}s_{2n+1}}$$

and let $n \to \infty$ to get the result

$$\beta = \alpha.$$

Show that

$$\frac{r_n}{s_n} = \langle x_0, x_1, \dots, x_n \rangle.$$

Use this to show that

$$\lim_{n \to \infty} \langle x_0, x_1, \dots, x_n \rangle$$

exists. We define $\langle x_0, x_1, \dots \rangle$ to be this limit. Show that for all $n > m$,

$$\langle x_0, x_1, \dots, x_m, \langle x_{m+1}, \dots, x_n \rangle \rangle = \langle x_0, x_1, \dots, x_n \rangle.$$

Let $n \to \infty$ in the above to show that

$$\langle x_0, x_1, \dots, x_m, \langle x_{m+1}, x_{m+2}, \dots \rangle \rangle = \langle x_0, x_1, x_2, \dots \rangle.$$

10. Suppose that $x_0, x_1, \dots$ are integers and that $x_n \geq 1$ if $n \geq 1$. Let

$$\alpha = \langle x_0, x_1, \dots \rangle$$

and let the continued fraction expansion of α be given by

$$\alpha = \langle a_0, a_1, a_2, \dots \rangle.$$

Show that $x_n = a_n$ for all n. (*Hint:* For $n \geq 0$, let

$$\beta_n = \langle x_n, x_{n+1}, \dots \rangle.$$

Show that for $n \geq 1$, β_n is positive and that

$$\beta_{n-1} = x_{n-1} + \frac{1}{\beta_n}.$$

Show as a result that for all $n \geq 0$,

$$x_n < \beta_n < x_n + 1.$$

Thus $[\beta_n] = x_n$. Starting with $\beta_0 = \alpha$, prove that $x_n = a_n$ for all n by induction.)

In any of the remaining problems, the reader should feel free to use the results of problems 9 and 10 if he desires.

Problems 11, 12, and 13 are related.

11. Let

$$\alpha = \langle 4,1,2,3,7,2,1,8,9,21,7,\ldots\rangle,$$

$$\beta = \langle 4,1,2,3,7,2,1,9,8,21,7,\ldots\rangle,$$

$$\gamma = \langle 4,1,2,3,7,2,1,8,10,20,7,\ldots\rangle,$$

where each of the numbers may or may not continue further and they may not continue in the same manner (but any further entries are positive integers). Which of the three numbers is largest and which is smallest?

12. Let the continued fraction expansions of α and β be

$$\alpha = \langle a_0, a_1, \ldots, a_n, \ldots\rangle,$$

$$\beta = \langle b_0, b_1, \ldots, b_n, \ldots\rangle,$$

where either expansion may terminate at or after a_n or b_n. Suppose $a_0 = b_0, a_1 = b_1, \ldots, a_{n-1} = b_{n-1}$ and $a_n > b_n$. One of the following is true, $\alpha \geq \beta$ or $\beta \geq \alpha$; which is it? Does your answer depend on n?

13. A certain number, α, rounded off to six decimal places is 3.141593 (π is such a number, but α may not be π). Thus the only thing you know about α is

$$3.1415925 \leq \alpha \leq 3.1415935.$$

Find the continued fraction expansion of α as far as possible and give the best possible bounds on the first doubtful a_n. Prove that $\frac{355}{113} = \langle 3,7,16\rangle = \langle 3,7,15,1\rangle$ is a convergent in the expansion of α and that

$$\left|\alpha - \frac{355}{113}\right| < \frac{1}{135 \cdot 113^2}.$$

*14. Let the continued fraction expansion of α be

$$\alpha = \langle a_0, a_1, a_2, \ldots\rangle.$$

Let $b_0, b_1, \ldots, b_n$ be integers with $b_j \geq 1$ if $1 \leq j \leq n$. Let

$$\begin{pmatrix} a & b \\ c & d \end{pmatrix} = \begin{pmatrix} 0 & 1 \\ 1 & b_n \end{pmatrix} \cdots \begin{pmatrix} 0 & 1 \\ 1 & b_1 \end{pmatrix}\begin{pmatrix} 0 & 1 \\ 1 & b_0 \end{pmatrix}$$

and let

$$\beta = \frac{a\alpha - b}{d - c\alpha}$$

(we assume that $d - c\alpha \neq 0$). Prove that $\beta > 1$ is a necessary and sufficient condition that

$$a_0 = b_0, a_1 = b_1, \ldots, a_n = b_n.$$

*15. Let p, q, P, and Q be nonnegative integers with

$$Q > q > 0,$$
$$qP - Qp = \pm 1.$$

Show that there is a unique sequence of integers $a_0, \ldots, a_n$ such that $a_k \geq 1$ if $1 \leq k \leq n$ and

$$\begin{pmatrix} q & p \\ Q & P \end{pmatrix} = \begin{pmatrix} 0 & 1 \\ 1 & a_n \end{pmatrix} \cdots \begin{pmatrix} 0 & 1 \\ 1 & a_1 \end{pmatrix} \begin{pmatrix} 0 & 1 \\ 1 & a_0 \end{pmatrix}.$$

[*Hints:* Show that the only solution to $Py - Qx = Pq - Qp$ with $Q > y \geq 0$ is $x = p$, $y = q$. Let the continued fraction expansion of P/Q be

$$\frac{P}{Q} = \langle b_0, b_1, \ldots, b_m \rangle = \langle b_0, b_1, \ldots, b_{m-1}, b_m - 1, 1 \rangle.$$

Show that $V_m = (Q,P)$ in the first expansion and $V_{m+1} = (Q,P)$ in the second expansion. Show that either in the first expansion det $M_{m+1} = qP - Qp$ or in the second expansion det $M_{m+2} = qP - Qp$ (the two determinants have opposite signs). Conclude that there are integers $a_0, \ldots, a_n$ and, in fact, either $n = m$ and $a_0 = b_0, \ldots, a_m = b_m$ or $n = m + 1$,

$$a_0 = b_0, \ldots, a_{m-1} = b_{m-1}, a_m = b_m - 1, a_{m+1} = 1.$$

To prove the uniqueness of the a_j's, suppose that

$$\begin{pmatrix} q & p \\ Q & P \end{pmatrix} = \begin{pmatrix} 0 & 1 \\ 1 & a_n \end{pmatrix} \cdots \begin{pmatrix} 0 & 1 \\ 1 & a_1 \end{pmatrix} \begin{pmatrix} 0 & 1 \\ 1 & a_0 \end{pmatrix}$$
$$= \begin{pmatrix} 0 & 1 \\ 1 & A_N \end{pmatrix} \cdots \begin{pmatrix} 0 & 1 \\ 1 & A_1 \end{pmatrix} \begin{pmatrix} 0 & 1 \\ 1 & A_0 \end{pmatrix},$$

where $a_j \geq 1$ and $A_j \geq 1$ if $j \geq 1$. Let

$$\alpha = \langle a_0, a_1, \ldots, a_n, 1,2,3,4,5,6,7, \ldots \rangle,$$

$$\alpha' = \langle A_0, A_1, \ldots, A_N, 1,2,3,4,5,6,7, \ldots \rangle.$$

Let $V_0, V_1, \ldots$ be the points found in the expansion of α and $V'_0, V'_1, \ldots$ be the points found in the expansion of α'. Show that for all $j \geq 0$,

$$V_{n+j-1} = V'_{N+j-1}.$$

[Use induction and note that $V_{n-1} = V_{N-1} = (q,p)$, $V_n = V_N = (Q,P)$. Use Theorem 7.9 to show that $\alpha = \alpha'$ and hence show that $N = n$ and

$$A_0 = a_0, A_1 = a_1, \ldots, A_n = a_n.]$$

*16. Starting with the equality

$$(244)^2 - 135(21)^2 = 1,$$

we find the matrix on the right side of equation (42) on page 245 to be

$$M = \begin{pmatrix} 21 & 244 \\ 475 & 5519 \end{pmatrix}.$$

Find a sequence of integers (see hints to problem 15) $a_0, a_1, \ldots, a_n$ with $a_k \geq 1$ for $1 \leq k \leq n$ such that

$$M = \begin{pmatrix} 0 & 1 \\ 1 & a_n \end{pmatrix} \cdots \begin{pmatrix} 0 & 1 \\ 1 & a_1 \end{pmatrix} \begin{pmatrix} 0 & 1 \\ 1 & a_0 \end{pmatrix}.$$

Let

$$\alpha = \langle a_0, \overline{a_1, \ldots, a_n} \rangle$$

(by problem 10, this is actually the continued fraction expansion of α). Use the method of Section 7.6 to show that

$$\alpha = \sqrt{135}.$$

*17. Suppose that d is a positive integer which is not a perfect square and suppose that x and y are positive integers such that

$$x^2 - dy^2 = 1.$$

Let $e = [\sqrt{d}]$ and let

$$M = \begin{pmatrix} y & x \\ x + ey & dy + ex \end{pmatrix}.$$

Use the result of problem 15 to show that there is a sequence of integers $a_0, \ldots, a_n$ with $a_k \geq 1$ if $1 \leq k \leq n$ such that

$$M = \begin{pmatrix} 0 & 1 \\ 1 & a_n \end{pmatrix} \cdots \begin{pmatrix} 0 & 1 \\ 1 & a_1 \end{pmatrix} \begin{pmatrix} 0 & 1 \\ 1 & a_0 \end{pmatrix}.$$

Let $\alpha = \langle a_0, \overline{a_1, \ldots, a_n} \rangle$ (by problem 10, this is the continued fraction expansion of α). Our goal is to show that

$$\alpha = \sqrt{d};$$

this will show that the periodic part of the continued fraction expansion of $\sqrt{d}$ does begin with a_1 and further that every solution to the Fermat–Pell equation is given by the method of Section 7.7. (It also follows that if the length of the period in the expansion of $\sqrt{d}$ is an even integer, then $x^2 - dy^2 = -1$ has no integral solutions.) To achieve our goal, it will be necessary to know a_0. Prove that $a_0 = e$. There are two cases here. If $y > 1$, then first show that

$$e < \frac{x}{y} < e + 1.$$

[Remember, $(x/y)^2 = d \pm (1/y^2)$ and $e^2 < d < (e+1)^2$.] If $y = 1$ and $x^2 - dy^2 = -1$, show that $x = e$, $d = e^2 + 1$, and

$$M = \begin{pmatrix} 0 & 1 \\ 1 & 2e \end{pmatrix} \begin{pmatrix} 0 & 1 \\ 1 & e \end{pmatrix};$$

if $y = 1$ and $x^2 - dy^2 = 1$, show that $x = e + 1$, $d = (e+1)^2 - 1$, and

$$M = \begin{pmatrix} 0 & 1 \\ 1 & 2e \end{pmatrix} \begin{pmatrix} 0 & 1 \\ 1 & 1 \end{pmatrix} \begin{pmatrix} 0 & 1 \\ 1 & e \end{pmatrix}.$$

Use the method of Section 7.6 to show that $\alpha = \sqrt{d}$. Show also that $a_n = 2e$. (When $y = 1$, you have done this above; when $y > 1$, show that

$$a_n = \left[\frac{x + ey}{y} \right] = 2e).$$

*18. Let d be a positive integer which is not a perfect square. By problem 17, the continued fraction expansion of $\sqrt{d}$ is

$$\sqrt{d} = \langle e, \overline{a_1, \ldots, a_{j-1}, 2e} \rangle,$$

where $e = [\sqrt{d}]$ and j is the length of the period. Let us suppose that $j \geq 2$, so that the period consists of more than just $2e$. [Thus besides the perfect squares, numbers d of the form $d = e^2 + 1$ ($\sqrt{d} = \langle e, \overline{2e} \rangle$) are not being considered.] By the method at the end of Section 7.7, we see that

$$\begin{pmatrix} 0 & 1 \\ 1 & 2e \end{pmatrix} \begin{pmatrix} 0 & 1 \\ 1 & a_{j-1} \end{pmatrix} \cdots \begin{pmatrix} 0 & 1 \\ 1 & a_1 \end{pmatrix} \begin{pmatrix} 0 & 1 \\ 1 & e \end{pmatrix} = \begin{pmatrix} y & x \\ x + ey & dy + ex \end{pmatrix},$$

where $x^2 - dy^2 = (-1)^j$. Let

$$A = \begin{pmatrix} 0 & 1 \\ 1 & a_{j-1} \end{pmatrix} \cdots \begin{pmatrix} 0 & 1 \\ 1 & a_2 \end{pmatrix} \begin{pmatrix} 0 & 1 \\ 1 & a_1 \end{pmatrix}.$$

Find A in terms of x, y, e, and d. Show that

$$A = A' \ (A \ \text{transpose}).$$

Use problem 15 to show that for $1 \leq k \leq j - 1$,

$$a_{j-k} = a_k.$$

This says that the period—with the $2e$ term removed—reads the same forward and backward.

*19. Let $a_0, a_1, \ldots, a_n$ be positive integers and let

$$M_{n+1} = \begin{pmatrix} q_{n-1} & p_{n-1} \\ q_n & p_n \end{pmatrix} = \begin{pmatrix} 0 & 1 \\ 1 & a_n \end{pmatrix} \cdots \begin{pmatrix} 0 & 1 \\ 1 & a_1 \end{pmatrix} \begin{pmatrix} 0 & 1 \\ 1 & a_0 \end{pmatrix}.$$

Show that

$$\frac{p_n}{p_{n-1}} = \langle a_n, a_{n-1}, \ldots, a_1, a_0 \rangle$$

and that

$$\frac{q_n}{q_{n-1}} = \langle a_n, a_{n-1}, \ldots, a_1 \rangle.$$

(*Hint:* Look at M'_{n+1}.)
Problems 20, 21, and 22 are related.

20. If an interval, I_1 of length λ, is contained in the interval I from 0 to 1, we shall say that λ is the probability that a randomly chosen real number in I is also in I_1. Let α be a randomly chosen number in I and let the continued fraction expansion of α be

$$\alpha = \langle 0, a_1, a_2, a_3, \ldots \rangle.$$

What is the probability that $a_1 \geq 1$? What is the probability that $a_1 \geq 2$? What is the probability that $a_1 = 1$? What is the probability that $a_1 = m$, where m is a given positive integer?

21. If the intervals $I_1, I_2, \ldots$ of lengths $\lambda_1, \lambda_2, \ldots$ are all contained in the unit interval I from 0 to 1, and if no two of the I_n's overlap, then we shall say that the probability that a randomly chosen real number from I is also in one of the I_n's is

$$\sum_{n=1}^{\infty} \lambda_n.$$

Let α be a randomly chosen real number in I and let the continued fraction expansion of α be

$$\alpha = \langle 0, a_1, a_2, a_3, \ldots \rangle.$$

Let m be a given positive integer. Show that the probability that $a_2 \geq m$ is

$$\sum_{a_1=1}^{\infty} \frac{1}{a_1(ma_1 + 1)}.$$

Check that

$$\frac{1}{a_1(ma_1 + 1)} = m\left(\frac{1}{ma_1} - \frac{1}{ma_1 + 1} \right).$$

Use this relation to show that the above infinite series "telescopes" when $m = 1$ and, in fact, the sum is 1. Why is this expected beforehand? Show that when $m = 2$, the series turns into

$$2(\tfrac{1}{2} - \tfrac{1}{3} + \tfrac{1}{4} - \tfrac{1}{5} + \tfrac{1}{6} - \tfrac{1}{7} + \cdots).$$

This has as its sum $2(1 - \ln 2) = .61 \ldots$. Thus the probability that $a_2 = 1$ is $1 - .61 \ldots = .38 \ldots$.

22. The notation is that of problem 21. Show that the probability that $a_3 \geq m$ is

$$\sum_{a_1=1}^{\infty} \sum_{a_2=1}^{\infty} \frac{1}{(a_1a_2 + 1)(a_1a_2m + m + a_1)}.$$

When $m = 1$, what should this sum turn out to be and why? Use the relation

$$\frac{1}{(a_1a_2 + 1)(a_1a_2m + m + a_1)} = \frac{a_2m + 1}{a_1(a_2m + 1) + m} - \frac{a_2}{a_1a_2 + 1}$$

to evaluate

$$\sum_{a_2=1}^{N} \frac{1}{(a_1 a_2 + 1)(a_1 a_2 m + m + a_1)}$$

when $m = 1$. Use this result to evaluate the sum of the double series. It should agree with your predicted answer.

23. Show that $\langle \tfrac{3}{2}, \overline{1,1,3} \rangle = \langle 2, \overline{16,4} \rangle$.

24. Suppose that x, y, and d are positive integers and

$$x^3 - dy^3 = n.$$

Show that if $y > 8|n|/(3d^{2/3})$, then x/y is a convergent in the continued fraction expansion of $d^{1/3}$. (*Hint:* Show that

$$\frac{x}{y} - d^{1/3} = \frac{x^3 - dy^3}{y^3 \left[\left(\dfrac{x}{y} + \dfrac{1}{2} d^{1/3} \right)^2 + \dfrac{3}{4} d^{2/3} \right]}.)$$

25. The continued fraction expansion of $2^{1/3}$ begins

$$2^{1/3} = \langle 1,3,1,5,1,1,4,1,1,8,1,14,1,10,2,1,4,12,2,3,2,\ldots \rangle.$$

Here $V_{19} = (1\ 070\ 524\ 477,\ 1\ 348\ 776\ 323)$.

Use problem 24 to help find all positive integral values of x and y with $y < 10^9$ such that

$$|x^3 - 2y^3| \le 10.$$

(*Note:* It is possible to do this problem without calculating any V_n past V_5. With a computer, it is perfectly feasible to do this problem with all positive $y < 10^{156}$. See miscellaneous exercise 10, Chapter 6, for the method of covering the infinite range $y \ge 10^{156}$.)

Chapter 8

QUADRATIC FIELDS

8.1. Introduction

One thing that is done time and again in solving Diophantine equations is to take some expression and factor it, preferably into linear factors. For example, in solving the equation

$$x^3 + y^3 = z^3,$$

we would like to write it in the form

$$(x + y)(x + ay)(x + by) = z^3$$

and then, hopefully, we could set each of $x + y$, $x + ay$, $x + by$ equal to cubes and solve the equation. Unfortunately,

$$a = \frac{-1 + \sqrt{-3}}{2}, \qquad b = \frac{-1 - \sqrt{-3}}{2},$$

and thus our factorization into linear factors has been too rash, and we are seemingly forced to beat a hasty retreat to the more awkward

$$(x + y)(x^2 - xy + y^2) = z^3.$$

Nevertheless, we would have solved the equation $x^3 + y^3 = z^3$ by exactly the method outlined above if the book had not been cut. Here we will give a brief illustration of the ideas involved by "solving" an equation considered in Chapter 5,

$$x^2 + y^2 = z^2.$$

We factor the left-hand side into linear factors and get

$$(x + y\sqrt{-1})(x - y\sqrt{-1}) = z^2.$$

Thus, by analogy with what we did in Chapter 5, we set $(x + y\sqrt{-1})$ and

$(x - y\sqrt{-1})$ equal to squares. But, squares of what? Well, since we are dealing with numbers of the form $u + v\sqrt{-1}$, we put

$$(x + y\sqrt{-1}) = (u + v\sqrt{-1})^2$$

or, multiplying out the right side,

$$x + y\sqrt{-1} = (u^2 - v^2) + (2uv)\sqrt{-1}.$$

Therefore,

$$x = u^2 - v^2, \qquad y = 2uv$$

and this gives

$$z = u^2 + v^2.$$

This is exactly the result given in Theorem 5.2. Amazing, is it not?

The reader will notice that we got the primitive triplets without making the assumption that $(x,y,z) = 1$. But this should not be too shocking; in order to set $(x + y\sqrt{-1})$ and $(x - y\sqrt{-1})$ equal to squares, we wish these numbers to be relatively prime—whatever that means. It is the purpose of this chapter to justify these outrageous manipulations and at the same time provide a bare introduction to one of the most thriving branches of modern number theory, the theory of algebraic numbers.

EXERCISE

1. Find solutions to the equation $x^2 + 2y^2 = w^2$ by using numbers of the form $x + y\sqrt{-2}$. Compare your answer with that of problem 1, Section 5.3.

8.2. Quadratic Fields and Quadratic Integers

It has become customary to let **Q** denote the set of rational numbers and **Z** the set of integers. Two numbers of **Q** can be added, subtracted, multiplied, and divided (providing that the divisor is not zero) and the result is again a number in **Q**. These are several of the properties that make **Q** a field.[1] We will now consider a larger collection of numbers in which addition, subtraction, multiplication, and division again lead to answers in the original collection.

Definition. Let d be a fixed rational number which is not the square of a rational number. We let $\mathbf{Q}(\sqrt{d})$ denote the set of numbers $a + b\sqrt{d}$, where

---

[1] The reader interested in seeing exactly what a field is should refer to Appendix C. We will not use anything from Appendix C in this chapter.

a and b are arbitrary rational numbers. We call $\mathbf{Q}(\sqrt{d})$ a **quadratic field**; if $d > 0$, we call it a **real quadratic field**: if $d < 0$, we call it a **complex quadratic field** or an **imaginary quadratic field**.

Thus for example, $3 + 4\sqrt{2}, \frac{9}{4} + \frac{7}{13}\sqrt{2}, 5 = 5 + 0\sqrt{2}, \sqrt{2} = 0 + 1\sqrt{2}$, and $\sqrt{32} = 0 + 4\sqrt{2}$ are all members of $\mathbf{Q}(\sqrt{2})$. Since any rational number a can be written in the form $a + 0\sqrt{d}$, the set of numbers $\mathbf{Q}(\sqrt{d})$ contains the set of rational numbers $\mathbf{Q}$. If d is itself the square of a rational number, then $a + b\sqrt{d}$ is rational when a and b are rational and thus $\mathbf{Q}(\sqrt{d})$ is just $\mathbf{Q}$ itself. It is for this reason that the definition of a quadratic field excludes the case of d being the square of a rational number; we get nothing new in this case. In this chapter, ordinary small Roman letters (with the possible exception of x, y, z) will always stand for rational numbers. Numbers in quadratic fields will be denoted by small Greek letters, $\alpha, \beta, \gamma, \delta, \ldots$. The letter d will always be restricted to a rational number such that $\sqrt{d}$ is not rational.

Theorem 8.1. $a + b\sqrt{d} = c + e\sqrt{d}$ if and only if $a = c$ and $b = e$. In particular, $a + b\sqrt{d} = 0$ if and only if $a = b = 0$.
(*Remark:* This theorem would be false if $\sqrt{d}$ were rational; it would also be false if a, b, c, and e were not restricted to $\mathbf{Q}$.)

Proof. Since $\sqrt{d}$ is irrational, the special case that $a + b\sqrt{d} = 0$ implies that $a = b = 0$ is a consequence of Theorem 6.4. Now, for the general case. If

$$a + b\sqrt{d} = c + e\sqrt{d},$$

then

$$(a - c) + (b - e)\sqrt{d} = 0,$$

where $(a - c)$ and $(b - e)$ are both in $\mathbf{Q}$. Therefore, $(a - c) = 0$ and $(b - e) = 0$ or $a = c$ and $b = e$.   ▲

Theorem 8.2. Let α and β be in $\mathbf{Q}(\sqrt{d})$. Then $\alpha + \beta$, $\alpha - \beta$, $\alpha\beta$, and, if $\beta \neq 0$, α/β are also in $\mathbf{Q}(\sqrt{d})$.

Proof. Let $\alpha = a + b\sqrt{d}$ and $\beta = c + e\sqrt{d}$ with a, b, c, e in $\mathbf{Q}$. Then

$$\alpha + \beta = (a + c) + (b + e)\sqrt{d},$$

$$\alpha - \beta = (a - c) + (b - e)\sqrt{d},$$

$$\alpha\beta = (ac + bed) + (ae + bc)\sqrt{d}$$

are all in $\mathbf{Q}(\sqrt{d})$ since the six numbers $(a + c)$, $(b + e)$, $(a - c)$, $(b - e)$, $(ac + bed)$, and $(ae + bc)$ are all rational. In addition, if $\beta \neq 0$, then not both of c and e are 0 and thus the number $c - e\sqrt{d} \neq 0$; therefore,

$$c^2 - e^2 d = (c + e\sqrt{d})(c - e\sqrt{d}) \neq 0,$$

since neither of the factors are zero. Therefore,

$$\frac{\alpha}{\beta} = \frac{(a + b\sqrt{d})(c - e\sqrt{d})}{(c + e\sqrt{d})(c - e\sqrt{d})} = \left(\frac{ac - bed}{c^2 - e^2 d}\right) + \left(\frac{bc - ae}{c^2 - e^2 d}\right)\sqrt{d}$$

is also in $\mathbf{Q}(\sqrt{d})$, since the numbers

$$\left(\frac{ac - bed}{c^2 - e^2 d}\right), \quad \left(\frac{bc - ae}{c^2 - e^2 d}\right)$$

are quotients of rationals with nonzero denominators and hence are rational.   ▲

Thus far, we have restricted d to being a rational number such that $\sqrt{d}$ is not rational. But we can further restrict d without losing any of our quadratic fields. For example,

$$a + b\sqrt{\frac{2}{3}} = a + \left(\frac{b}{3}\right)\sqrt{6}$$

and thus $\mathbf{Q}(\sqrt{\frac{2}{3}}) = \mathbf{Q}(\sqrt{6})$; that is, $\mathbf{Q}(\sqrt{\frac{2}{3}})$ and $\mathbf{Q}(\sqrt{6})$ consist of exactly the same numbers. In exactly the same way

$$\mathbf{Q}\left(\sqrt{\frac{r}{s}}\right) = \mathbf{Q}(\sqrt{rs}),$$

and thus we need only consider integral values of d. But we can go even further. For example,

$$a + b\sqrt{12} = a + (2b)\sqrt{3}$$

and thus $\mathbf{Q}(\sqrt{12}) = \mathbf{Q}(\sqrt{3})$. In like manner, $\mathbf{Q}(\sqrt{r^2 s}) = \mathbf{Q}(\sqrt{s})$. Thus we need only consider integral values of d that have no square factors greater than one (such a number is said to be square-free). *For the rest of this chapter, d will denote an integer other than 0 or 1 without square factors greater than one.* (Since $d \neq 0$ or 1 and d has no square factors greater than one, d is not a perfect square and thus $\sqrt{d}$ is irrational.) The first few positive admissible values of d thus restricted are $2, 3, 5, 6, 7, 10, 11, 13, \ldots$. The first few negative admissible values of d thus restricted are $-1, -2, -3, -5, -6, -7, -10,$ $-11, -13, \ldots$.

The numbers of $\mathbf{Q}(\sqrt{d})$ are solutions of quadratic equations with integral coefficients. The number $a + b\sqrt{d}$ is a root of the equation

$$x^2 - 2ax + (a^2 - b^2d) = [x - (a + b\sqrt{d})][x - (a - b\sqrt{d})] = 0,$$

which has rational coefficients; if we multiply the equation through by a common denominator for the rational numbers $(2a)$ and $(a^2 - b^2d)$, then we get a quadratic equation with integral coefficients. The two roots of this equation, $a + b\sqrt{d}$ and $a - b\sqrt{d}$, are said to be conjugates of each other.

Definition. If $\alpha = a + b\sqrt{d}$, then we define the **conjugate** of α to be the number $\bar{\alpha} = a - b\sqrt{d}$.

If $d < 0$, this definition should be familiar to the reader as the definition of the ordinary complex conjugate of a complex number. As a result, the following theorem should not be too surprising.

Theorem 8.3. If α and β are in $\mathbf{Q}(\sqrt{d})$, then $\overline{(\bar{\alpha})} = \alpha$, $\overline{(\alpha + \beta)} = \bar{\alpha} + \bar{\beta}$, $\overline{(\alpha - \beta)} = \bar{\alpha} - \bar{\beta}$, $\overline{(\alpha\beta)} = \bar{\alpha}\bar{\beta}$, and if $\beta \neq 0$, then $\bar{\beta} \neq 0$ and $\overline{(\alpha/\beta)} = \bar{\alpha}/\bar{\beta}$. Further, $\alpha = \bar{\alpha}$ if and only if α is rational.

Proof. Let $\alpha = a + b\sqrt{d}$ and $\beta = c + e\sqrt{d}$. Then

$$\overline{(\bar{\alpha})} = \overline{(a - b\sqrt{d})} = a + b\sqrt{d} = \alpha,$$

$$\overline{(\alpha + \beta)} = \overline{[(a + c) + (b + e)\sqrt{d}]} = (a + c) - (b + e)\sqrt{d} = \bar{\alpha} + \bar{\beta},$$

$$\overline{(\alpha - \beta)} = \overline{[(a - c) + (b - e)\sqrt{d}]} = (a - c) - (b - e)\sqrt{d} = \bar{\alpha} - \bar{\beta},$$

$$\overline{(\alpha\beta)} = \overline{[(ac + bed) + (ae + bc)\sqrt{d}]} = (ac + bed) - (ae + bc)\sqrt{d} = \bar{\alpha}\bar{\beta}.$$

If $\alpha = \bar{\alpha}$, then $a + b\sqrt{d} = a - b\sqrt{d}$ and thus (by Theorem 8.1) $b = -b$, whence $b = 0$ and $\alpha = a$ is rational. Conversely, if $\alpha = a + 0\sqrt{d}$ is rational, then $\bar{\alpha} = a - 0\sqrt{d} = \alpha$. We can use the parts of Theorem 8.3 proved thus far to simplify the arithmetic in the remaining part. If $\beta \neq 0$, then not both

c and e are 0 and thus $\bar{\beta} = c - e\sqrt{d} \neq 0$. Therefore, since $1/(\beta\bar{\beta}) = 1/(c^2 - e^2 d)$ is rational,

$$\overline{\left(\frac{\alpha}{\beta}\right)} = \overline{\left(\frac{1}{\beta\bar{\beta}} \cdot \alpha \cdot \bar{\beta}\right)} = \overline{\left(\frac{1}{\beta\bar{\beta}}\right)} \cdot \bar{\alpha} \cdot \overline{(\bar{\beta})} = \frac{1}{\beta\bar{\beta}} \cdot \bar{\alpha} \cdot \beta = \frac{\bar{\alpha}}{\bar{\beta}}. \qquad \blacktriangle$$

We return for a minute to the quadratic equation satisfied by a number α of $\mathbf{Q}(\sqrt{d})$. We have seen that α satisfies a quadratic equation with integral coefficients, say,

$$(1) \qquad\qquad\qquad ax^2 + bx + c = 0.$$

Since α satisfies this equation and since $a, b,$ and c are rational,

$$a(\bar{\alpha})^2 + b\bar{\alpha} + c = \bar{a}(\overline{\alpha^2}) + \bar{b}\bar{\alpha} + \bar{c} = \overline{(a\alpha^2)} + \overline{(b\alpha)} + \bar{c}$$

$$= \overline{(a\alpha^2 + b\alpha + c)} = \overline{(0)} = 0.$$

In other words, if α is a root of (1), then so is $\bar{\alpha}$. If α is irrational, then $\alpha \neq \bar{\alpha}$ and we have thus found the two roots of (1). Since roots of quadratic equations correspond to factors, when α is irrational,

$$ax^2 + bx + c = a(x - \alpha)(x - \bar{\alpha}).$$

Thus when α is irrational, there are infinitely many quadratic equations with integral coefficients satisfied by α, but they are all multiples of one another. By dividing equation (1) through by the greatest common divisor of $a, b,$ and c and then multiplying by -1 if necessary, we get a unique quadratic equation for α.

Definition. If α is an irrational number in $\mathbf{Q}(\sqrt{d})$, then the equation $ax^2 + bx + c = 0$ is called the **defining equation** for α if α satisfies the equation and $a, b,$ and c are integers, $(a,b,c) = 1$, and $a > 0$.

For example, the number $(3 + \sqrt{17})/4$ satisfies the equations

$$8x^2 - 12x - 4 = 0, \quad -10x^2 + 15x + 5 = 0, \quad -2x^2 + 3x + 1 = 0.$$

They are all multiples of the equation $2x^2 - 3x - 1 = 0$, which is the defining equation for $(3 + \sqrt{17})/4$.

As an example of conjugation and defining equations, let us show that an irrational number can be in at most one quadratic field (in other words, the intersection of two different quadratic fields is $\mathbf{Q}$). Suppose that the irrational number

$$\alpha = a + b\sqrt{d} = a_1 + b_1\sqrt{d_1}$$

is in $\mathbf{Q}(\sqrt{d})$ and $\mathbf{Q}(\sqrt{d_1})$. The other root of the defining equation for α is, on the one hand, $a - b\sqrt{d}$ and, on the other hand, $a_1 - b_1\sqrt{d_1}$. In other words, $\bar{\alpha}$ is the same in both fields:

$$\bar{\alpha} = a - b\sqrt{d} = a_1 - b_1\sqrt{d_1}.$$

Hence

$$b\sqrt{d} = \frac{\alpha - \bar{\alpha}}{2} = b_1\sqrt{d_1},$$

and a slight alteration of this gives ($b \neq 0$ since α is irrational)

$$\sqrt{dd_1} = \frac{b_1 d_1}{b}.$$

Thus $\sqrt{dd_1}$ is rational and, by Theorem 2.13, $\sqrt{dd_1}$ is an integer. But since d and d_1 are square-free, the unique factorization theorem guarantees that the only way that dd_1 could be a perfect square is that d and d_1 both have the same sign and are composed of the same prime factors. Thus $d = d_1$, as desired.

The combination $\alpha\bar{\alpha}$ is used sufficiently often that it is convenient to give it a name.

Definition. If α is in $\mathbf{Q}(\sqrt{d})$, we define the **norm** of α to be the number $\mathbf{N}(\alpha) = \alpha\bar{\alpha}$.

Theorem 8.4. $\mathbf{N}(a) = a^2$. If α is in $\mathbf{Q}(\sqrt{d})$, then $\mathbf{N}(\alpha)$ is rational, $\mathbf{N}(\alpha) = 0$ if and only if $\alpha = 0$, and if $d < 0$, then $\mathbf{N}(\alpha) \geq 0$. If β is also in $\mathbf{Q}(\sqrt{d})$, then

$$\mathbf{N}(\alpha\beta) = \mathbf{N}(\alpha)\mathbf{N}(\beta)$$

and, if $\beta \neq 0$,

$$\mathbf{N}\left(\frac{\alpha}{\beta}\right) = \frac{\mathbf{N}(\alpha)}{\mathbf{N}(\beta)}.$$

Proof. Since a is rational, $\bar{a} = a$ and thus $\mathbf{N}(a) = a\bar{a} = a^2$. If $\alpha = a + b\sqrt{d}$, then

$$\mathbf{N}(\alpha) = \alpha\bar{\alpha} = (a + b\sqrt{d})(a - b\sqrt{d}) = a^2 - b^2 d = a^2 + (-d)b^2,$$

which is a rational number, and, if $d < 0$, then $a^2 \geq 0$, $(-d) \geq 0$, $b^2 \geq 0$, and thus $N(\alpha) \geq 0$. The fact that $N(\alpha)$ is rational could have been proved by noting that

$$\overline{N(\alpha)} = \overline{(\alpha\bar\alpha)} = \bar\alpha\alpha = \alpha\bar\alpha = N(\alpha),$$

so that $N(\alpha)$ is rational by Theorem 8.3. Clearly $N(0) = 0$ and, if $\alpha \neq 0$, then $\bar\alpha \neq 0$ and thus $N(\alpha) = \alpha\bar\alpha \neq 0$. Therefore, $N(\alpha) = 0$ if and only if $\alpha = 0$. Next,

$$N(\alpha\beta) = (\alpha\beta)(\overline{\alpha\beta}) = \alpha\beta\bar\alpha\bar\beta = (\alpha\bar\alpha)(\beta\bar\beta) = N(\alpha)N(\beta)$$

and, if $\beta \neq 0$, then

$$N\left(\frac{\alpha}{\beta}\right) = \left(\frac{\alpha}{\beta}\right)\overline{\left(\frac{\alpha}{\beta}\right)} = \frac{\alpha}{\beta}\cdot\frac{\bar\alpha}{\bar\beta} = \frac{N(\alpha)}{N(\beta)}. \qquad \blacktriangle$$

The identity $N(\alpha)N(\beta) = N(\alpha\beta)$ may seem trivial, but if we put $\alpha = a + b\sqrt{d}$, $\beta = c + e\sqrt{d}$, then it leads to the identity

(2) $$(a^2 - db^2)(c^2 - de^2) = (ac + bed)^2 - d(ae + bc)^2.$$

The special case of $d = -1$ says that if two numbers can be written as the sum of two squares, then so can their product. Thus the identity $N(\alpha\beta) = N(\alpha)N(\beta)$ has more content than one might suspect at first glance.

Now that we know something about arithmetic in quadratic fields, it is time to decide just which elements of $Q(\sqrt{d})$ we wish to call integers. To begin with, a rational number should be an integer in $Q(\sqrt{d})$ if and only if it is already an integer. Otherwise, the word "integer" would create too much confusion to be tolerated. But since $Q(\sqrt{d})$ consists of numbers other than Q, it seems reasonable to require that the integers of $Q(\sqrt{d})$ consist of more than just the ordinary integers. But what? If we think back over the properties of the ordinary integers, we find a theorem that says that if x^n is an integer and x is rational, then x is an integer. This has been a very useful property and it seems like one that we should desire to keep. In our case, it means that if x^n is an integer and x is in $Q(\sqrt{d})$, then x should be an integer. In particular, since d is an integer and $\sqrt{d}$ is a root of $x^2 = d$, we will define $\sqrt{d}$ to be an integer. We want sums and products of integers to be integers. If a and b are ordinary integers, then we wish $b\sqrt{d}$ and as a result $a + b\sqrt{d}$ to be integers also. It would be nice if we could stop here, but there are many advantages to going further. For example, the number $(-\frac{1}{2}) + (\frac{1}{2})\sqrt{-3}$ is a root of the equation $x^3 = 1$ and hence it seems desirable to make $(-\frac{1}{2}) + (\frac{1}{2})\sqrt{-3}$ an integer even though it is not of the form (ordinary integer) + (ordinary integer) $\sqrt{-3}$.

We have thus far been defining integers by the equations they satisfy. What is it about the equations that distinguishes an integer from a noninteger? The defining equation for $\sqrt{d}$ is $x^2 - d = 0$. The defining equation for $(-\frac{1}{2}) + (\frac{1}{2})\sqrt{-3}$ is $x^2 + x + 1 = 0$. The defining equation for $5 + 7\sqrt{2}$ (which we have agreed shall be an integer) is $x^2 - 10x - 73 = 0$. What, if anything, do these equations have in common? They all begin with x^2, not $2x^2$ or $3x^2$ or $71x^2$, but just x^2. Could this be the key to our definition? Let us test the idea on rational numbers. If a/b is a rational number with a and b integers, $(a,b) = 1$ and $b > 0$, then the defining equation for a/b is $bx - a = 0$. The integers are exactly those numbers with $b = 1$, in other words, any rational such that the coefficient of x in its defining equation is one. Thus, by analogy with the ordinary integers and with what we have already seen, the following definition seems to be in order:

Definition. A number α of $\mathbf{Q}(\sqrt{d})$ shall be called a **quadratic integer**, or just integer for short, if either α is rational and is in $\mathbf{Z}$ or if α is irrational and the coefficient of x^2 in the defining equation for α is 1. The numbers in $\mathbf{Z}$ will be called **rational integers**.

Hence, by definition, a quadratic integer that is rational is a rational integer; we have introduced no new integers among the numbers of $\mathbf{Q}$. Note also that since the defining equations are the same, $\bar{\alpha}$ is an integer whenever α is. We have attempted to motivate this definition somewhat, but in the end, it will stand or fall on its usefulness and not on its motivation. You may ask: "Why should $(-1 + \sqrt{-3})/2$ (defining equation $x^2 + x + 1 = 0$) be an integer and not $(-3 + 6\sqrt{-3})/2$ (defining equation $4x^2 + 12x + 117 = 0$)? That doesn't seem natural." One might also ask: "Why should -1 be an integer?" The second question does not occur to anyone anymore since they have been brought up to believe that -1 is an integer. The first question would not occur to anyone either if he had been brought up on our definition. A definition should not be condemned just because it is unnatural. One should not pass judgment on it until one sees if it leads to new results and clarifies old ones or rather, as in some of the "new math," it leads to excess verbiage that does little except obfuscate the clear concepts at the foundations.

The integers of $\mathbf{Q}(\sqrt{d})$ satisfy the fundamental properties that the sum, difference, and product of two integers is again an integer. Before we prove this, however, it is useful to learn how to recognize an integer without having to find its defining equation.

Theorem 8.5. If $d \not\equiv 1(\bmod 4)$, then the integers of $\mathbf{Q}(\sqrt{d})$ are exactly those numbers of the form $a + b\sqrt{d}$, where a and b are rational integers. If $d \equiv 1(\bmod 4)$, then the integers of $\mathbf{Q}(\sqrt{d})$ are those numbers of the form $(a + b\sqrt{d})/2$, where a and b are rational integers, both even or both odd.

[*Remark:* It follows that if a and b are rational integers, then $a + b\sqrt{d}$ is always an integer; if $d \equiv 1 \pmod 4$, this is because such a number can be expressed as $(2a + 2b\sqrt{d})/2$, where $2a$ and $2b$ are both even.]

Corollary. If $d \equiv 1 \pmod 4$, a number of $\mathbf{Q}(\sqrt{d})$ is an integer if and only if it can be written as $a + b[(1 + \sqrt{d})/2]$, where a and b are rational integers.

Proof of Theorem 8.5. If $\alpha = a + b\sqrt{d}$, where a and b are rational, then α is rational if and only if $b = 0$. Thus α is a rational integer if and only if a is in $\mathbf{Z}$ and $b = 0$ and then it is in the form

$$\alpha = a + b\sqrt{d} \qquad (a \text{ and } b \text{ in } \mathbf{Z})$$

$$= \frac{2a + 2b\sqrt{d}}{2} \qquad (2a \text{ and } 2b \text{ both even numbers in } \mathbf{Z}).$$

For the rest of the proof we suppose that $\alpha = a + b\sqrt{d}$ is irrational. If a and b are in $\mathbf{Z}$, then α satisfies the equation

$$x^2 - 2ax + (a^2 - b^2 d) = 0,$$

where $(-2a)$ and $(a^2 - b^2 d)$ are in $\mathbf{Z}$. In this equation, the coefficients obviously have no common factors other than ± 1 (since 1 is a coefficient), and thus this is the defining equation for α. Therefore, α is an integer. If $d \equiv 1 \pmod 4$ and $\alpha = (a + b\sqrt{d})/2$, where a and b are both odd numbers in $\mathbf{Z}$ (a and b both even reduces α to the previous case), then α is a root of the equation

$$x^2 - ax + \left(\frac{a^2 - b^2 d}{4}\right) = 0,$$

where $-a$ and $(a^2 - b^2 d)/4$ are in $\mathbf{Z}$. The latter number is in $\mathbf{Z}$, since if a and b are both odd, then

$$a^2 - b^2 d \equiv 1 - 1 \cdot 1 \equiv 0 \pmod 4$$

and there is a factor of 4 in the numerator. Thus α is an integer.

Conversely, suppose that the irrational number α in $\mathbf{Q}(\sqrt{d})$ is a root of the equation

$$x^2 + bx + c = 0,$$

where b and c are in $\mathbf{Z}$. Thus α is an integer in $\mathbf{Q}(\sqrt{d})$. There are two cases to consider here. Suppose first that b is even and let $b = 2a$. Then α is one of the two numbers

$$-a \pm \sqrt{a^2 - c}.$$

If we put $a^2 - c = e^2 d'$, where d' has no square factors greater than 1 (and d' is not 0 or 1, since α is not rational), then α is one of the two numbers $-a \pm e\sqrt{d'} = (-2a \pm 2e\sqrt{d'})/2$ in $\mathbf{Q}(\sqrt{d'})$, where $-a$ and $\pm e$ are in $\mathbf{Z}$, $-2a$ and $\pm 2e$ are both even numbers in $\mathbf{Z}$. In the second case, we suppose that b is odd and then α is one of the two numbers

$$\frac{-b \pm \sqrt{b^2 - 4c}}{2}.$$

Since b is odd,

$$b^2 - 4c \equiv 1 - 4c \equiv 1 (\mathrm{mod}\ 4).$$

Let us put $b^2 - 4c = e^2 d'$, where d' has no square factors > 1 (and $d' \neq 0$ or 1 since α is irrational). Then e is odd since $b^2 - 4c$ is odd and thus

$$d' \equiv 1 \cdot d' \equiv e^2 d' \equiv b^2 - 4c \equiv 1 (\mathrm{mod}\ 4).$$

Hence α is one of the two numbers

$$\frac{-b \pm e\sqrt{d'}}{2}$$

in $\mathbf{Q}(\sqrt{d'})$, where $d' \equiv 1 (\mathrm{mod}\ 4)$ and $-b$ and $\pm e$ are both odd. In each of these cases, $d' = d$, as we have already shown on page 263. ▲

Proof of the Corollary to Theorem 8.5. We naturally use Theorem 8.5. If a and b are in $\mathbf{Z}$, then

$$a + b\left(\frac{1 + \sqrt{d}}{2}\right) = \frac{(2a + b) + b\sqrt{d}}{2},$$

where $2a + b \equiv b(\mathrm{mod}\ 2)$ and thus $(2a + b)$ and b are both even or both odd. Also if a and b are in $\mathbf{Z}$, both even or both odd, then

$$\frac{a + b\sqrt{d}}{2} = \left(\frac{a - b}{2}\right) + b\left(\frac{1 + \sqrt{d}}{2}\right),$$

where $(a - b)/2$ and b are in $\mathbf{Z}$, the former since $a \equiv b(\mathrm{mod}\ 2)$, $a - b \equiv 0(\mathrm{mod}\ 2)$, and so $2|(a - b)$. ▲

Theorem 8.6. If α and β are integers in $\mathbf{Q}(\sqrt{d})$, then $\alpha + \beta$, $\alpha - \beta$, and $\alpha\beta$ are also integers in $\mathbf{Q}(\sqrt{d})$.

Proof. We can prove the theorem for both cases $d \not\equiv 1 \pmod{4}$ and $d \equiv 1 \pmod{4}$ by slightly complicating the notation. Let

$$\omega = \begin{cases} \sqrt{d} & \text{if } d \not\equiv 1 \pmod{4}, \\ \dfrac{1 + \sqrt{d}}{2} & \text{if } d \equiv 1 \pmod{4}. \end{cases}$$

Theorem 8.5 and its corollary show that for a number γ of $\mathbf{Q}(\sqrt{d})$ to be an integer, it is necessary and sufficient that γ can be written

(3) $\gamma = $ (rational integer) $+$ (rational integer)ω.

Since α and β are integers in $\mathbf{Q}(\sqrt{d})$, we can write

$$\alpha = a + b\omega, \qquad \beta = c + e\omega,$$

where a, b, c, e are in $\mathbf{Z}$. Thus

$$\alpha + \beta = (a + c) + (b + e)\omega,$$
$$\alpha - \beta = (a - c) + (b - e)\omega$$

are both in the form of (3) and hence are integers. We may write

$$\omega^2 = s\omega + t,$$

where s and t are in $\mathbf{Z}$; in fact, if $d \not\equiv 1 \pmod{4}$, then $s = 0$, $t = d$, and if $d \equiv 1 \pmod{4}$, then $s = 1$, $t = (d - 1)/4$ with t being an integer since $d \equiv 1 \pmod{4}$. Thus

$$\alpha\beta = ac + (ae + bc)\omega + be\omega^2$$
$$= ac + (ae + bc)\omega + be(s\omega + t)$$
$$= (ac + bet) + (ae + bc + bes)\omega,$$

which is in the form of (3), and thus $\alpha\beta$ is an integer.   ▲

We know that the norm of a number in $\mathbf{Q}(\sqrt{d})$ is a rational number. Can we say more about the norm of an integer? The answer is yes, and the result will be extremely useful in later sections; indeed, the concept of norm would be much less valuable without the following theorem.

Theorem 8.7. If α is an integer in $\mathbf{Q}(\sqrt{d})$, then $\mathbf{N}(\alpha)$ is a rational integer.

Proof. When α is an integer, $\bar{\alpha}$ is also an integer (this follows from either the definition of integer or from Theorem 8.5), and thus by Theorem 8.6,

$N(\alpha) = \alpha\bar{\alpha}$ is an integer. But $N(\alpha)$ is also rational, and hence $N(\alpha)$ is a rational integer. ▲

Theorem 8.7 could also be proved for irrational α (it is obvious for rational α) by observing that the defining equation for α must be of the form

$$x^2 + bx + c = 0,$$

where b and c are in $\mathbf{Z}$, and since the two roots of this equation are α and $\bar{\alpha}$, the left side must factor as

$$x^2 + bx + c = (x - \alpha)(x - \bar{\alpha}).$$

If we multiply out the left-hand side of the above equation, then we see that

$$\alpha + \bar{\alpha} = -b, \qquad \alpha\bar{\alpha} = c.$$

Thus $\alpha + \bar{\alpha}$ and $N(\alpha) = \alpha\bar{\alpha}$ are rational integers. Theorem 8.7 may also be proved directly from Theorem 8.5.

EXERCISES

1. Which of the following numbers are quadratic integers in some quadratic field: $9 + \sqrt{7}$, $1 - 2\sqrt{-7}$, 10, $\frac{1}{2}$,

$$\frac{1 + \sqrt{3}}{2}, \quad \frac{9 + 15\sqrt{-7}}{2}, \quad \frac{9 + 15\sqrt{-7}}{6}, \quad \frac{10 + \sqrt{-108}}{4}, \quad \frac{3 + 2\sqrt{6}}{1 - \sqrt{6}}?$$

2. In each of the following two equations, one root is $3 + 2\sqrt{-2}$. What is the other root? $3x^2 - 18x + 51 = 0$, and

$$(1 + 3\sqrt{-2})x^2 + (2 - 13\sqrt{-2})x + (13 + 20\sqrt{-2}) = 0.$$

3. If we let $\alpha = 3 + 4\sqrt{-1}$, then the equation $N(\alpha) = 5^2$ is another way of expressing the fact that 3,4,5 is a Pythagorean triplet. Using the identity $N(\alpha^n) = [N(\alpha)]^n$, with $n = 2$ and 3, find two other triplets. Let $\beta = 5 + 12\sqrt{-1}$ so that $N(\beta) = 13^2$. Find a triplet with hypotenuse 65. Using $\gamma = 12 + 5\sqrt{-1}$, find another triplet with hypotenuse 65.

4. Show that the roots of the equation $x^2 = 37 + 12\sqrt{7}$ are in $\mathbf{Q}(\sqrt{7})$. Are the roots integers in $\mathbf{Q}(\sqrt{7})$?

5. Give three examples of nonintegers whose norms are rational integers. Does this contradict Theorem 8.7?

6. If α is an integer, show that $\alpha^2 + (\bar{\alpha})^2$ is a rational integer.

7. Show that if α is in $\mathbf{Q}(\sqrt{d})$, then α can be written as the ratio of two integers in $\mathbf{Q}(\sqrt{d})$. (*Hint:* The denominator can even be taken from $\mathbf{Z}$.)

8. It was claimed on page 262 that if α is an irrational root of the equation $ax^2 + bx + c = 0$, then

$$ax^2 + bx + c = a(x - \alpha)(x - \bar{\alpha}).$$

Let

$$a(x - \alpha)(x - \bar{\alpha}) = ax^2 + ex + f.$$

Show that e and f are rational and that

$$(ax^2 + bx + c) - (ax^2 + ex + f) = (b - e)x + (c - f)$$

is 0 when $x = \alpha$. Use this to prove that $e = b$ and $f = c$.
9. Prove Theorem 8.7 directly by using Theorem 8.5.

8.3. Divisibility and Factorization into Primes

We now begin to build up some of the theory of divisibility of integers of $\mathbf{Q}(\sqrt{d})$ analogous to what was done for the rational integers in Chapter 1.

Definition. If α and β are integers in $\mathbf{Q}(\sqrt{d})$, with $\alpha \neq 0$, we say that α **divides** β and write $\alpha|\beta$ if there is an integer γ in $\mathbf{Q}(\sqrt{d})$ such that $\beta = \alpha\gamma$ (in other words, $\alpha|\beta$ if β/α is an integer). Whenever we write $\alpha|\beta$, it is assumed that α and β are integers in $\mathbf{Q}(\sqrt{d})$ with $\alpha \neq 0$ (and, of course, that β/α is an integer).

We recall that if α and β are rational integers, we already had a definition of $\alpha|\beta$. Fortunately, in this case the two definitions are the same: β/α should be a quadratic integer in the new definition and a rational integer in the old; in either case β/α is a rational number and, by definition, a rational number is a quadratic integer if and only if it is a rational integer. Our definition of divisibility leads to the analog of Theorem 1.2, and the proof is also the same.

Theorem 8.8. If $\alpha|\beta$ and $\alpha|\gamma$, then $\bar{\alpha}|\bar{\beta}$ and for any integers δ and ε in $\mathbf{Q}(\sqrt{d})$, $\alpha|(\beta\delta + \gamma\varepsilon)$. [Special cases include $\alpha|(\beta + \gamma)$ with $\delta = \varepsilon = 1$, $\alpha|(\beta - \gamma)$ with $\delta = 1$ and $\varepsilon = -1$, $\alpha|\beta\delta$ with $\varepsilon = 0$.] If $\alpha|\beta$ and $\beta|\gamma$, then $\alpha|\gamma$.

Proof. By definition, there are integers ζ and η such that $\beta = \alpha\zeta$, $\gamma = \alpha\eta$. Then $\bar{\beta} = \bar{\alpha}\bar{\zeta}$ and

$$\beta\delta + \gamma\varepsilon = \alpha\zeta\delta + \alpha\eta\varepsilon = \alpha(\zeta\delta + \eta\varepsilon).$$

Since ζ is an integer, $\bar{\zeta}$ is an integer; since $\delta, \eta, \varepsilon$ are also integers, Theorem 8.6

says that $\zeta\delta + \eta\varepsilon$ is an integer. Thus, by definition, $\bar{\alpha}|\bar{\beta}$ and $\alpha|(\beta\delta + \gamma\varepsilon)$. If $\alpha|\beta$ and $\beta|\gamma$, then there are integers ζ and η such that $\beta = \alpha\zeta$, $\gamma = \beta\eta$. Therefore, $\gamma = \alpha\zeta\eta$ and hence $\alpha|\gamma$.   ▲

Now we are ready to look at the problem of factoring the integers of $\mathbf{Q}(\sqrt{d})$ into other integers. Notice that we didn't say, "factoring the positive integers of $\mathbf{Q}(\sqrt{d})$ into positive integers," which is all that we considered in $\mathbf{Z}$. One reason for this is that when $d < 0$, the numbers of $\mathbf{Q}(\sqrt{d})$ are complex numbers and thus are neither positive nor negative, these being concepts which were defined for real numbers. The more creative student may well ask: "You have just made a definition that makes certain complex numbers integers; why don't you define positive and negative for complex numbers?" This is a good question and one worth answering. Any definition of positive and negative should have at least the following properties:

1. Every nonzero complex number shall be either positive or negative but not both.

2. If α is positive, then $-\alpha$ shall be negative; if α is negative, then $-\alpha$ shall be positive.

3. If α and β are positive, then $\alpha\beta$ is positive.

Without these minimal properties, the words positive and negative would be too much in conflict with their usual meaning to be of any use. Unfortunately, when $d < 0$, these properties are self-contradictory; it is impossible to make a definition of positive and negative which has all three of these properties. Since this fact is of no importance in the rest of this book, we leave it for the reader as an exercise at the end of this section. Since we will be factoring all the numbers of $\mathbf{Q}(\sqrt{d})$, it may be worthwhile to consider what happens in $\mathbf{Z}$ when we factor numbers of $\mathbf{Z}$ into products of numbers in $\mathbf{Z}$ and not just factor positive integers into products of positive integers.

As an example, we factor the number 12 in several manners:

$$12 = 2 \cdot 2 \cdot 3,$$
$$12 = 2 \cdot 3 \cdot 2,$$
$$12 = 1 \cdot 1 \cdot 1 \cdot 1 \cdot 2 \cdot 2 \cdot 3,$$
$$12 = (-2)(2)(-3),$$
$$12 = (-1)(-2)(-2)(-3),$$
$$12 = (-1)(-1)(1)(-1)(-1)(-2)(3)(-2).$$

The first three factorizations are familiar; the first two factor 12 into the same primes except for order, and the third illustrates why it was advantageous not to call 1 a prime number. The last two factorizations show that it is equally advantageous not to call -1 a prime. The two integers, 1 and -1, share the property that any number of them (actually only an even number of minus ones) can be introduced into a factorization. It is thus convenient to consider the two integers 1 and -1 as neither primes nor composite numbers but something else. This something else is called a **unit**; 1 and -1 are the units of **Q**. Except for 0 and the units ± 1, the other integers of **Q** are called primes or composite numbers according to the way they factor. Any integer can always be factored into a given unit such as (-1) and another integer, for example $2 = (-1)\cdot(-2)$. Thus a rational prime is one in which whenever it is written as the product of two rational integers, one of those integers is a unit. What about unique factorization? In the old sense, we no longer have it. For instance, the first and fourth factorizations of 12 above factor 12 into products of different primes. Yet they are not really that different. One factorization can be turned into the other by appropriate use of the unit -1. The primes 2 and -2 are associated by the fact that either equals the other multiplied by the unit -1. If p is a rational prime, then $-p$ is also; any factorization containing p as one of the factors can be turned into a factorization containing $-p$ as one of the factors and vice versa. The numbers p and $-p$ are called **associated primes** or just **associates**. The unique factorization theorem for positive integers can now be replaced by the slightly more complicated **unique factorization theorem for rational integers**:

Theorem 8.9. If n is a nonzero rational integer, then any two factorizations of n into primes are the same except possibly for the order of the factors and primes being replaced by their associates.

We will not use this theorem in this book and hence we leave its proof to the reader as an exercise at the end of the section. The factorizations of 12 show that the extra complications in the theorem are necessary.

If the complications of units and associates are necessary in **Z** to give unique factorization, we can certainly expect these complications to appear in $\mathbf{Q}(\sqrt{d})$. If ε and δ are integers in $\mathbf{Q}(\sqrt{d})$ and $\varepsilon\delta = 1$, then ε and δ can be added to any factorization as often as desired. For example, if α, β, and γ are integers in $\mathbf{Q}(\sqrt{d})$ with $\gamma = \alpha\beta$, then we also have

$$\gamma = \varepsilon\delta\alpha\beta = (\varepsilon)(\varepsilon)(\delta)(\alpha)(\delta\beta),$$

and so on. Thus ε and δ play the same role in $\mathbf{Q}(\sqrt{d})$ as -1 does in **Q**. Since $\varepsilon\delta = 1$ with ε and δ being integers is another way of saying $\varepsilon|1$, we make the following definition:

Definition. An integer ε in $\mathbf{Q}(\sqrt{d})$ is called a **unit** if $\varepsilon|1$. In particular, 1 and -1 are always units in $\mathbf{Q}(\sqrt{d})$.

The following facts about units will be useful.

Theorem 8.10. If ε_1 and ε_2 are units in $\mathbf{Q}(\sqrt{d})$, then $\bar{\varepsilon}_1$, $\varepsilon_1\varepsilon_2$, $\varepsilon_1/\varepsilon_2$ (in particular $1/\varepsilon_2$) are also units in $\mathbf{Q}(\sqrt{d})$. Further, an integer ε of $\mathbf{Q}(\sqrt{d})$ is a unit if and only if $\mathbf{N}(\varepsilon) = \pm 1$.
[*Remark:* The restriction that ε be an integer is necessary. For example, $\mathbf{N}(\frac{3}{5} + \frac{4}{5}\sqrt{-1}) = 1$ but $\frac{3}{5} + \frac{4}{5}\sqrt{-1}$ is not an integer in $\mathbf{Q}(\sqrt{-1})$ and hence not a unit.]

Proof. Since ε_1 and ε_2 are units, they are integers and there are integers δ_1 and δ_2 in $\mathbf{Q}(\sqrt{d})$ such that $\varepsilon_1\delta_1 = \varepsilon_2\delta_2 = 1$. Then $\bar{\varepsilon}_1$, $\varepsilon_1\varepsilon_2$, $\bar{\delta}_1$, and $\delta_1\delta_2$ are integers and

$$\bar{\varepsilon}_1\bar{\delta}_1 = \overline{(\varepsilon_1\delta_1)} = \bar{1} = 1,$$

$$(\varepsilon_1\varepsilon_2)(\delta_1\delta_2) = (\varepsilon_1\delta_1)(\varepsilon_2\delta_2) = 1,$$

so that $\bar{\varepsilon}_1$ and $\varepsilon_1\varepsilon_2$ are units. Further,

$$\frac{\varepsilon_1}{\varepsilon_2} = \frac{\varepsilon_1\delta_2}{\varepsilon_2\delta_2} = \varepsilon_1\delta_2$$

is an integer and $(\varepsilon_1/\varepsilon_2)(\varepsilon_2\delta_1) = 1$ with $\varepsilon_2\delta_1$ also being an integer. Thus $\varepsilon_1/\varepsilon_2$ is also a unit. Finally, if ε is an integer and $\mathbf{N}(\varepsilon) = \pm 1$, then $\bar{\varepsilon}$ and $-\bar{\varepsilon}$ are also integers and either $\mathbf{N}(\varepsilon) = 1$, so that

$$\varepsilon\bar{\varepsilon} = \mathbf{N}(\varepsilon) = 1$$

or $\mathbf{N}(\varepsilon) = -1$, so that

$$\varepsilon(-\bar{\varepsilon}) = -\mathbf{N}(\varepsilon) = 1.$$

In either case, we see that ε is a unit. Conversely, suppose ε is a unit. Then there is an integer δ such that $\varepsilon\delta = 1$ and thus

$$\mathbf{N}(\varepsilon)\mathbf{N}(\delta) = \mathbf{N}(\varepsilon\delta) = \mathbf{N}(1) = 1.$$

Since $\mathbf{N}(\varepsilon)$ and $\mathbf{N}(\delta)$ are rational integers and their product is 1, either $\mathbf{N}(\varepsilon) = \mathbf{N}(\delta) = 1$ or $\mathbf{N}(\varepsilon) = \mathbf{N}(\delta) = -1$. In any event, $\mathbf{N}(\varepsilon) = \pm 1$. ▲

Now that we know how to recognize a unit, the next question to be asked is: What are the units in $\mathbf{Q}(\sqrt{d})$? We will answer this question only for the complex quadratic fields and settle here for a knowledge of how many units the real quadratic fields have.

Theorem 8.11. If $d < 0$, $d \neq -1$, $d \neq -3$, then $\mathbf{Q}(\sqrt{d})$ has exactly two units, namely ± 1. Also $\mathbf{Q}(\sqrt{-1})$ has exactly four units, ± 1 and $\pm\sqrt{-1}$; $\mathbf{Q}(\sqrt{-3})$ has exactly six units, ± 1, $\pm(-1 + \sqrt{-3})/2$, $\pm(-1 - \sqrt{-3})/2$. If $d > 0$, then $\mathbf{Q}(\sqrt{d})$ has infinitely many units.

Proof. Suppose that $d < 0$ and $d \not\equiv 1 \pmod 4$. Then the integers of $\mathbf{Q}(\sqrt{d})$ are of the form $\alpha = a + b\sqrt{d}$, where a and b are in $\mathbf{Z}$. If α is a unit, then $N(\alpha) = \pm 1$ and, in fact, $N(\alpha) = 1$, since norms are not negative when $d < 0$ (Theorem 8.4). We recall that

$$N(\alpha) = a^2 - b^2 d = a^2 + b^2(-d).$$

For $d \leq -2$ (and therefore $-d \geq 2$), we have $N(\alpha) \geq a^2 + 2b^2$, and if $b \neq 0$ (so that $b^2 \geq 1$), then

$$N(\alpha) \geq a^2 + 2 \cdot 1 \geq 2$$

and hence α is not a unit. Thus if α is a unit and $d \leq -2$, then $b = 0$ and $N(\alpha) = a^2 = 1$. Therefore, $a = \pm 1$ and therefore $\alpha = \pm 1$. Thus for $d \leq -2$, $d \not\equiv 1 \pmod 4$, the only units of $\mathbf{Q}(\sqrt{d})$, are ± 1. Suppose that $d = -1$. Then $1 = N(\alpha) = a^2 + b^2$, assuming that α is a unit. It is quickly seen that $a^2 + b^2 > 1$ unless $|a| \leq 1$, $|b| \leq 1$ and, of these choices, $a^2 + b^2 = 1$ only for $a = \pm 1, b = 0$ or $a = 0, b = \pm 1$. This gives ± 1 and $\pm\sqrt{-1}$ as the only units in $\mathbf{Q}(\sqrt{-1})$.

Now suppose that $d < 0$ and $d \equiv 1 \pmod 4$. In this case, the integers α of $\mathbf{Q}(\sqrt{d})$ are of the form

$$\alpha = \frac{a + b\sqrt{d}}{2},$$

where a and b are both odd or both even integers in $\mathbf{Z}$ and here

$$N(\alpha) = \frac{a^2 + b^2(-d)}{4}.$$

Again since $d < 0$, α is a unit if and only if $N(\alpha) = 1$; that is,

$$a^2 + b^2(-d) = 4.$$

For $d \leq -7$ (and thus $-d \geq 7$) and $b \neq 0$,

$$a^2 + b^2(-d) \geq a^2 + 7b^2 \geq a^2 + 7 \cdot 1 \geq 7 > 4,$$

and hence α a unit implies that $b = 0$. In this case, $4 = N(a) = a^2$, so that $a = \pm 2$ and $\alpha = \pm 1$. Thus if $d \leq -7$ and $d \equiv 1 \pmod 4$, the only units of $\mathbf{Q}(\sqrt{d})$ are ± 1. If $d < 0$, $d \equiv 1 \pmod 4$ and $-7 < d$, then $d = -3$. So now assume that $d = -3$. In this case α a unit implies

$$a^2 + 3b^2 = 4.$$

If $|b| \geq 2$, then $a^2 + 3b^2 \geq 12$ and thus if $a^2 + 3b^2 = 4$, then the only choices for b are $b = \pm 1, 0$. If $b = 0$, then $a = \pm 2$ and $\alpha = \pm 1$. If $b = 1$, then $a = \pm 1$ and $\alpha = (\pm 1 + \sqrt{-3})/2$. If $b = -1$, then $a = \pm 1$ and $\alpha = (\pm 1 - \sqrt{-3})/2$. Thus, the only units of $\mathbf{Q}(\sqrt{-3})$ are $\pm 1, (\pm 1 + \sqrt{-3})/2, (\pm 1 - \sqrt{-3})/2$, and these are the six units of the theorem.

Now let us suppose that $d > 0$. We will show that there are infinitely many units of the form

$$\alpha = a + b\sqrt{d},$$

where a and b are rational integers. This is, of course, what we must do if $d \not\equiv 1(\bmod 4)$, but when $d \equiv 1(\bmod 4)$, we are throwing away potential units of the form $(c + e\sqrt{d})/2$, where c and e are both odd rational integers. In fact, we will show that there are infinitely many such units α with $\mathbf{N}(\alpha) = 1$; we thereby ignore whatever units there may be of norm -1. So suppose $\mathbf{N}(\alpha) = 1$. This happens if and only if

$$a^2 - db^2 = 1.$$

But this is the Fermat–Pell equation. Since $\sqrt{d}$ is irrational, Theorem 7.26 says that there are infinitely many solutions to this equation in integers a and b. Each of these solutions leads to a unit of norm $+1$; hence $\mathbf{Q}(\sqrt{d})$ has infinitely many units. ▲

Thus when $d < 0$, the unit situation in $\mathbf{Q}(\sqrt{d})$ is exactly the same as in $\mathbf{Q}$ except when $d = -1$ and -3, when the situation is slightly complicated. On the other hand, the situation in $\mathbf{Q}(\sqrt{d})$ with $d > 0$ is considerably more complicated. But it is still a finite process to determine if two factorizations of a number differ only by unit factors. For example, in $\mathbf{Q}(\sqrt{6})$ we have

$$5 = (1 + \sqrt{6})(-1 + \sqrt{6}) = (71 + 29\sqrt{6})(-71 + 29\sqrt{6}).$$

Once we prove Theorem 8.12, the reader may return and see that the numbers $(1 + \sqrt{6}), (-1 + \sqrt{6}), (71 + 29\sqrt{6}), (-71 + 29\sqrt{6})$ are all primes in $\mathbf{Q}(\sqrt{6})$. Does this destroy the hope for unique factorization in $\mathbf{Q}(\sqrt{6})$, or is it just a more complicated example of something like $6 = (2)(3) = (-2)(-3)$? To see if one of the two factorizations can be turned into the other by insertion of unit factors, we wish to know if either $(71 + 29\sqrt{6})/(1 - \sqrt{6})$ or $(-71 + 29\sqrt{6})/(1 + \sqrt{6})$ is a unit. In the first case,

$$\frac{71 + 29\sqrt{6}}{1 + \sqrt{6}} = \frac{103 + 42\sqrt{6}}{5}$$

is not even an integer and hence not a unit. In the second case,

$$\frac{-71 + 29\sqrt{6}}{1 + \sqrt{6}} = 49 - 20\sqrt{6}$$

is an integer and thus may be a unit. [We will see immediately from the definition of primes that if one prime in $\mathbf{Q}(\sqrt{d})$ divides another prime, the quotient must be a unit.] To see for sure, we look at

$$N(49 - 20\sqrt{6}) = 49^2 - 6 \cdot 20^2 = 1$$

and hence $49 - 20\sqrt{6}$ is a unit. In like manner

$$\frac{71 + 29\sqrt{6}}{-1 + \sqrt{6}} = 49 + 20\sqrt{6}$$

is also a unit. (Incidentally, this division need not be performed; if the numerator and denominator are multiplied by -1, then the division to be done involves dividing the conjugates of the earlier division and hence, by Theorem 8.3, the answer is the conjugate of the earlier answer.] Therefore, since $(49 + 20\sqrt{6})(49 - 20\sqrt{6}) = 1$,

$$\begin{aligned} 5 &= (1 + \sqrt{6})(-1 + \sqrt{6}) \\ &= [(1 + \sqrt{6})(49 - 20\sqrt{6})][(-1 + \sqrt{6})(49 + 20\sqrt{6})] \\ &= (-71 + 29\sqrt{6})(71 + 29\sqrt{6}), \end{aligned}$$

and indeed the first factorization of 5 can be turned into the second (except for order) merely by inserting the proper unit factors. Clearly this is much harder to tell when there are units other than ± 1 available, unless of course the reader is willing to use an "obvious" (but not yet proved) theorem on unique factorization for $\mathbf{Q}(\sqrt{6})$ analogous to Theorem 8.9 for $\mathbf{Q}$. And, in view of this last example, is such a theorem really obvious?

We are now ready to consider factoring the nonzero, nonunit integers of $\mathbf{Q}(\sqrt{d})$ into primes. We first note that if α is an integer, $N(\alpha) = 0$ implies $\alpha = 0$, and $|N(\alpha)| = 1$ implies α is a unit. Thus an integer α is nonzero and a nonunit if and only if $|N(\alpha)| \geq 2$. This fact will be useful to remember.

Definition. An integer π in $\mathbf{Q}(\sqrt{d})$ which is neither zero nor a unit is called a **prime** in $\mathbf{Q}(\sqrt{d})$ if for every decomposition of π into a product of two integers, say $\pi = \alpha\beta$, either α or β is a unit.[2] To distinguish the primes

[2] In this chapter, we will have no occasion to refer to the number $3.14159\ldots$ which does not belong to any quadratic field. The letter π is the Greek letter corresponding to p and it is standard practice to use it to denote a prime.

of $\mathbf{Q}(\sqrt{d})$ from the primes of $\mathbf{Q}(\pm 2, \pm 3, \pm 5, \pm 7, \ldots)$ we will now call the primes of $\mathbf{Q}$ **rational primes**. A nonzero, nonunit, nonprime integer of $\mathbf{Q}(\sqrt{d})$ is called **composite**. In other words, a composite number is the product of two nonzero, nonunit integers of $\mathbf{Q}(\sqrt{d})$.

It is very important to distinguish between "prime" and "rational prime." A rational prime is not necessarily a prime in $\mathbf{Q}(\sqrt{d})$. In fact, we have just seen such an example. We had the factorization

$$5 = (1 + \sqrt{6})(-1 + \sqrt{6})$$

and $N(1 + \sqrt{6}) = N(-1 + \sqrt{6}) = -5$, so that neither $(1 + \sqrt{6})$ nor $(-1 + \sqrt{6})$ are units and hence the rational prime 5 is not a prime in $\mathbf{Q}(\sqrt{6})$. The following theorem will enable the reader to distinguish many primes (but not all) at a glance.

Theorem 8.12. If α is an integer in $\mathbf{Q}(\sqrt{d})$ and $N(\alpha)$ is a rational prime, then α is a prime.

Proof. Since $N(\alpha)$ is a rational prime, α is not zero or a unit. Suppose $\alpha = \beta\gamma$, where β and γ are integers in $\mathbf{Q}(\sqrt{d})$. Then

$$N(\alpha) = N(\beta)N(\gamma),$$

where $N(\beta)$ and $N(\gamma)$ are rational integers. Thus by the definition of a rational prime, one of $N(\beta)$ and $N(\gamma)$ is a rational unit. The rational units are ± 1, and thus either $N(\beta)$ or $N(\gamma)$ is ± 1. Hence either β or γ is a unit and therefore, by definition, α is a prime. ▲

For example, $N[(3 + \sqrt{-163})/2] = 43$ and thus $(3 + \sqrt{-163})/2$ is a prime in $\mathbf{Q}(\sqrt{-163})$; $N(-1 + \sqrt{6}) = -5$ and thus $-1 + \sqrt{6}$ is a prime in $\mathbf{Q}(\sqrt{6})$; $N[(11 + 2\sqrt{-1})/5] = 5$ but $(11 + 2\sqrt{-1})/5$ is not an integer and hence not a prime.

This is also a good spot to give an example of a prime whose norm is not a rational prime. We will show that 7 is a prime in $\mathbf{Q}(\sqrt{6})$ in spite of the fact that $N(7) = 49$ is not a rational prime. Suppose that 7 is not a prime in $\mathbf{Q}(\sqrt{6})$. Then there are integers α and β in $\mathbf{Q}(\sqrt{6})$ with $7 = \alpha\beta$ and neither α nor β being a unit. As a result, $|N(\alpha)| \geq 2$, $|N(\beta)| \geq 2$ and since

$$|N(\alpha)| \cdot |N(\beta)| = |N(\alpha)N(\beta)| = |N(7)| = 49,$$

with $|N(\alpha)|$ and $|N(\beta)|$ being positive rational integers, we must have

$$|N(\alpha)| = |N(\beta)| = 7.$$

To show that 7 is a prime in $\mathbf{Q}(\sqrt{7})$, it thus suffices to show that there are no integers in $\mathbf{Q}(\sqrt{6})$ of norm ± 7. Let γ be an integer in $\mathbf{Q}(\sqrt{6})$; then

$\gamma = a + b\sqrt{6}$, where a and b are in **Z**. Suppose $\mathbf{N}(\gamma) = \pm 7$. Then

(4) $a^2 - 6b^2 = \pm 7.$

Modulo 7 this equation reads

(5) $a^2 + b^2 \equiv a^2 - 6b^2 \equiv \pm 7 \equiv 0 (\text{mod } 7).$

By Theorem 3.30, it follows from (5) that $a \equiv b \equiv 0 (\text{mod } 7)$. Thus $7|a$, $7|b$, $49|a^2$, $49|b^2$ and hence $49|(a^2 - 6b^2)$. This is a contradiction since $a^2 - 6b^2 = \pm 7$. As a result, there are no (rational) integral solutions to equation (4), and hence there are no integers in $\mathbf{Q}(\sqrt{6})$ of norm ± 7. Therefore, 7 is a prime in $\mathbf{Q}(\sqrt{6})$.

We have one remaining concept left to generalize to $\mathbf{Q}(\sqrt{d})$.

Definition. If α and β are nonzero integers in $\mathbf{Q}(\sqrt{d})$ such that $\alpha = \beta\varepsilon$, where ε is a unit, then α is said to be an **associate** of β. In other words, α is an associate of β if and only if α/β is a unit.

Under this definition, any integer is an associate of itself. The main properties of associates are given in the following theorem.

Theorem 8.13. If α and β are integers in $\mathbf{Q}(\sqrt{d})$, then α is an associate of β if and only if β is an associate of α (and then we can say that α and β are **associates** or **associated**). Also α and β are associates if and only if $\alpha|\beta$ and $\beta|\alpha$. If α and β are associates and $\gamma|\alpha$, then $\gamma|\beta$ and if $\alpha|\delta$, then $\beta|\delta$. If α is a prime, then every associate of α is a prime; if α is a composite, then every associate of α is composite.

Proof. If ε is a unit, then $1/\varepsilon$ is also a unit. Thus if $\alpha = \beta\varepsilon$, then $\beta = \alpha(1/\varepsilon)$. Therefore, by definition, if α is an associate of β, then β is an associate of α. Likewise if β is an associate of α, then α is an associate of β. If the integers α and β are associates, then there is a unit ε such that $\alpha = \beta\varepsilon$ and $\beta = \alpha(1/\varepsilon)$. Since ε and $1/\varepsilon$ are units and hence integers, $\beta|\alpha$ and $\alpha|\beta$ by the definition of divisibility. On the other hand, if $\alpha|\beta$ and $\beta|\alpha$, then by definition, $\beta = \alpha\varepsilon$ and $\alpha = \beta\delta$ with ε and δ being integers. Therefore $\varepsilon\delta = (\beta/\alpha)(\alpha/\beta) = 1$ and thus ε and δ are units. Hence α and β are associates. If α and β are associates and $\gamma|\alpha$, then since $\alpha|\beta$, it follows from Theorem 8.8 that $\gamma|\beta$. If α and β are associates, so that $\beta|\alpha$ and $\alpha|\delta$, then, by Theorem 8.8, $\beta|\delta$.

Theorem 8.10 shows us that all the associates of units are units and hence no associate of a nonunit is a unit. Suppose that α and β are nonzero, nonunit associated integers in $\mathbf{Q}(\sqrt{d})$. Then α is either prime or composite, and the same holds for β. We will show that both are prime or both are composite.

Suppose that one is prime and one composite; since it makes no difference, let α be prime and β be composite. Then we may write $\beta = \gamma\delta$, where neither γ nor δ is a unit. Since α and β are associates, there is a unit ε such that $\alpha = \beta\varepsilon$. Thus $\alpha = (\gamma)(\delta\varepsilon)$. Now γ is a nonunit. Also $(\delta\varepsilon)$ is an associate of a nonunit and hence, as we have just noted, is itself a nonunit. Thus we have factored α into two nonunits, which contradicts the assumption that α is prime. Hence α and β are both prime or both composite, as was to be shown. ▲

We conclude this section by showing that factorization into primes is possible (incidentally, there are systems where factorization into primes is not possible) and we save the problem of unique factorization for the next section.

Theorem 8.14. If $\alpha, \alpha_1, \alpha_2, \ldots, \alpha_n$ are nonzero integers in $\mathbf{Q}(\sqrt{d})$ and

$$\alpha = \alpha_1\alpha_2\cdots\alpha_n \qquad \text{with } n > \log_2(|N(\alpha)|),$$

then at least one of the numbers $\alpha_1, \ldots, \alpha_n$ is a unit. It follows that if α is not a unit, then it can be represented as a product of a finite number of primes in $\mathbf{Q}(\sqrt{d})$.

Proof. Suppose none of the α_j's are units. Then for each j in the range $1 \leq j \leq n$, we have $|N(\alpha_j)| \geq 2$. Thus

$$\begin{aligned}
|N(\alpha)| &= |N(\alpha_1)N(\alpha_2)\cdots N(\alpha_n)| \\
&= |N(\alpha_1)| \cdot |N(\alpha_2)| \cdot \cdots \cdot |N(\alpha_n)| \\
&\geq 2 \cdot 2 \cdot \cdots \cdot 2 \\
&= 2^n.
\end{aligned}$$

We now look at the logarithms (to the base 2) of both sides of the inequality, $|N(\alpha)| \geq 2^n$, and see that $\log_2(|N(\alpha)|) \geq \log_2(2^n) = n$. This contradicts the fact that $n > \log_2(|N(\alpha)|)$ and hence at least one of the numbers $\alpha_1, \ldots, \alpha_n$ is a unit.

Now suppose α is a nonzero, nonunit integer in $\mathbf{Q}(\sqrt{d})$. We will show that α can be written as a product of one or more primes. If α is a prime, we are done. If α is composite, then we can set $\alpha = \alpha_1\beta_1$, where neither α_1 nor β_1 are units. If α_1 and β_1 are primes, we are done. If α_1 and β_1 are not both primes, then one of them is composite and we can assume that α_1 and β_1 are named so that β_1 is composite. Then $\beta_1 = \alpha_2\beta_2$, where α_2 and β_2 are nonunits. Thus $\alpha = \alpha_1\alpha_2\beta_2$ gives α as a product of three nonunits. If α_1, α_2 and β_2 are primes, we are done; if not, one of them is composite and, by renaming the numbers, we may assume that β_2 is composite. Then $\beta_2 = \alpha_3\beta_3$, where α_3 and β_3 are nonunit integers. Thus $\alpha = \alpha_1\alpha_2\alpha_3\beta_3$ gives α as a product of four nonunits. We continue this way as long as at least one of the factors is composite. At the nth step, we get

(6) $$\alpha = \alpha_1\alpha_2\cdots\alpha_{n-1}\beta_{n-1}$$

with none of the numbers on the right-hand side being units. For some $n \leq \log_2(|N(\alpha)|)$ this process gives $\alpha_1, \alpha_2, \ldots, \alpha_{n-1}$ and β_{n-1} all primes. If not, the process is repeated until we reach an $n > \log_2(|N(\alpha)|)$. But then there are n factors on the right, and the first half of our theorem says that at least one of the factors on the right is a unit. The process outlined above, if carried n steps, would have given all nonunits on the right. Thus the factoring process must have ended with an earlier n and, when it ends, it gives a representation of α as a finite product of primes.   ▲

We have actually proved more than the second half of the theorem. We have shown that if we start with α and factor it into two nonunits, and if we keep factoring any composite numbers that appear into two nonunits, that after a maximum of $\log_2(|N(\alpha)|)$ factorizations (no matter how we proceeded), we will have α represented as a product of primes.

EXERCISES

1. Show that it is impossible to make a definition of positive and negative in $Q(\sqrt{-1})$ that satisfies the three properties listed on page 271. Do this by showing that since $\alpha \neq 0$ implies α or $-\alpha$ is positive, then α^2 is always positive. Use this to show that both 1 and -1 are positive.

2. Prove Theorem 8.9 by using Theorem 2.9 and the facts that the absolute value of a nonzero rational integer is a positive rational integer and the absolute value of a product is the product of the absolute values.

3. Which of the following numbers are units:

$$2 + \sqrt{7}, \ 19 + 231\sqrt{-71}, \ \frac{7 + \sqrt{53}}{2}, \ \frac{39 + 5\sqrt{61}}{2}, \ \frac{13 + 3\sqrt{17}}{4}?$$

4. Prove that if the product of two integers in $Q(\sqrt{d})$ is a unit, then each is a unit.

5. If π_1 and π_2 are primes in $Q(\sqrt{d})$ and $\pi_1|\pi_2$, show that π_1 and π_2 are associates.

6. Show that each of the following numbers are primes in $Q(\sqrt{-5})$: $3 + 2\sqrt{-5}, 37, 1 + 2\sqrt{-5}$.

7. Factor the number $33 + 11\sqrt{-7}$ into primes in $Q(\sqrt{-7})$. [*Note:* $-7 \equiv 1 \pmod 4$.]

8. Show that 2 is the product of two associated primes in $Q(\sqrt{6})$. Show also that 3 is the product of two associated primes in $Q(\sqrt{6})$.

9. Factor the number $\sqrt{14}$ into the product of two primes in $\mathbf{Q}(\sqrt{14})$.
10. Show that if a and b are both odd rational integers, then either $4|(a + b)$ or $4|(a - b)$. Use this to help you show that if a and b are both odd rational integers, then either

$$\left(\frac{1 + \sqrt{-7}}{2}\right)\left|\left(\frac{a + b\sqrt{-7}}{2}\right)\right. \quad \text{or} \quad \left(\frac{1 + \sqrt{-7}}{2}\right)\left|\left(\frac{a - b\sqrt{-7}}{2}\right)\right.$$

(*Hint:* Perform the divisions.)
11. Suppose α and β are integers in $\mathbf{Q}(\sqrt{d})$. Prove that if $\mathbf{N}(\alpha) = \mathbf{N}(\beta) = 1$, then $\alpha|\beta$. Prove or give a counterexample to the more general conjecture that if $\mathbf{N}(\alpha) = \mathbf{N}(\beta)$, then $\alpha|\beta$.
12. Show that there is no integer in $\mathbf{Q}(\sqrt{7})$ of norm 3. Show that in spite of this, 3 is not a prime in $\mathbf{Q}(\sqrt{7})$.
13. Suppose ε is a unit in $\mathbf{Q}(\sqrt{d})$ and $\sqrt{\varepsilon}$ is an integer in $\mathbf{Q}(\sqrt{d})$. Show that $\sqrt{\varepsilon}$ is a unit.

8.4. Unique Factorization and Euclidean Domains

The discussion in the previous section leading up to Theorem 8.9 shows what form a unique factorization theorem for $\mathbf{Q}(\sqrt{d})$ should take.

Definition. Suppose $\mathbf{Q}(\sqrt{d})$ has the following property: If α is any non-zero, nonunit integer in $\mathbf{Q}(\sqrt{d})$ and we have the two factorizations

$$\alpha = \varepsilon\pi_1\pi_2\cdots\pi_r, \qquad \alpha = \varepsilon'\pi'_1\pi'_2\cdots\pi'_s,$$

where ε and ε' are units and $\pi_1,\ldots,\pi_r,\pi'_1,\ldots,\pi'_s$ are primes not necessarily distinct, then $r = s$ and the primes $\pi'_1,\pi'_2,\ldots,\pi'_s$ can be given new subscripts in such a way that π_j and π'_j are associates for $j = 1, 2,\ldots,r$. Then we say that the set of integers of $\mathbf{Q}(\sqrt{d})$ is a **unique factorization domain** (**UFD** for short).

Other authors say that $\mathbf{Q}(\sqrt{d})$ has the **unique factorization property** and still others say $\mathbf{Q}(\sqrt{d})$ is a **simple field**. The word "domain" (short for "integral domain") in the above definition is a technical word which refers to a set of numbers which satisfies all the field axioms of Appendix C except possibly axiom 9, which is replaced by a less restrictive axiom on cancellation; thus sums, differences, and products of numbers of the set are in the set, but quotients are not necessarily in the set. The technical meaning of the word "domain" is unimportant here. Unique factorization has been used time and again in the last six chapters. We temporarily delay consideration of

whether or not the integers of $\mathbf{Q}(\sqrt{d})$ do form a UFD and prove some
analogues of some of the most applicable theorems of the second chapter.

Theorem 8.15. The integers of $\mathbf{Q}(\sqrt{d})$ form a UFD if and only if $\mathbf{Q}(\sqrt{d})$
has the following property: If $\pi | \alpha\beta$, where π is a prime and α and β are
integers in $\mathbf{Q}(\sqrt{d})$, then $\pi | \alpha$ or $\pi | \beta$.

Proof. Suppose first that the integers of $\mathbf{Q}(\sqrt{d})$ form a UFD and then
suppose that $\pi | \alpha\beta$, where π is a prime and α and β are integers in $\mathbf{Q}(\sqrt{d})$.
By definition of $\pi | \alpha\beta$, there is an integer γ such that $\alpha\beta = \pi\gamma$. By Theorem
8.14, there are primes $\pi_1, \pi_2, \ldots, \pi_n$ and a unit ε in $\mathbf{Q}(\sqrt{d})$ such that

$$\gamma = \varepsilon\pi_1\pi_2 \cdots \pi_n$$

and then

$$\alpha\beta = \varepsilon\pi\pi_1\pi_2 \cdots \pi_n$$

is a factorization of $\alpha\beta$ into primes. Likewise, there are primes

$$\pi'_1, \ldots, \pi'_r, \pi''_1, \ldots, \pi''_s$$

and units $\varepsilon_1, \varepsilon_2$ such that
$$\alpha = \varepsilon_1\pi'_1 \cdots \pi'_r \qquad \beta = \varepsilon_2\pi''_1 \cdots \pi''_s$$

so that

$$\alpha\beta = (\varepsilon_1\varepsilon_2)\pi'_1 \cdots \pi'_r\pi''_1 \cdots \pi''_s.$$

The number $\alpha\beta$ is not a unit since it is divisible by the prime π, but α may be a
unit, β may be a unit (but not both α and β are units) and γ may be a unit.
This is the interpretation to be given either $r = 0$, $s = 0$, or $n = 0$ and is the
reason for using the units ε_1, ε_2, and ε; for example, if γ is a unit, we let
$n = 0$ and $\gamma = \varepsilon$; if γ is not a unit, then $n \geq 1$ and we may take $\varepsilon = 1$ if we
wish. We now have the two factorizations of the nonunit $\alpha\beta$ into primes:

$$\alpha\beta = \varepsilon\pi\pi_1 \cdots \pi_n = (\varepsilon_1\varepsilon_2)\pi'_1 \cdots \pi'_r\pi''_1 \cdots \pi''_s.$$

By the definition of a UFD, one of the primes on the right-hand side is an
associate of π and hence divisible by π (the quotient being a unit). If one of
the π''s is divisible by π, then, by Theorem 8.8, $\pi | \alpha$; if one of the π'''s is divis-
ible by π, then $\pi | \beta$. Thus π divides either α or β (and possibly both).

Now we suppose that $\mathbf{Q}(\sqrt{d})$ has the property that if $\pi | \alpha\beta$, where π is a
prime and α and β are integers in $\mathbf{Q}(\sqrt{d})$, then $\pi | \alpha$ or $\pi | \beta$. Suppose α is a
nonunit integer in $\mathbf{Q}(\sqrt{d})$ and

(7) $$\alpha = \varepsilon\pi_1\pi_2 \cdots \pi_r = \varepsilon'\pi'_1\pi'_2 \cdots \pi'_s,$$

where ε and ε' are units and $\pi_1, \pi_2, \ldots, \pi_r, \pi'_1, \ldots, \pi'_s$ are primes. Either $r \leq s$ or $s \leq r$, and since it makes no difference, we will let $r \leq s$. We now show that one of the π''s is an associate of π_1. Since $\pi_1 | \alpha$, and hence all associates of α, we see that

$$\pi_1 | (\pi'_1 \cdots \pi'_{s-1}) \pi'_s.$$

Thus either $\pi_1 | (\pi'_1 \cdots \pi'_{s-1})$ or $\pi_1 | \pi'_s$. If $\pi_1 | \pi'_s$, then π'_s / π_1 is a unit, since otherwise π'_s would be composite, and therefore π'_s is an associate of π_1. If π'_s is not an associate of π_1, then $\pi_1 | (\pi'_1 \cdots \pi'_{s-2}) \pi'_{s-1}$. In like manner, if, in addition to not being an associate of π'_s, π_1 is not an associate of π'_{s-1}, then $\pi_1 | (\pi'_1 \cdots \pi'_{s-3}) \pi'_{s-2}$. We continue in this way until we reach the point that either π_1 is an associate of one of the primes $\pi'_2, \pi'_3, \ldots, \pi'_s$ or $\pi_1 | \pi'_1$, and in the latter case, π'_1 and π_1 are associates. Thus, as was to be shown, one of the π''s is an associate of π_1 and by renumbering the π''s, we may assume that π'_1 and π_1 are associates. Thus $\pi'_1 = \varepsilon_1 \pi_1$, where ε is a unit, and hence it follows from equation (7) that

$$(8) \qquad \varepsilon \pi_2 \cdots \pi_r = (\varepsilon' \varepsilon_1) \pi'_2 \cdots \pi'_s,$$

where $\varepsilon' \varepsilon_1$ is also a unit.

We now repeat the above process with π_2 instead of π_1 and we find that one of the primes $\pi'_2, \ldots, \pi'_s$ is an associate of π_2 and, by renumbering the π''s, we may assume that π'_2 and π_2 are associates. Thus $\pi'_2 = \varepsilon_2 \pi_2$, where ε_2 is a unit and hence it follows from equation (8) that

$$\varepsilon \pi_3 \cdots \pi_r = (\varepsilon' \varepsilon_1 \varepsilon_2) \pi'_3 \cdots \pi'_s,$$

where $\varepsilon' \varepsilon_1 \varepsilon_2$ is a unit. We repeat this whole process with π_3 and then π_4 and so on until we at last come to the point that π'_j is an associate of π_j for each j in the range $1 \leq j \leq r - 1$ and

$$(9) \qquad \varepsilon \pi_r = (\varepsilon' \varepsilon_1 \varepsilon_2 \cdots \varepsilon_{r-1}) \pi'_r \cdots \pi'_s,$$

where $(\varepsilon' \varepsilon_1 \cdots \varepsilon_{r-1})$ is a unit. The number on the left of (9) is a prime and hence the number on the right of (9) is also. The number on the right of (9) is composite for $s > r$ and therefore $s = r$. Hence $\varepsilon \pi_r = \varepsilon'' \pi'_r$, $\pi'_r = \varepsilon / \varepsilon'' \pi_r$, and $\varepsilon / \varepsilon''$ is a unit. Thus π'_r is an associate of π_r. This shows that the integers of $Q(\sqrt{d})$ form a UFD. ▲

One frequently uses Theorem 8.15 to justify the following statement. If $\alpha_1, \ldots, \alpha_n$ are integers in a UFD and π is a prime such that $\pi | \alpha_1 \cdots \alpha_n$, then $\pi | \alpha_j$ for some j in the range $1 \leq j \leq n$. Theorem 8.15 really is only this statement for $n = 2$, but the statement for general n follows from this in the same way that we showed in the proof of Theorem 8.15 that π_1 and π'_j were associated for some j. It will also be useful to prove an analogue of Theorem 2.12.

Theorem 8.16. Suppose that $\mathbf{Q}(\sqrt{d})$ has the unique factorization property. If α, β, and γ are integers and ε is a unit in $\mathbf{Q}(\sqrt{d})$ such that α and β have no common integral factors other than units and $\alpha\beta = \varepsilon\gamma^n$, where n is a positive rational integer, then there are units ε' and ε'' and integers δ and ζ in $\mathbf{Q}(\sqrt{d})$ such that $\alpha = \varepsilon'\delta^n$ and $\beta = \varepsilon''\zeta^n$.

Proof. If γ is a unit, then $\alpha\beta$ is a unit and hence both α and β are units. In this case, the theorem is trivial: We put $\varepsilon' = \alpha$, $\varepsilon'' = \beta$, $\delta = \zeta = 1$. If $\gamma = 0$, then one of α and β is 0; since anything divides 0, the only way that units divide α and β is that the other of α and β is a unit; the theorem is trivial in this case also, with one of δ and ζ being 0 and the other 1. Thus we may assume that γ is not zero and not a unit. We can therefore set

$$\gamma = \pi_1\pi_2\cdots\pi_r$$

where $\pi_1,\ldots,\pi_r$ are primes some of which may be associates. It is sufficient to show that α is a unit times an nth power, the proof for β is identical.

If α is a unit, then set $\varepsilon' = \alpha$ and $\delta = 1$. Thus we shall assume that α is not a unit; it is also not zero, since $\gamma \neq 0$. Hence we can also break α up into a product of primes, say

$$\alpha = \pi_1'\pi_2'\cdots\pi_s'$$

and then

(10) $$\pi_1'\pi_2'\cdots\pi_s'\beta = \varepsilon\pi_1^n\pi_2^n\cdots\pi_r^n.$$

The unique factorization property says that π_1' is an associate of one of the π_j's and they can be renumbered so that π_1' is an associate of π_1. Now if any associate of π_1 divides β, then $\pi_1|\beta$ and then $\pi_1'|\beta$ so that π_1' divides both α and β. By hypothesis, this cannot happen. Thus π_1 or its associates must show up n times among the primes $\pi_1',\ldots,\pi_s'$ which can be renumbered so that $\pi_1',\pi_2',\ldots,\pi_n'$ are associates of π_1. This also means that $s \geq n$. Hence there are units $\varepsilon_1,\varepsilon_2,\ldots,\varepsilon_n$ such that

$$\pi_1' = \varepsilon_1\pi_1, \pi_2' = \varepsilon_2\pi_1,\ldots,\pi_n' = \varepsilon_n\pi_1$$

and thus

$$\pi_1'\pi_2'\cdots\pi_n' = (\varepsilon_1\varepsilon_2\cdots\varepsilon_n)\pi_1^n.$$

If $s = n$, then we are finished, since the left-hand side of the above equation is then α. If $s > n$, then we divide both sides of (10) by π_1^n and get

(11) $$(\varepsilon_1\varepsilon_2\cdots\varepsilon_n)\pi_{n+1}'\pi_{n+2}'\cdots\pi_s'\beta = \varepsilon\pi_2^n\pi_3^n\cdots\pi_r^n.$$

We now repeat the above process. By the unique factorization property, one of the π_j's is an associate of π'_{n+1}, and they can be renumbered so that π_2 is an associate of π'_{n+1}. No associate of π_2 divides β, since then π'_{n+1} also divides β as well as α. Thus π_2 or its associates must show up n times among the primes $\pi'_{n+1}, \ldots, \pi'_s$ and these may be renumbered so that $\pi'_{n+1}, \ldots, \pi'_{2n}$ are associates of π_2. It follows from this that $s \geq 2n$. It also follows that there are units $\varepsilon_{n+1}, \ldots, \varepsilon_{2n}$ such that

$$\pi'_{n+1} = \varepsilon_{n+1}\pi_2, \ldots, \pi'_{2n} = \varepsilon_{2n}\pi_2$$

and hence

$$\pi'_{n+1} \cdots \pi'_{2n} = (\varepsilon_{n+1} \cdots \varepsilon_{2n})\pi_2^n.$$

If $s = 2n$, then $\alpha = (\varepsilon_1\varepsilon_2 \cdots \varepsilon_{2n})(\pi_1\pi_2)^n$ and we are done; if $s > 2n$, then we divide both sides of (11) by π_2^n and repeat the process a third time. Since there are a finite number of primes in the factorization of α, the repetitions of this process must eventually come to an end; when we have finally gone through this process the last time (say on the kth repetition), we will have found $s = kn$ and we will have renumbered the π''s and the π's and found units $\varepsilon_1, \ldots, \varepsilon_{kn}$ such that

$$\alpha = \pi'_1\pi'_2 \cdots \pi'_{kn} = (\varepsilon_1\varepsilon_2 \cdots \varepsilon_{kn})(\pi_1\pi_2 \cdots \pi_k)^n.$$

This is the form required by the theorem with

$$\varepsilon' = \varepsilon_1\varepsilon_2 \cdots \varepsilon_{kn}, \qquad \delta = \pi_1\pi_2 \cdots \pi_k. \qquad \blacktriangle$$

There is one more theorem which will be useful very soon; its proof does not depend on the unique factorization property in $\mathbf{Q}(\sqrt{d})$.

Theorem 8.17. Let a and b be rational integers not both zero, and let $(a,b) = c$. If α is an integer in $\mathbf{Q}(\sqrt{d})$ such that $\alpha|a$ and $\alpha|b$, then $\alpha|c$. In particular, if a and b are relatively prime [as defined in Chapter 2, we have not made a definition of relatively prime in $\mathbf{Q}(\sqrt{d})$], then the only common divisors of a and b in $\mathbf{Q}(\sqrt{d})$ are units.

Proof. If $(a,b) = c$, then by Theorem 2.2, there are rational integers r and s such that $ar + bs = c$. By Theorem 8.8, if $\alpha|a$ and $\alpha|b$, then $\alpha|(ar + bs)$, in other words, $\alpha|c$. If a and b are relatively prime, then $c = 1$ and hence $\alpha|c$ implies that α is a unit. $\blacktriangle$

It may be useful before going on to give an example of the application of these theorems. We will use them in an attempt to find all rational integer solutions to the equation

(12) $$x^2 + 47 = y^3.$$

Some of these theorems depend on unique factorization and this is something we have not yet shown. Thus for the duration of this example, we make the assumption that the set of integers of $\mathbf{Q}(\sqrt{-47})$ is a UFD. Before going over to $\mathbf{Q}(\sqrt{-47})$, it is useful to establish that $(x,47) = 1$. Since 47 is a rational prime, either $(x,47) = 1$ or $(x,47) = 47$. In the latter case, $47|x$ and then $47|(x^2 + 47)$ or $47|y^3$ and hence $47|y$. But then $47^2|x^2$, $47^2|y^3$ (47^3 even divides y^3) and therefore $47^2|(y^3 - x^2)$ or $47^2|47$. This is false and hence $(x,47) = 1$. We consider two separate cases. In the first case we assume x is odd, and in the second case we assume x is even.

If x is odd, then y is even and hence we have

$$(x + \sqrt{-47})(x - \sqrt{-47}) = 8\left(\frac{y}{2}\right)^3$$

or

(13)
$$\left(\frac{x + \sqrt{-47}}{2}\right)\left(\frac{x - \sqrt{-47}}{2}\right) = 2\left(\frac{y}{2}\right)^3.$$

Since x is odd, y is even, and $-47 \equiv 1 \pmod 4$, it follows that $(x + \sqrt{-47})/2$, $(x - \sqrt{-47})/2$, and $y/2$ are all integers in $\mathbf{Q}(\sqrt{-47})$. We pause to show that 2 is a prime in $\mathbf{Q}(\sqrt{-47})$; it will follow easily from this that equation (13) is impossible if x is odd. If 2 is not a prime, then $2 = \alpha\beta$ with neither α nor β being a unit. Thus

$$\mathbf{N}(\alpha)\mathbf{N}(\beta) = \mathbf{N}(2) = 4,$$

and since norms are positive in complex quadratic fields and since $\mathbf{N}(\alpha) = 1$ or $\mathbf{N}(\beta) = 1$ implies α or β is a unit, we must have

$$\mathbf{N}(\alpha) = \mathbf{N}(\beta) = 2.$$

If $\alpha = (a + b\sqrt{-47})/2$ with a and b both odd or both even numbers in $\mathbf{Z}$, then

$$2 = \mathbf{N}(\alpha) = \frac{a^2 + 47b^2}{4}$$

or

$$a^2 + 47b^2 = 8.$$

This is impossible, since if $b \neq 0$, then the left side is too big and, if $b = 0$, then $a = \pm\sqrt{8}$ is not a rational integer. Thus there are no integers in $\mathbf{Q}(\sqrt{-47})$ of norm 2 and hence 2 is a prime in $\mathbf{Q}(\sqrt{-47})$. It follows from equation (13) that

$$2 \left| \left(\frac{x + \sqrt{-47}}{2} \right) \left(\frac{x - \sqrt{-47}}{2} \right) \right.$$

and hence, by Theorem 8.15, either

$$2 \left| \left(\frac{x + \sqrt{-47}}{2} \right) \right. \qquad \text{or} \qquad 2 \left| \left(\frac{x - \sqrt{-47}}{2} \right) \right..$$

Thus either $(x + \sqrt{-47})/4$ is an integer or $(x - \sqrt{-47})/4$ is an integer. But this is patently false, since these numbers are of the form

$$\frac{\dfrac{x}{2} + \dfrac{1}{2}\sqrt{-47}}{2}, \qquad \frac{\dfrac{x}{2} - \dfrac{1}{2}\sqrt{-47}}{2}$$

and $x/2$ and $\pm\frac{1}{2}$ are not even both integers, let alone both being odd or even. Hence there are no solutions to (12) with x odd.

In the second case, we assume that x is even and it follows that y is odd. Our equation is then

$$(14) \qquad\qquad (x + \sqrt{-47})(x - \sqrt{-47}) = y^3.$$

We pause to show that $(x + \sqrt{-47})$ and $(x - \sqrt{-47})$ have no common factors in $\mathbf{Q}(\sqrt{-47})$ other than units. We will then apply Theorem 8.16. If $\alpha|(x + \sqrt{-47})$ and $\alpha|(x - \sqrt{-47})$, then

$$\alpha|[(x + \sqrt{-47}) + (x - \sqrt{-47})], \qquad \alpha|(x + \sqrt{-47})(x - \sqrt{-47}).$$

In other words,

$$\alpha|2x, \qquad \alpha|(x^2 + 47).$$

Since $x^2 + 47$ is odd, any rational divisor of $(x^2 + 47)$ is odd. Thus $d = (2x, x^2 + 47)$ is odd. Thus, since $d|2x$ and $(d,2) = 1$, $d|x$. But $d|(x^2 + 47)$ and hence $d|[(x^2 + 47) - x^2]$, or $d|47$. Since $d|x$ and $d|47$, and since we have seen that $(x,47) = 1$, we see that $d = 1$; that is,

$$(2x, x^2 + 47) = 1.$$

Therefore, by Theorem 8.17, if $\alpha|(x + \sqrt{-47})$ and $\alpha|(x - \sqrt{-47})$, then α is a unit. It follows from this, equation (14), and Theorem 8.16 that there is a unit ε and integer α such that

$$x + \sqrt{-47} = \varepsilon\alpha^3.$$

The units of $\mathbf{Q}(\sqrt{-47})$ are ± 1; if $\varepsilon = 1$, then

$$x + \sqrt{-47} = \alpha^3,$$

and if $\varepsilon = -1$, then

$$x + \sqrt{-47} = (-\alpha)^3.$$

Hence in this instance, $x + \sqrt{-47}$ actually is a cube and we may set

$$x + \sqrt{-47} = \left(\frac{a + b\sqrt{-47}}{2}\right)^3,$$

where a and b are rational integers, both odd or both even. (It frequently happens that one can get rid of the unit that arises from Theorem 8.16, but it also often happens that one cannot get rid of it, particularly when $d > 0$.) If we multiply out the right-hand side of the above and collect terms, we see that

$$8x + 8\sqrt{-47} = (a^3 - 141ab^2) + (3a^2b - 47b^3)\sqrt{-47}$$

and thus, by Theorem 8.1,

$$8x = a^3 - 141ab^2,$$

$$8 = 3a^2b - 47b^3 = b(3a^2 - 47b^2).$$

The second of these equations is momentarily the most important. If a and b are both odd, then $(b,8) = 1$ and thus $8|(3a^2 - 47b^2)$. But this contradicts the fact that for a and b both odd,

$$3a^2 - 47b^2 \equiv 3 \cdot 1 - 47 \cdot 1 \equiv 4 \pmod 8.$$

Hence a and b are both even, say $a = 2r$ and $b = 2s$, and thus

$$x + \sqrt{-47} = (r + s\sqrt{-47})^3,$$

$$x = r^3 - 141rs^2,$$

$$1 = s(3r^2 - 47s^2).$$

Hence $s|1$ and thus $s = \pm 1$. If $s = -1$, then $1 = -(3r^2 - 47)$ and $3r^2 = 46$, which is impossible since $3 \nmid 46$. Thus $s = 1$ and then $1 = 3r^2 - 47$, $r^2 = 16$, $r = \pm 4$. As a result,

$$x = r(r^2 - 141s) = \pm 4(-125) = \pm 500.$$

We could find y from this directly by means of equation (12), but there is a less computational way. The reason that we never used Theorem 8.16 to set $(x - \sqrt{-47}) = \varepsilon_1\beta^3$ is that when we know what happens to $x + \sqrt{-47}$, we then know what happens to $x - \sqrt{-47}$ by conjugation. In this case, $x + \sqrt{-47} = (r + s\sqrt{-47})^3$ and thus

$$x - \sqrt{-47} = \overline{(x + \sqrt{-47})} = \overline{(r + s\sqrt{-47})^3} = \left(\overline{r + s\sqrt{-47}}\right)^3$$

$$= (r - s\sqrt{-47})^3.$$

As a result,

$$y^3 = (x + \sqrt{-47})(x - \sqrt{-47}) = [(r + s\sqrt{-47})(r - s\sqrt{-47})]^3$$
$$= (r^2 + 47s^2)^3,$$

and since a real number has only one real cube root,

$$y = r^2 + 47s^2 = 63.$$

Thus we have found that the only solutions to equation (12) are $x = 500$, $y = 63$ and $x = -500$, $y = 63$. This is a difficult example, but it would have been far more difficult to find these solutions without using $\mathbf{Q}(\sqrt{-47})$. There is a fly in the ointment, however. We seem to have missed the solutions $x = \pm 13$ and $y = 6$. The first reaction must be that there is an error in the proof, but this is not true. Well, then, there must be somewhere that we made an unwarranted assumption. There was an assumption made, namely that $\mathbf{Q}(\sqrt{-47})$ has the unique factorization property. But surely it could not be this! Or could it? Let us see. The solution $x = 13$, $y = 6$ comes under the heading of case 1 (x is odd). If we put this solution into equation (13), then we see that

$$\left(\frac{13 + \sqrt{-47}}{2}\right)\left(\frac{13 - \sqrt{-47}}{2}\right) = 2 \cdot 3^3,$$

which is true, and indeed this is an equation involving integers, as was claimed. By the definition of divisibility,

$$2 \left| \left(\frac{3 + \sqrt{-47}}{2}\right)\left(\frac{13 - \sqrt{-47}}{2}\right)\right..$$

We have shown that 2 is a prime in $\mathbf{Q}(\sqrt{-47})$. But still, $(13 + \sqrt{-47})/4$ and $(13 - \sqrt{-47})/4$ are not integers and hence 2 does not divide either $(13 + \sqrt{-47})/2$ or $(13 - \sqrt{-47})/2$. Thus we have found an actual instance in which a prime divides a product of two integers although it divides neither integer separately! As shown in Theorem 8.15, this is a property that is equivalent to unique factorization, and hence $\mathbf{Q}(\sqrt{-47})$ does not have the unique factorization property!

Perhaps this disaster is due to a poor definition of integer. For instance, perhaps one of the numbers $(13 \pm \sqrt{-47})/4$ ought to be an integer. Unfortunately, this leads to too many difficulties to be of any value. Another, more subtle, way out of the troubles arising above would be that since 2 does not divide either of the numbers $(13 \pm \sqrt{-47})/2$, perhaps 2 should not be a prime. This is more in the spirit of what was actually done to restore unique factorization to $\mathbf{Q}(\sqrt{-47})$. We shall say a little more about this in Section 8.6.

Now that we have seen that $\mathbf{Q}(\sqrt{-47})$ does not have the unique factorization property, it may well be asked if any $\mathbf{Q}(\sqrt{d})$ has the unique factorization property. The reader, having just seen the "obvious" totally destroyed, cannot be blamed for being pessimistic, but fortunately there are quite a few values of d such that $\mathbf{Q}(\sqrt{d})$ has the unique factorization property. We proved the unique factorization theorem for the positive rational integers by means of the Euclidean algorithm. A generalization of this algorithm enables one to do the same thing for several different quadratic fields [we will see shortly that $\mathbf{Q}(\sqrt{-1})$ is such a field]. In honor of Euclid, the fields for which this generalization works are called Euclidean fields.

Definition. A quadratic field, $\mathbf{Q}(\sqrt{d})$, is called a **Euclidean field** if it has the following property: Given integers α and β in $\mathbf{Q}(\sqrt{d})$ with $\beta \neq 0$, there are integers γ and δ in $\mathbf{Q}(\sqrt{d})$ such that

$$\alpha = \gamma\beta + \delta, \qquad |\mathbf{N}(\delta)| < |\mathbf{N}(\beta)|.$$

(If $d < 0$, then the absolute–value signs can be dispensed with, since norms are then nonnegative.)

In Euclidean fields we may develop an algorithm for finding the "greatest" common divisor; this in turn can be used to prove the unique factorization property. The proofs are all similar to those of Chapter 2.

Theorem 8.18. If $\mathbf{Q}(\sqrt{d})$ is a Euclidean field and α and β are integers in $\mathbf{Q}(\sqrt{d})$, not both zero, then there is an integer δ in $\mathbf{Q}(\sqrt{d})$ such that
1. $\delta|\alpha$ and $\delta|\beta$.
2. If $\gamma|\alpha$ and $\gamma|\beta$, then $\gamma|\delta$.
An integer δ' has the above two properties if and only if it is an associate of δ. Further, if an integer δ has the properties 1 and 2, then it is a linear

combination of α and β, in other words, there exist integers ξ and η in $\mathbf{Q}(\sqrt{d})$ such that

3. $\delta = \alpha\xi + \beta\eta$.

Proof. Since α and β are not both zero, we may assume that $\beta \neq 0$. By the definition of a Euclidean field, there are integers γ_1 and β_1 such that

$$\alpha = \gamma_1\beta + \beta_1, \qquad |N(\beta)| > |N(\beta_1)|.$$

If β_1 is zero, we stop here. If $\beta_1 \neq 0$, then there are integers γ_2 and β_2 such that

$$\beta = \gamma_2\beta_1 + \beta_2, \qquad |N(\beta_1)| > |N(\beta_2)|.$$

If β_2 is zero, we stop here. Otherwise there are integers γ_3 and β_3 such that

$$\beta_1 = \gamma_3\beta_2 + \beta_3, \qquad |N(\beta_2)| > |N(\beta_3)|.$$

We continue this process as long as we do not get some $\beta_n = 0$. In this way, we get a sequence of integers $\beta, \beta_1, \beta_2, \ldots$ such that $|N(\beta)| > |N(\beta_1)| > |N(\beta_2)| > \cdots$ and the numbers $|N(\beta_j)|$ are rational integers which are greater than or equal to 0. A sequence of decreasing nonnegative rational integers must be a finite sequence. As a result, there is a last β_j in the sequence, say β_n. If $\beta_n \neq 0$, then there would be a β_{n+1} and hence $\beta_n = 0$. Thus we have a sequence of equations

$$\alpha = \gamma_1\beta + \beta_1$$
$$\beta = \gamma_2\beta_1 + \beta_2$$
$$\beta_1 = \gamma_3\beta_2 + \beta_3$$
$$\vdots$$
$$\beta_{n-3} = \gamma_{n-1}\beta_{n-2} + \beta_{n-1}$$
$$\beta_{n-2} = \gamma_n\beta_{n-1} + 0.$$

The number β_{n-1} will be the δ in the theorem. By the last equation, $\beta_{n-1}|\beta_{n-2}$; by the equation before it, $\beta_{n-1}|\beta_{n-3}$; by the previous equation, $\beta_{n-1}|\beta_{n-4}$; and so on. We finally get up to the second of these equations with $\beta_{n-1}|\beta_2$ and $\beta_{n-1}|\beta_1$ and hence $\beta_{n-1}|\beta$, and then by the first equation $\beta_{n-1}|\alpha$. This is property 1 of the theorem. Again, $\beta_{n-1} = \beta_{n-3} - \gamma_{n-1}\beta_{n-2}$ and thus

$$\beta_{n-1} = \beta_{n-3} - \gamma_{n-1}(\beta_{n-4} - \gamma_{n-2}\beta_{n-3})$$
$$= -\gamma_{n-1}\beta_{n-4} + (1 + \gamma_{n-1}\gamma_{n-2})\beta_{n-3}$$

is a linear combination of β_{n-4} and β_{n-3}. We proceed up the ladder until we finally reach the top, at which point we get

$$\beta_{n-1} = \alpha\zeta + \beta\eta$$

for some integers ζ and η. Hence any divisor of α and β also divides β_{n-1}, and this is property 2 of the theorem. It is clear that any associate of β_{n-1} has properties 1 and 2. Further, if δ' also has these properties, then by using property 2 for both β_{n-1} and δ', $\delta'|\beta_{n-1}$ and $\beta_{n-1}|\delta'$. Hence by Theorem 8.13, δ' and β_{n-1} are associates. If ε is a unit, then

$$\varepsilon\beta_{n-1} = \alpha(\varepsilon\zeta) + \beta(\varepsilon\eta),$$

and hence any associate of β_{n-1} can be written in the form of property 3 of the theorem. By what we have just shown, this means that any δ with properties 1 and 2 also has property 3.   ▲

Theorem 8.19. A Euclidean quadratic field has the unique factorization property.

Proof. Suppose $\mathbf{Q}(\sqrt{d})$ is Euclidean and let $\pi|\alpha\beta$, where π is a prime and α and β are integers in $\mathbf{Q}(\sqrt{d})$. We will show that either $\pi|\alpha$ or $\pi|\beta$ and then use Theorem 8.15. Suppose that $\pi\nmid\alpha$. Then no associate of π divides α. Also any divisor of π is either an associate of π or a unit. These two facts together show that a common divisor of α and π must be a unit. Since any unit divides 1, we see that the number 1 has the first two properties of the number δ of Theorem 8.18, namely $1|\alpha$, $1|\pi$, and if $\gamma|\alpha$ and $\gamma|\pi$ (and hence γ is a unit), then $\gamma|1$. Therefore, by Theorem 8.18, there exist integers ζ and η such that

$$\alpha\zeta + \pi\eta = 1.$$

Therefore,

$$(\alpha\beta)\zeta + \pi\beta\eta = \beta$$

and since $\pi|\alpha\beta$, $\pi|(\alpha\beta\zeta + \pi\beta\eta)$. In other words, $\pi|\beta$. Thus either $\pi|\alpha$ or $\pi|\beta$ and therefore, by Theorem 8.15, $\mathbf{Q}(\sqrt{d})$ has the unique factorization property.   ▲

Now is a good time to give some examples of Euclidean fields.

Theorem 8.20. If $d = -11, -7, -3, -2, -1, 2, 3, 5$, then $\mathbf{Q}(\sqrt{d})$ is Euclidean.

Proof. We will split the theorem up into two parts: in the first part, we take those $d \not\equiv 1 \pmod 4$, in the second part, we take those $d \equiv 1 \pmod 4$. We now assume that d is either -2, -1, 2, or 3. Let α and β be integers in $\mathbf{Q}(\sqrt{d})$ such that $\beta \neq 0$. Let $\alpha/\beta = x + y\sqrt{d}$, where x and y are rational but not necessarily integers. Since any rational number is between two consecutive integers and within $\frac{1}{2}$ of the nearest integer, there are rational integers r and s such that $|x - r| \leq \frac{1}{2}$, $|y - s| \leq \frac{1}{2}$. Let

$$\gamma = r + s\sqrt{d}, \qquad \delta = \beta[(x - r) + (y - s)\sqrt{d}]$$

so that

$$\alpha = \beta(x + y\sqrt{d}) = \beta\gamma + \delta.$$

Since r and s are rational integers, γ is an integer and, since $\delta = \alpha - \beta\gamma$, δ is an integer. Also

$$|\mathbf{N}(\delta)| = |\mathbf{N}(\beta)| \cdot |\mathbf{N}[(x - r) + (y - s)\sqrt{d}]| = |\mathbf{N}(\beta)| \cdot |[(x - r)^2 - d(y - s^2)]|$$

and here we have

$$|(x - r)^2 - d(y - s)^2| \leq |x - r|^2 + |-d| |y - s|^2 \leq (\tfrac{1}{2})^2 + 3 \cdot (\tfrac{1}{2})^2 = 1.$$

The only time there could possibly be equality is when $|x - r| = |y - s| = \frac{1}{2}$ and $d = 3$ and then $|(x - r)^2 - d(y - s)^2| = |\tfrac{1}{4} - 3 \cdot \tfrac{1}{4}| = \tfrac{1}{2} < 1$. Therefore, we always have

$$|(x - r)^2 - d(y - s)^2| < 1.$$

Thus

$$|\mathbf{N}(\delta)| = |\mathbf{N}(\beta)| \cdot |(x - r)^2 - d(y - s)^2| < |\mathbf{N}(\beta)| \cdot 1 = \mathbf{N}(\beta).$$

Hence $\mathbf{Q}(\sqrt{d})$ is Euclidean.

Now suppose that d is either -11, -7, -3, or 5. Let α and β be integers in $\mathbf{Q}(\sqrt{d})$ with $\beta \neq 0$. Put $\alpha/\beta = x + y\sqrt{d}$, where x and y are rational. The number $2y$ is between consecutive rational integers and the nearest integer to $2y$ is within $\frac{1}{2}$ of $2y$; that is, there is a rational integer s such that $|2y - s| \leq \frac{1}{2}$ and thus

$$\left| y - \frac{s}{2} \right| \leq \frac{1}{4}.$$

Similarly, there is a rational integer r within $\frac{1}{2}$ of $x - (s/2)$ and thus

$$\left| \left(x - \frac{s}{2} \right) - r \right| \leq \frac{1}{2}.$$

Let $\gamma = r + s[(1 + \sqrt{d})/2]$, which is an integer by the corollary to Theorem 8.5, and let $\delta = \beta\{[x - r - (s/2)] + [y - (s/2)]\sqrt{d}\}$. Then

$$\alpha = \beta(x + y\sqrt{d}) = \beta\gamma + \delta,$$

so that, in particular, $\delta = \alpha - \beta\gamma$ is also an integer. Also

$$|N(\delta)| = |N(\beta)| \cdot |[x - r - (s/2)]^2 - d[y - (s/2)]^2|$$

and here

$$\left|\left(x - r - \frac{s}{2}\right)^2 - d\left(y - \frac{s}{2}\right)^2\right| \leq \left|x - r - \frac{s}{2}\right|^2 + |d| \cdot \left|y - \frac{s}{2}\right|^2$$

$$\leq \left(\frac{1}{2}\right)^2 + 11\left(\frac{1}{4}\right)^2 < 1.$$

Thus

$$|N(\delta)| < |N(\beta)|$$

and $Q(\sqrt{d})$ is Euclidean.   ▲

There are other Euclidean fields. Surprisingly enough, the problem of determining all the Euclidean fields is quite difficult, and it is only in recent times that they have been completely determined. We present the result here for the reader's edification; the proof is beyond the scope of this text.

Theorem 8.21. $Q(\sqrt{d})$ is Euclidean if and only if d is one of the twenty-one integers $-11, -7, -3, -2, -1, 2, 3, 5, 6, 7, 11, 13, 17, 19, 21, 29, 33, 37, 41, 57,$ and 73.

It is relatively easy to show that a particular field is or is not Euclidean [although it was once mistakenly thought that $Q(\sqrt{97})$ was Euclidean]. The great difficulty in this theorem is to provide bounds for d outside of which no field can possibly be Euclidean. When $d < 0$, the problem is actually quite easy; we have shown that the five fields listed are Euclidean in Theorem 8.20. The fact that there are no other Euclidean fields with $d < 0$ is the subject of Problem 8 at the end of the section. When $d > 0$, things are much harder; the difficult part was settled in 1950 by H. Davenport, who showed that if $d > 16384 (= 2^{14})$, then $Q(\sqrt{d})$ is not Euclidean. After this, there remained only finitely many (several thousand!) individual fields to be checked out. This was done in the same year; Theorem 8.21 was announced in 1950 by Chatland and Davenport.

Amazingly enough, the twenty-one Euclidean fields of Theorem 8.21 are not the only quadratic fields with the unique factorization property. We cannot list them all here because they are not yet all known; in fact, it is not even known if there are infinitely many such fields. However, in the case of the complex quadratic fields, the problem has just been solved. We state the result without proof.

Theorem 8.22. If $d < 0$, then $\mathbf{Q}(\sqrt{d})$ has the unique factorization property if and only if d is one of the nine numbers -1, -2, -3, -7, -11, -19, -43, -67, and -163.

This theorem has a long and honorable history. It can be found stated in other terminology as a special case of a conjecture made by Gauss in his famous work, *Disquisitiones Arithmeticae* (written in Latin but now available in English translation). As late as 1934, it was not known whether or not there were infinitely many complex quadratic fields with the unique factorization property, although it was shown in 1933 that the only such fields with d in the range

$$-1 \geq d \geq -5 \cdot 10^9$$

were the nine given in Theorem 8.22. In 1934 Heilbronn and Linfoot proved the remarkable theorem that there are at most ten such fields. Thus the situation was that nine complex quadratic fields with UFD's were known and there may have been a tenth but definitely not an eleventh. By 1966, the bounds had been improved to where it was known that there are only nine such fields in the range

(15) $-1 \geq d \geq -10^{9000000}$

(still only a finite range; there are infinitely many d's outside it). Finally, at the end of 1966, the problem was settled; in fact, it was essentially settled by two different people using two different methods at the same time.

The whole problem is essentially to find a negative number, d_0, for which one could prove that if $d < d_0$, then $\mathbf{Q}(\sqrt{d})$ does not have the unique factorization property. If d_0 turns out to be in the range (15), then Theorem 8.22 would be completely proved. If d_0 turns out to be outside the range (15), then it would still be necessary to check the finitely many d's between d_0 and $-10^{9000000}$. H. M. Stark did this and found the value $d_0 = -200$, thereby proving Theorem 8.22. At the same time, A. Baker found a completely different method for calculating a number d_0, but thus far no one has actually calculated Baker's value of d_0.[3] The papers of Heilbronn and

---

[3] It now looks as though Baker's method will lead to a value of $d_0 \approx -10^{500}$, which is safely in (15).

Linfoot, Stark, and Baker all made use of the theory of functions of a complex variable whereby knowledge about integers is gotten from such continuous processes as differentiation and integration.

The problem of which real quadratic fields have the unique factorization property is still wide open. Although most number theorists conjecture that there are infinitely many such fields, even this has not yet been established. We content ourselves here with a partial result given without proof.

Theorem 8.23. There are exactly 38 real quadratic fields, $\mathbf{Q}(\sqrt{d})$, having the unique factorization property with d in the range $2 \leq d < 100$. These are given by

$$d = 2, 3, 5, 6, 7, 11, 13, 14, 17, 19, 21, 22, 23, 29, 31, 33, 37,$$
$$38, 41, 43, 46, 47, 53, 57, 59, 61, 62, 67, 69, 71, 73, 77,$$
$$83, 86, 89, 93, 94, 97.$$

Incidentally, there are 60 real quadratic fields with $2 \leq d < 100$.

In spite of the great difficulty of the proof of Theorem 8.22, the part having to do with $d \not\equiv 1 \pmod 4$ is quite easy and will be given here. In fact, the same method of proof can be made to provide further motivation for our definition of an integer in $\mathbf{Q}(\sqrt{d})$ when $d \equiv 1 \pmod 4$. A more natural definition of an integer would seemingly be that it should be a number of the form $a + b\sqrt{d}$, where a and b are in $\mathbf{Z}$. This is indeed what the integers of $\mathbf{Q}(\sqrt{d})$ are when $d \not\equiv 1 \pmod 4$; why not simplify things by letting this be the definition always? The answer is that under this definition, there would *never* be unique factorization when $d \equiv 1 \pmod 4$! Under our definition, we at least have unique factorization sometimes (and possibly for infinitely many fields). Before proving all this, it is convenient to give a definition.

Definition. Let $\mathbf{Z}[\sqrt{d}]$ denote the set of numbers of the form $a + b\sqrt{d}$, where a and b are rational integers.

The numbers of $\mathbf{Z}[\sqrt{d}]$ have the property that sums, differences, and products of such numbers are again in $\mathbf{Z}[\sqrt{d}]$. If α and β are in $\mathbf{Z}[\sqrt{d}]$ and $\alpha \neq 0$, then we say $\alpha \mid \beta$ in $\mathbf{Z}[\sqrt{d}]$ if β/α is in $\mathbf{Z}[\sqrt{d}]$. For example, $2 \mid (1 + \sqrt{5})$ in $\mathbf{Q}(\sqrt{5})$ but not in $\mathbf{Z}[\sqrt{5}]$. The norm of a number of $\mathbf{Z}[\sqrt{d}]$ is a rational integer. We may define a unit in $\mathbf{Z}[\sqrt{d}]$ to be a number in $\mathbf{Z}[\sqrt{d}]$ which divides 1 in $\mathbf{Z}[\sqrt{d}]$. Here also, a number of $\mathbf{Z}[\sqrt{d}]$ is a unit if and only if its norm is ± 1. We make definitions similar to those made earlier for primes and associates in $\mathbf{Z}[\sqrt{d}]$. Again, there are infinitely many units in $\mathbf{Z}[\sqrt{d}]$ if $d > 0$ and again any nonzero, nonunit number of $\mathbf{Z}[\sqrt{d}]$ may be factored into a finite product of primes in $\mathbf{Z}[\sqrt{d}]$. The proofs of all these facts are exactly the same as given before; in fact, the earlier proofs were given for $\mathbf{Z}[\sqrt{d}]$ when $d \not\equiv 1 \pmod 4$. We define $\mathbf{Z}[\sqrt{d}]$ to be a unique factorization domain (UFD) if factorization into primes is unique up to order and associates. Theorem 8.15 and its proof are valid for $\mathbf{Z}[\sqrt{d}]$; Theorem 8.15 will be crucial in what follows. We first state a preliminary theorem which will yield our desired results.

Theorem 8.24. If $\mathbf{Z}[\sqrt{d}]$ is a UFD, then 2 is not a prime in $\mathbf{Z}[\sqrt{d}]$.

Proof. Either d or $(d-1)$ is even and hence $2|d(d-1)$. Since

$$(d + \sqrt{d})(d - \sqrt{d}) = d^2 - d = d(d-1),$$

we see that

$$2|(d + \sqrt{d})(d - \sqrt{d}).$$

But in $\mathbf{Z}[\sqrt{d}]$, $2 \nmid (d + \sqrt{d})$ and $2 \nmid (d - \sqrt{d})$, since neither $(d/2) + \frac{1}{2}\sqrt{d}$ nor $(d/2) - \frac{1}{2}\sqrt{d}$ is in $\mathbf{Z}[\sqrt{d}]$. Thus 2 divides a product of two numbers in $\mathbf{Z}[\sqrt{d}]$, although 2 divides neither of the numbers individually. By Theorem 8.15, if $\mathbf{Z}[\sqrt{d}]$ is a UFD, then 2 is not a prime. ▲

Theorem 8.25. If $d < 0$, then $\mathbf{Z}[\sqrt{d}]$ is a UFD if and only if $d = -1$ or $d = -2$. [Therefore, if $d < 0$ and $d \not\equiv 1 \pmod 4$, then $\mathbf{Q}(\sqrt{d})$ has the unique factorization property if and only if $d = -1$ or -2.] If $d \equiv 1 \pmod 4$, then $\mathbf{Z}[\sqrt{d}]$ is never a UFD.

Proof. We will show that if $d \leq -3$ or if $d \equiv 1 \pmod 4$, then 2 is a prime in $\mathbf{Z}[\sqrt{d}]$, and then we will use Theorem 8.24. We have already shown that $\mathbf{Z}[\sqrt{-1}]$ and $\mathbf{Z}[\sqrt{-2}]$ are UFD's. So when considering $d < 0$, we may restrict our attention to those $d \leq -3$. Suppose that 2 is not a prime in $\mathbf{Z}[\sqrt{d}]$. Then there are numbers α and β in $\mathbf{Z}[\sqrt{d}]$ such that

$$2 = \alpha\beta, \qquad |N(\alpha)| > 1, \qquad |N(\beta)| > 1.$$

Therefore, $N(\alpha)N(\beta) = 4$ and since $|N(\alpha)|$ and $|N(\beta)|$ are rational integers greater than one,

$$|N(\alpha)| = |N(\beta)| = 2.$$

Thus if 2 is not a prime in $\mathbf{Z}[\sqrt{d}]$, then there is a number

$$\alpha = a + b\sqrt{d} \ (a \text{ and } b \text{ in } \mathbf{Z}),$$

such that

(16) $$\mathbf{N}(\alpha) = a^2 - db^2 = \pm 2.$$

If $d \leq -3$ and $b \neq 0$, then

$$a^2 - db^2 = a^2 + (-d)b^2 \geq 0 + 3 \cdot 1 > \pm 2,$$

while if $b = 0$, then

$$a^2 - db^2 = a^2 \neq \pm 2$$

when a is in $\mathbf{Z}$. Thus when $d \leq -3$, there are no numbers in $\mathbf{Z}[\sqrt{d}]$ of norm ± 2, and hence 2 is a prime in $\mathbf{Z}[\sqrt{d}]$. Therefore, by Theorem 8.24, if $d \leq -3$, then $\mathbf{Z}[\sqrt{d}]$ is not a UFD.

Now suppose that $d \equiv 1 \pmod 4$. Modulo 4, equation (16) reduces to

$$a^2 - b^2 \equiv a^2 - db^2 \equiv \pm 2 \equiv 2 \pmod 4.$$

But the squares (mod 4) are 0 and 1, and therefore $a^2 - b^2$ is congruent to either -1, 0, or 1 (mod 4). Thus the congruence

$$a^2 - b^2 \equiv 2 \pmod 4$$

has no solutions in rational integers and therefore equation (16) is impossible. Hence 2 is a prime in $\mathbf{Z}[\sqrt{d}]$ and, by Theorem 8.24, $\mathbf{Z}[\sqrt{d}]$ is not a UFD.   ▲

EXERCISES
 1. Use the factorizations $6 = 2 \cdot 3 = (1 + \sqrt{-5})(1 - \sqrt{-5})$ to show directly that $\mathbf{Q}(\sqrt{-5})$ does not have the unique factorization property.
 2. Use congruences (mod 5) to show that there is no integer in $\mathbf{Q}(\sqrt{10})$ of norm ± 2. Show as a result that 2 is a prime in $\mathbf{Q}(\sqrt{10})$ and then show that $\mathbf{Q}(\sqrt{10})$ does not have the unique factorization property.
 3. Show directly that $\mathbf{Q}(\sqrt{10})$ does not have the unique factorization property by considering the factorizations

$$6 = 2 \cdot 3 = (4 + \sqrt{10})(4 - \sqrt{10}).$$

 4. Do the factorizations $14 = 2 \cdot 7 = (217 + 56\sqrt{15})(62 - 16\sqrt{15})$ show that $\mathbf{Q}(\sqrt{15})$ does not have the unique factorization property?

5. Show that 21 has two essentially different factorizations into primes in $\mathbf{Q}(\sqrt{-5})$.
6. Prove that $\mathbf{Q}(\sqrt{13})$ is Euclidean.
7. Find the "greatest common divisor" of $-25 + 47\sqrt{-1}$ and $34 + 32\sqrt{-1}$ by means of the generalized Euclidean algorithm used in the proof of Theorem 8.18.
8. Use $\alpha = 1 + \sqrt{d}$ $[d \le -5, d \not\equiv 1(\text{mod } 4)]$, $\alpha = (1 + \sqrt{d})/2$ $[d \le -15, d \equiv 1(\text{mod } 4)]$, $\beta = 2$ to show that $\mathbf{Q}(\sqrt{d})$ is not Euclidean.
9. Suppose $5|d$. Show that 2 is a prime in $\mathbf{Q}(\sqrt{d})$. Conclude that $\mathbf{Q}(\sqrt{d})$ does not have the unique factorization property if $d \not\equiv 1(\text{mod } 4)$ and $5|d$.
10. In the proof of Theorem 8.24, we had the equality

$$d(d - 1) = (d - \sqrt{d})(d + \sqrt{d}).$$

Why do not these two different factorizations of $d^2 - d$ show that $\mathbf{Z}[\sqrt{d}]$ is not a UFD?
11. We have seen that

$$\left(\frac{13 + \sqrt{-47}}{2}\right)\left(\frac{13 - \sqrt{-47}}{2}\right) = 2 \cdot 3^3 = 54,$$

where 2 is a prime in $\mathbf{Q}(\sqrt{-47})$. Show that $2, 3, (13 + \sqrt{-47})/2$, and $(13 - \sqrt{-47})/2$ are all primes in $\mathbf{Q}(\sqrt{-47})$, no two of which are associates. Thus we have presented two different factorizations of 54 into primes in $\mathbf{Q}(\sqrt{-47})$.

8.5. Applications of Quadratic Fields to Diophantine Equations

We have already seen one example of such an application in Section 8.4 when we examined the equation

$$x^2 + 47 = y^3.$$

That example serves as a model of the procedure to be followed when we have unique factorization; it also shows that one can sometimes find solutions by assuming unique factorization even when we do not actually have it. In this section, we give two more examples using $\mathbf{Q}(\sqrt{-1})$. Here we have unique factorization in $\mathbf{Z}[i]$, where we write $\sqrt{-1} = i$.

Our first example will be to give and justify the argument sketched in Section 8.1 for finding primitive Pythagorean triples. Let

$$x^2 + y^2 = z^2, \qquad x > 0, \qquad y > 0, \qquad z > 0, \qquad (x,y,z) = 1.$$

The condition $(x,y,z) = 1$ may be replaced by the condition $(x,y) = 1$; this is because if $d|x$ and $d|y$, then $d|z$. As was shown in Section 5.3, one of x and y is even, the other odd; also z is odd. The above equation may be written

(17) $$(x + yi)(x - yi) = z^2.$$

Our first task is to show that $(x + yi)$ and $(x - yi)$ have no common factors in $\mathbf{Z}[i]$ other than units. Suppose

$$\alpha|(x + yi), \qquad \alpha|(x - yi).$$

Then

$$\alpha|[(x + yi) + (x - iy)] \qquad \text{or} \qquad \alpha|2x$$

and

$$\alpha|[-i(x + iy) + i(x - iy)] \qquad \text{or} \qquad \alpha|2y.$$

Therefore by Theorem 8.17, $\alpha|(2x,2y)$ and thus $\alpha|2$, since $(2x,2y) = 2(x,y) = 2$. Since $2 = -i(1 + i)^2$, where $1 + i$ is a prime by Theorem 8.12, and since $\mathbf{Z}[i]$ is a UFD, the divisors of 2 fall into three groups: 2 and its associates, $1 + i$ and its associates, and units. Thus it suffices to show that neither 2 nor $(1 + i)$ divide both $(x + iy)$ and $(x - iy)$. In fact, since $(1 + i)|2$, it is sufficient to show that $(1 + i)$ does not divide both $x + iy$ and $x - iy$. If $(1 + i)$ divides both $(x + iy)$ and $(x - iy)$, then $(1 + i)|(x + iy)(x - iy)$ or $(1 + i)|z^2$. But

$$\frac{z^2}{1 + i} = \frac{z^2}{2} - \frac{z^2}{2}i,$$

which is not in $\mathbf{Z}[i]$ since z^2 is odd. Hence the only divisors of both $x + iy$ and $x - iy$ in $\mathbf{Z}[i]$ are units.

We can now apply Theorem 8.16 to equation (17). The result is that there is a unit ε and an integer α such that

$$x + iy = \varepsilon\alpha^2.$$

The units of $\mathbf{Z}[i]$ are ± 1 and $\pm i$. Further, since

$$(-1)\alpha^2 = (1)(i\alpha)^2, \qquad (-i)\alpha^2 = (i)(i\alpha)^2,$$

we may restrict ourselves to two cases:
Case 1. $x + iy = \alpha^2$.
Case 2. $x + iy = i\alpha^2$.
We take case 1 first. Put $\alpha = u + iv$ and then

$$x + iy = (u + iv)^2 = (u^2 - v^2) + (2uv)i.$$

Therefore,

$$x = u^2 - v^2, \qquad y = 2uv.$$

In this case $x - iy = \overline{(x + iy)} = \overline{(\alpha^2)} = (\bar{\alpha})^2$, and so

$$z^2 = \alpha^2(\bar{\alpha})^2 = (\alpha\bar{\alpha})^2 = (u^2 + v^2)^2,$$

whence

$$z = u^2 + v^2,$$

the positive root being taken since $z > 0$.

In case 2, we write $\alpha = u - vi$. [This is in the general form (integer of $\mathbf{Z}$) + i(integer of $\mathbf{Z}$); it was written this way just to help the final answer look more familiar.] Then

$$x + iy = i(u - vi)^2 = (2uv) + (u^2 - v^2)i,$$

and, as a result,

$$x = 2uv, \qquad y = u^2 - v^2.$$

Here again,

$$z = u^2 + v^2.$$

Thus case 1 corresponds to x being odd and y even, while case 2 corresponds to x being even and y odd. The remaining restrictions $x > 0$, $y > 0$, and $(x,y) = 1$ lead to $u^2 > v^2$, u and v have the same sign, and $(u,v) = 1$, u and v not both odd. Since $\alpha^2 = (-\alpha)^2$, we could have even assumed that $u > 0$ and then $v > 0$ also. Thus we have been led to the result of Theorem 5.2. We showed in the proof of Theorem 5.2 that the above restrictions on u and v always give primitive triangles; we will not repeat this here.

It will be useful to know more about the primes of $\mathbf{Q}(i)$ before considering our next example.

Theorem 8.26. Let p be a positive rational prime. If $p \equiv 3 \pmod 4$, then p is a prime in $\mathbf{Q}(i)$. If $p = 2$ or if $p \equiv 1 \pmod 4$, then p is not a prime in $\mathbf{Q}(i)$ and, in fact, there is a prime π in $\mathbf{Q}(i)$ such that $\mathbf{N}(\pi) = p$. Further, if π is a prime in $\mathbf{Q}(i)$, then either π is an associate of a rational prime $\equiv 3 \pmod 4$ or $\mathbf{N}(\pi)$ is a rational prime $\equiv 1 \pmod 4$ or $\mathbf{N}(\pi) = 2$.

Proof. We first take the case of $p \equiv 3 \pmod 4$. If p is not a prime in $\mathbf{Q}(i)$, then there exist nonunit integers α and β such that $\alpha\beta = p$. Thus

$$\mathbf{N}(\alpha)\mathbf{N}(\beta) = \mathbf{N}(p) = p^2,$$

and since norms are nonnegative in $\mathbf{Q}(i)$, $\mathbf{N}(\alpha) > 1$, $\mathbf{N}(\beta) > 1$ so that

$$\mathbf{N}(\alpha) = \mathbf{N}(\beta) = p.$$

If $\alpha = a + bi$, then a and b are rational integers and

$$a^2 + b^2 = \mathbf{N}(\alpha) = p.$$

Therefore,

$$a^2 + b^2 \equiv 0(\mathrm{mod}\ p),$$

and, by Theorem 3.30 this means that $a \equiv b \equiv 0(\mathrm{mod}\ p)$. Hence $p|a$, $p|b$, and thus $p^2|(a^2 + b^2)$. But this is impossible since $a^2 + b^2 = p$, and therefore p is a prime in $\mathbf{Q}(i)$.

The case of $p = 2$ is simple since

$$\mathbf{N}(1 + i) = 2$$

and, by Theorem 8.12, $1 + i$ is a prime in $\mathbf{Q}(i)$. We now consider the remaining case, $p \equiv 1(\mathrm{mod}\ 4)$. By Theorem 3.29, there is a rational integer a such that

$$a^2 + 1 \equiv 0(\mathrm{mod}\ p).$$

Hence there is an integer b such that

$$a^2 + 1 = pb,$$

or

$$(a + i)(a - i) = pb.$$

If p is a prime, then since $\mathbf{Z}[i]$ is a UFD, either $p|(a + i)$ or $p|(a - i)$. But this is impossible since neither $(a/p) + (1/p)i$ nor $(a/p) - (1/p)i$ are integers. Therefore, p is not a prime in $\mathbf{Q}(i)$ and hence there are nonunit integers π_1 and π_2 such that $\pi_1\pi_2 = p$. Therefore,

$$\mathbf{N}(\pi_1)\mathbf{N}(\pi_2) = p^2$$

and since $\mathbf{N}(\pi_1) > 1$, $\mathbf{N}(\pi_2) > 1$, we must have

$$\mathbf{N}(\pi_1) = \mathbf{N}(\pi_2) = p.$$

By Theorem 8.12, both π_1 and π_2 are primes in $\mathbf{Q}(i)$.

We now prove the converse—that every prime in $\mathbf{Q}(i)$ is an associate of one of the three types given above. Let π be a prime in $\mathbf{Q}(i)$. Then $\mathbf{N}(\pi)$ is a positive rational integer greater than 1 and is therefore a product of rational primes,

$$\pi\bar{\pi} = \mathbf{N}(\pi) = p_1 p_2 \cdots p_n,$$

where $p_1, p_2, \ldots, p_n$ are rational primes. Thus $\pi | p_1 p_2 \cdots p_n$ and, since $\mathbf{Z}[i]$ is a UFD, $\pi | p_j$ for some j. If $p_j \equiv 3 \pmod 4$, then p_j is a prime in $\mathbf{Q}(i)$ and thus p_j / π is a unit and π is an associate of p_j. If $p_j \not\equiv 3 \pmod 4$, then p_j is not a prime and as we have seen earlier, $\mathbf{N}(\pi) = p_j$.   ▲

We now use the previous theorem to help us settle the problem of what rational integers can be written as the sum of two squares (we allow one of the squares to be 0 so that $4 = 2^2 + 0^2$ is acceptable).

Theorem 8.27. Let n be a fixed positive rational integer. The Diophantine equation $x^2 + y^2 = n$ with unknowns x and y has a solution in rational integers if and only if n can be written in the form $n = m^2 k$, where m and k are positive rational integers and k has no positive rational prime divisors $\equiv 3 \pmod 4$. (Thus, for example, 45 is the sum of two squares while 27 is not.)

Proof. Suppose $n = m^2 k$, where m and k are positive integers such that if p is a positive prime dividing k, then $p \not\equiv 3 \pmod 4$. If $k = 1$, then $n = m^2 + 0^2$. If $k > 1$, then we may set

$$k = p_1 p_2 \cdots p_r,$$

where each p_j is a prime either equal to 2 or $\equiv 1 \pmod 4$. Hence by Theorem 8.26, there are primes $\pi_1, \pi_2, \ldots, \pi_r$ in $\mathbf{Q}(i)$ such that for each j in the range $1 \le j \le r$,

$$\mathbf{N}(\pi_j) = p_j.$$

Let

$$a + bi = m \pi_1 \pi_2 \cdots \pi_r.$$

Then

$$\begin{aligned}
a^2 + b^2 &= \mathbf{N}(a + bi) \\
&= \mathbf{N}(m)\mathbf{N}(\pi_1)\mathbf{N}(\pi_2) \cdots \mathbf{N}(\pi_r) \\
&= m^2 p_1 p_2 \cdots p_r \\
&= m^2 k \\
&= n,
\end{aligned}$$

and hence the equation $x^2 + y^2 = n$ has solutions.

Conversely, suppose that there are rational integers a and b such that $a^2 + b^2 = n$ or

$$\mathbf{N}(a + bi) = n.$$

If $a + bi$ is a unit, then $n = 1$, which can be put in the form $1^2 \cdot 1$ demanded by the theorem. If $a + bi$ is not a unit, then we can factor $a + bi$ into a product of primes,

$$a + bi = \pi_1 \pi_2 \ldots \pi_r.$$

By Theorem 8.26, we can assume (after renumbering the π's if necessary) that $\pi_1, \pi_2, \ldots, \pi_s$ are associates of rational primes $p_1, p_2, \ldots, p_s$ all congruent to 3(mod 4), while $\pi_{s+1}, \pi_{s+2}, \ldots, \pi_r$ have norms $p_{s+1}, p_{s+2}, \ldots, p_r$ which are rational primes either equal to 2 or congruent to 1(mod 4). Let

$$m = p_1 p_2 \cdots p_s, \qquad k = p_{s+1} p_{s+2} \cdots p_r;$$

then we see that

$$
\begin{aligned}
n &= \mathbf{N}(a + bi) \\
  &= \mathbf{N}(\pi_1)\mathbf{N}(\pi_2) \cdots \mathbf{N}(\pi_s)\mathbf{N}(\pi_{s+1})\mathbf{N}(\pi_{s+2}) \cdots \mathbf{N}(\pi_r) \\
  &= p_1^2 p_2^2 \cdots p_s^2 p_{s+1} p_{s+2} \cdots p_r \\
  &= m^2 k.
\end{aligned}
$$

Further, the prime divisors of k are either 2 or $\equiv$ 1(mod 4) and thus no prime congruent to 3(mod 4) divides k.   ▲

It is an immediate corollary to Theorem 8.27 that 2 and every rational prime $\equiv$ 1(mod 4) can be broken up into the sum of two squares, and no prime $\equiv$ 3(mod 4) can be split up into the sum of two squares.

EXERCISES

1. Find a nontrivial infinite family of solutions to the equation

$$x^2 + y^2 = z^3.$$

(By trivial, we mean something whereby you take a solution such as $x = 2, y = 11, z = 5$ and then give the infinite family $2a^3, 11a^3, 5a^2$).

2. Solve the equation $x^2 + 2 = y^3$.
*3. Solve the equation $x^2 + 4 = y^3$.
4. Solve the equation $x^2 + 11 = y^3$.
5. In the second part of the proof of Theorem 8.27, we had

$$a + bi = \pi_1 \pi_2 \cdots \pi_r.$$

What happens in the proof if all the π's are associates of rational primes (so that $s = r$)? What happens in the proof if none of the π's are associates of rational primes (so that $s = 0$)?

6. Find all solutions to the equation

$$x^2 + y^2 = 2z^2.$$

7. It can be shown that if p is an odd prime, then the congruence $x^2 \equiv -2 \pmod p$ has a solution if $p \equiv 1$ or $3 \pmod 8$ and does not have a solution if $p \equiv 5$ or $7 \pmod 8$. Use this fact to show that an odd prime p can be written in the form $a^2 + 2b^2$ if and only if $p \equiv 1$ or $3 \pmod 8$.

8. Suppose n cannot be written as the sum of two integral squares. Show that the equation $x^2 + y^2 = n$, where x and y are merely rational, is still impossible.

9. Can $\sqrt{665}$ be the hypotenuse of a right triangle with integral legs?

10. Show that if π is a prime in $\mathbf{Q}(i)$ such that $\mathbf{N}(\pi)$ is a rational prime, then π and $\bar{\pi}$ are associates if and only if $\mathbf{N}(\pi) = 2$.

8.6. Historical Comments

A large part of algebraic number theory is due to attempts to solve one equation,

$$x^n + y^n = z^n.$$

As early as Euler's time, people were "solving" Diophantine equations such as

$$x^2 + 2 = y^3$$

by using quadratic fields. However, no one wondered about unique factorization. Indeed, rigorous proofs as we know them today were largely unknown 200 years ago. In this sense, Gauss was one of the first really modern mathematicians. It was Gauss who showed that $\mathbf{Z}[\sqrt{-1}]$ is a UFD, and in fact it was he who invented the generalization of the Euclidean algorithm presented in Section 8.4. It is in honor of this achievement that the integers of $\mathbf{Q}(\sqrt{-1})$ are called the **Gaussian integers**.

Thus it was that in the 1840s, the concept of unique factorization was recognized, but it was still fairly well believed that it would automatically happen. In 1843, Kummer (1810–1893) believed, incorrectly, that he had settled Fermat's Last Theorem. Kummer used fields gotten from the rationals by adding pth roots of unity to $\mathbf{Q}$ (where p is an odd prime). Let ρ be a primitive pth root of unity ($\rho^p = 1$ but $\rho^n \neq 1$ if $1 \leq n < p$). Let $\mathbf{Q}(\rho)$ denote the set of all numbers of the form

$$a_0 + a_1\rho + a_2\rho^2 + \cdots + a_{p-2}\rho^{p-2},$$

where $a_0, a_1, \ldots, a_{p-2}$ are rational numbers. Such a number is called an integer of $\mathbf{Q}(\rho)$ if $a_0, a_1, \ldots, a_{p-2}$ are rational integers. Kummer knew that sums, differences, products, and quotients of numbers in $\mathbf{Q}(\rho)$ are again in $\mathbf{Q}(\rho)$ and that sums, differences, and products of integers in $\mathbf{Q}(\rho)$ are integers in $\mathbf{Q}(\rho)$. He also knew that in $\mathbf{Q}(\rho)$, the equation

$$x^p + y^p = z^p$$

could be factored as

$$(x + y)(x + \rho y)(x + \rho^2 y) \cdots (x + \rho^{p-1} y) = z^p.$$

He then proceeded to show that this equation has no solution with x, y, z being nonzero rational integers. Unfortunately, Kummer's proof needed the fact that the integers of $\mathbf{Q}(\rho)$ have the unique factorization property, and this is not always true.[4] When Kummer realized this, he set about finding a way of restoring unique factorization to $\mathbf{Q}(\rho)$.

We have seen earlier that $\mathbf{Z}[\sqrt{-7}]$ does not have the unique factorization property. But when we also agree to let $a + b\sqrt{-7}$ be an integer when a and b are both halves of odd rational integers, then the integers of $\mathbf{Q}(\sqrt{-7})$ do have the unique factorization property. This suggests the idea of creating more integers in $\mathbf{Q}(\rho)$ so as to restore unique factorization. This is what Kummer did except that his new integers were not in the original field $\mathbf{Q}(\rho)$. Kummer's new integers, called ideal numbers, are of the form

$$\sqrt[r]{a_0 + a_1\rho + \cdots + a_{p-2}\rho^{p-2}},$$

where $a_0, a_1, \ldots, a_{p-2}$ are rational integers and r is a positive rational integer. The number r is not allowed to be just anything, but rather is restricted to certain admissible values according to the choice of

$$\alpha = a_0 + a_1\rho + \cdots + a_{p-2}\rho^{p-2}.$$

It turns out that there is an integer h, called the **class number** of the field, which depends only on the given field and is such that for any given α, all admissible values of r divide h. When $\mathbf{Q}(\rho)$ has the unique factorization property, the value $r = 1$ is clearly all that is needed to "restore" unique factorization. This is reflected in the fact that the class number, h, equals 1 if and only if $\mathbf{Q}(\rho)$ has the unique factorization property.

When Kummer returned to Fermat's last theorem, he found that he could settle the problem for more values of p than before but still not for all p. He found a proof that holds for all p which do not divide h, the class number

---

[4] It is true for $p = 3, 5, 7, 11, 13, 17, 19$ but fails for $p = 23$. There are infinitely many p for which unique factorization fails.

of $\mathbf{Q}(\rho)$. If $p\nmid h$, p is called a **regular prime** while if $p|h$, p is called an **irregular prime**. The only irregular primes less than 100 are $p = 37, 59, 67$.[5] Special methods have been devised for particular irregular primes, and this is how the present result that Fermat's last theorem is true for all p from 3 to 4001 has been obtained.

Although Kummer did not extend his results to other fields, his results are extendable. As an example, the class number of the field $\mathbf{Q}(\sqrt{-47})$ is 5. We saw in Section 8.4 that although 2 is a prime in $\mathbf{Q}(\sqrt{-47})$ and

$$(18) \qquad \left(\frac{13 + \sqrt{-47}}{2}\right)\left(\frac{13 - \sqrt{-47}}{2}\right) = 2 \cdot 3^3,$$

2 does not divide either factor on the left. In fact, (18) gives two different factorizations of 54 into primes in $\mathbf{Q}(\sqrt{-47})$ (see problem 11, Section 8.4). But considered in terms of ideal numbers, none of the numbers $2, 3$, $(13 + \sqrt{-47})/2$, and $(13 - \sqrt{-47})/2$ are primes, and in fact they factor into prime ideal numbers as follows:

$$2 = \sqrt[5]{\frac{9 + \sqrt{-47}}{2}} \cdot \sqrt[5]{\frac{9 - \sqrt{-47}}{2}},$$

$$3 = \sqrt[5]{14 + \sqrt{-47}} \cdot \sqrt[5]{14 - \sqrt{-47}},$$

$$\frac{13 + \sqrt{-47}}{2} = -\sqrt[5]{\frac{9 + \sqrt{-47}}{2}} \cdot (\sqrt[5]{14 - \sqrt{-47}})^3,$$

$$\frac{13 - \sqrt{-47}}{2} = -\sqrt[5]{\frac{9 - \sqrt{-47}}{2}} \cdot (\sqrt[5]{14 + \sqrt{-47}})^3.$$

(Interpret these as both sides to the fifth power are equal.) In terms of ideal numbers, both sides of (18) give the same factorization of 54:

$$54 = \sqrt[5]{\frac{9 + \sqrt{-47}}{2}} \cdot \sqrt[5]{\frac{9 - \sqrt{-47}}{2}} \cdot (\sqrt[5]{14 + \sqrt{-47}})^3 \cdot (\sqrt[5]{14 - \sqrt{-47}})^3.$$

[5] Thus, even though $\mathbf{Q}(\rho)$ does not have the unique factorization property when $p = 23$, Kummer's results settle Fermat's last theorem for $p = 23$ anyway.

Thus it is that unique factorization is restored to $\mathbf{Q}(\sqrt{-47})$ by introducing new numbers of the form $\sqrt[5]{\alpha}$, where α is in $\mathbf{Q}(\sqrt{-47})$. Not all α are used in obtaining these new numbers, for example, $\sqrt[5]{2}$ is not allowed to be used. Just how we determine which are the chosen α's and how we determine the class number unfortunately cannot be included here.

Dedekind (1831–1916) was the first to define ideals for all algebraic field extensions of the rationals. His definition is completely different from Kummer's, and it takes a considerable amount of work to show that they are equivalent. The Dedekind approach is the one that is used in modern algebra and is the one most likely to be seen by the reader elsewhere.

EXERCISES
1. Show that when $p = 3$, the $\mathbf{Q}(\rho)$ of this section is nothing more than $\mathbf{Q}(\sqrt{-3})$.
2. When $p = 5$, show that the sum, difference, and product of two numbers in $\mathbf{Q}(\rho)$ is again in $\mathbf{Q}(\rho)$. (*Hint:* $\rho^5 = 1$ and

$$\rho^4 + \rho^3 + \rho^2 + \rho + 1 = \frac{\rho^5 - 1}{\rho - 1} = \frac{0}{\rho - 1} = 0.)$$

3. It can be shown that any nonzero member of $\mathbf{Q}(\rho)$, say α, satisfies an equation of the form

$$a_n\alpha^n + a_{n-1}\alpha^{n-1} + \cdots + a_1\alpha + a_0 = 0,$$

where $a_n, a_{n-1}, \ldots, a_0$ are integers and $a_0 \neq 0$. Assuming that sums and products of numbers in $\mathbf{Q}(\rho)$ are also in $\mathbf{Q}(\rho)$, divide both sides of the above equation by αa_0 and show that $1/\alpha$ is also in $\mathbf{Q}(\rho)$. If β is in $\mathbf{Q}(\rho)$, show that β/α is in $\mathbf{Q}(\rho)$.
4. In $\mathbf{Q}(\sqrt{-5})$ we have the two factorizations of 21 into primes (see problem 5, Section 8.4)

$$21 = 3 \cdot 7 = (4 + \sqrt{-5})(4 - \sqrt{-5}).$$

In terms of ideal numbers, 3, 7, $4 + \sqrt{-5}$, and $4 - \sqrt{-5}$ can all be factored. The class number of $\mathbf{Q}(\sqrt{-5})$ is 2, and thus we expect that these numbers can be put in the form $\sqrt{\alpha}\sqrt{\beta}$, where α and β are in $\mathbf{Q}(\sqrt{-5})$. In fact,

$$3 = \sqrt{-2 + \sqrt{-5}} \cdot \sqrt{-2 - \sqrt{-5}},$$

$$7 = \sqrt{2 + 3\sqrt{-5}} \cdot \sqrt{2 - 3\sqrt{-5}}.$$

Show that $4 + \sqrt{-5}$ and $4 - \sqrt{-5}$ may be written as a product of the same ideal numbers and that both factorizations of 21 above are simply

$$21 = \sqrt{-2 + \sqrt{-5}} \cdot \sqrt{-2 - \sqrt{-5}} \cdot \sqrt{2 + 3\sqrt{-5}} \cdot \sqrt{2 - 3\sqrt{-5}}$$

except for the order of the factors.

MISCELLANEOUS PROBLEMS

1. Suppose that $\mathbf{Q}(\sqrt{d})$ has the unique factorization property and $d \equiv 1 \pmod 8$. Show that

$$2 \left| \left(\frac{1 + \sqrt{d}}{2} \right) \left(\frac{1 - \sqrt{d}}{2} \right) \right.$$

although 2 divides neither factor. Thus 2 is not a prime in $\mathbf{Q}(\sqrt{d})$. Show that if π is a prime divisor of 2 in $\mathbf{Q}(\sqrt{d})$, then π is of the form $(a + b\sqrt{d})/2$ where a and b are both odd. Show that if e and f are both odd, then either π or $\bar{\pi}$ divides $(e + f\sqrt{d})/2$. Use this to show that any prime in $\mathbf{Q}(\sqrt{d})$ which does not divide 2 is in $\mathbf{Z}[\sqrt{d}]$. Thus, in this case, the lack of unique factorization in $\mathbf{Z}[\sqrt{d}]$ is due entirely to the number 2.
2. Show that if $d > 0$, there is no such thing as the smallest positive integer in $\mathbf{Q}(\sqrt{d})$.
3. Show that if $d > 0$ and $\varepsilon = a + b\sqrt{d}$ is a unit $\mathbf{Q}(\sqrt{d})$ (where a and b may be halves of odd rational integers) and $\varepsilon > 1$, then $a > 0$, $b > 0$. (*Hint:* Look at $|\varepsilon\bar{\varepsilon}|$.) Use this to show that there is a smallest unit greater than 1 in $\mathbf{Q}(\sqrt{d})$. This unit is called the **fundamental unit** of the field and is denoted by ε_0. Show that every unit of $\mathbf{Q}(\sqrt{d})$ is of the form $\pm\varepsilon_0^n$, where n is in $\mathbf{Z}$.
4. Let $\alpha = a + b\sqrt{d}$ and let M_α be a matrix with rational entries,

$$M_\alpha = \begin{pmatrix} a & bd \\ b & a \end{pmatrix}.$$

Show that $\alpha(1, \sqrt{d}) = (1, \sqrt{d})M_\alpha$ and that M_α is uniquely determined by this equation. Use this to prove that $M_{\alpha+\beta} = M_\alpha + M_\beta$, $M_{\alpha\beta} = M_\alpha M_\beta$ (it follows that $M_\alpha M_\beta = M_\beta M_\alpha$). Show that $\mathbf{N}(\alpha) = \det M_\alpha$. Let $f(x) = \det(M_\alpha - xI)$, so that $f(x)$ is a quadratic polynomial with rational coefficients, say $f(x) = ex^2 + fx + g$. Show that $f(\alpha) = 0$ and use the correspondence between α and M_α to show that

$$eM_\alpha^2 + fM_\alpha + gI = (eI)M_\alpha^2 + (fI)M_\alpha + gI = 0$$

(the zero matrix). It is possible to study a certain set, S, of 2×2 matrices (the set of all M_α) in place of $\mathbf{Q}(\sqrt{d})$ since S and $\mathbf{Q}(\sqrt{d})$ have the same properties and there is a correspondence between the matrices of S and the numbers of $\mathbf{Q}(d)$ which is preserved under addition and multiplication.

5. Use $|N(\pi_1\pi_2 \cdots \pi_r)| + 1$ to show that $\mathbf{Q}(\sqrt{d})$ has infinitely many primes.

6. Show that every integer of $\mathbf{Q}(\sqrt{-3})$ has an associate in $\mathbf{Z}[\sqrt{-3}]$. This is interesting because the integers of $\mathbf{Q}(\sqrt{-3})$ form a UFD while $\mathbf{Z}[\sqrt{-3}]$ is not a UFD.

7. Suppose that $\mathbf{Q}(\sqrt{d})$ has the property that if α and β are integers in $\mathbf{Q}(\sqrt{d})$ with no common nonunit divisors, then there are integers γ and δ in $\mathbf{Q}(\sqrt{d})$ such that $\alpha\gamma + \beta\delta = 1$. Prove that $\mathbf{Q}(\sqrt{d})$ has the unique factorization property. The converse is also true, but the proof is considerably more difficult.

Problems 8–10 are related.

8. Let $\alpha = x + y[(1 + \sqrt{d})/2]$, where x and y are in $\mathbf{Q}(\sqrt{d})$. Show that $N(\alpha) = x^2 + xy + [(1 - d)/4]y^2$. Suppose that $d < 0$ and $d \equiv 1 \pmod 4$. Prove that if α is an integer in $\mathbf{Q}(\sqrt{d})$, α not rational, then $N(\alpha) \geq (1 - d)/4$.

9. Let $d < 0, d \equiv 1 \pmod 4$. Show that if $\mathbf{Q}(\sqrt{d})$ has the unique factorization property, then $-d$ is a positive rational prime. Do not use any of the unproved theorems in Section 8.4. (*Hints:* Let $-d = ab$, where $a > 1$, $b > 1$, and show that $\sqrt{d}\,|\,ab$, although $\sqrt{d}\nmid a$, $\sqrt{d}\nmid b$. Use problem 8 to show that in spite of this, $\sqrt{d}$ is a prime in $\mathbf{Q}(\sqrt{d})$. Note that 3, 7, 11 are primes and if $d \leq -15$, then $[(1 - d)/4]^2 > -d$.)

10. Let $d < 0$, $d \equiv 1 \pmod 4$, and suppose that $\mathbf{Q}(\sqrt{d})$ has the unique factorization property. Let $\alpha = x + y[(1 + \sqrt{d})/2]$ be an integer in $\mathbf{Q}(\sqrt{d})$, $(x,y) = 1$, $y \neq 0$. Show that if $N(\alpha) < [(1 - d)/4]^2$, then $N(\alpha)$ is a rational prime. (*Hint:* Use problem 8.) The special case of $d \leq -15$, $x = -1$, $y = 2$ is problem 9. The special case of $y = 1$ says that $x^2 + x + [(1 - d)/4]$ is a prime for $0 \leq x < [(1 - d)/4] - 1$. This special case was first given in 1913 by Rabinovitch, who proved the converse, that if $x^2 + x + [(1 - d)/4]$ is a prime for $0 \leq x < [(1 - d)/4] - 1$, then $\mathbf{Q}(\sqrt{d})$ has the unique factorization property. When $d = -163$, this gives the polynomial $x^2 + x + 41$ mentioned in Section 1.1.

Problems 11–22 are related.

11. Suppose that $\mathbf{Q}(\sqrt{d})$ has the unique factorization property. Show that if p is an odd rational prime such that the congruence

$$x^2 \equiv d \pmod p$$

has solutions, then p is not a prime in $\mathbf{Q}(\sqrt{d})$, and, in fact, there is a prime π such that $\mathbf{N}(\pi) = \pm p$. Where does your proof use the fact that p is odd? Show that if p is a rational prime such that the congruence

$$x^2 \equiv d(\bmod p)$$

has no solutions, then p is a prime in $\mathbf{Q}(\sqrt{d})$. Does this last statement hold without the unique factorization property?

12. Let p be an odd prime and a an integer such that $p \nmid a$. If the congruence

$$x^2 \equiv a(\bmod p)$$

has solutions, then we say that a is a square (mod p); otherwise, we say that a is a nonsquare (mod p). Show that the product of two squares (mod p) is a square (mod p), the product of a square (mod p) and a nonsquare (mod p) is a nonsquare (mod p), and the product of two nonsquares (mod p) is a square (mod p). (*Hint:* Look at a primitive root of p and its powers.)

13. Show that if the congruence

$$x^2 \equiv -2(\bmod p)$$

has a solution, where p is an odd rational prime, then there are rational integers a and b such that

$$a^2 + 2b^2 = p.$$

Prove that this last equation is impossible if $p \equiv 5,7(\bmod 8)$. Show as a result that the congruence equation

$$x^2 \equiv 2(\bmod p)$$

is solvable if $p \equiv 7(\bmod 8)$ and is not solvable if $p \equiv 5(\bmod 8)$.

14. Suppose that $\mathbf{Q}(\sqrt{p})$ has the unique factorization property where p is a rational prime $\equiv 3(\bmod 4)$. Show that there are rational integers a and b such that

$$a^2 - pb^2 = \pm 2,$$

and in fact the right side is $+2$ if $p \equiv 7(\bmod 8)$ and -2 if $p \equiv 3(\bmod 8)$. Conclude that the equation

$$x^2 \equiv 2(\bmod p)$$

is solvable if $p \equiv 7(\bmod 8)$ and is not solvable if $p \equiv 3(\bmod 8)$. Note that when $p \equiv 7(\bmod 8)$, the result here agrees with the result of problem 13 except that here there is an extra (and in view of problem 13, unnecessary) restriction on p.

15. Suppose that $\mathbf{Q}(\sqrt{p})$ has the unique factorization property where p is a rational prime $\equiv 1 \pmod 8$. Show that

$$2 \left| \left(\frac{1 + \sqrt{p}}{2} \right) \left(\frac{1 - \sqrt{p}}{2} \right) \right.$$

although 2 divides neither factor and hence show that 2 is not a prime in $\mathbf{Q}(\sqrt{p})$. Show that there are integers a and b such that

$$a^2 - pb^2 = \pm 8$$

and use this to show that the equation

$$x^2 \equiv 2 \pmod p$$

has solutions. Note that this was also proved in miscellaneous exercise 16 of Chapter 3 for all $p \equiv 1 \pmod 8$.

16. Suppose that p and q are rational primes, $p \equiv 3 \pmod 4$, and that the congruence

$$x^2 \equiv p \pmod q$$

is solvable. Show that if $\mathbf{Q}(\sqrt{p})$ has the unique factorization property, then there are rational integers a and b such that

$$a^2 - pb^2 = \begin{cases} q & \text{if } q \equiv 1 \pmod 4, \\ -q & \text{if } q \equiv 3 \pmod 4 \end{cases}$$

and use this to show that the congruence

$$x^2 \equiv q \pmod p$$

is solvable if $q \equiv 1 \pmod 4$ and insolvable if $q \equiv 3 \pmod 4$.

17. Suppose that p and q are both rational primes $\equiv 3 \pmod 4$ and the congruence

$$x^2 \equiv p \pmod q$$

is insolvable. Show that if $\mathbf{Q}(\sqrt{-p})$ has the unique factorization property, then there are rational integers a and b such that

$$a^2 + pb^2 = 4q.$$

Use this to show that the congruence

$$x^2 \equiv q \pmod p$$

is solvable. The unique factorization condition here is rather severe; by Theorem 8.22, p is restricted to one of the seven primes 3, 7, 11, 19, 43, 67, 163.

18. Suppose that p and q are rational primes, $p \equiv 3 \pmod 4$, $q \equiv 3 \pmod 8$, and the congruence

$$x^2 \equiv p \pmod q$$

is insolvable. Show that if $\mathbf{Q}(\sqrt{q})$ and $\mathbf{Q}(\sqrt{2p})$ have the unique factorization property, then there are rational integers a and b such that

$$a^2 - 2pb^2 = q$$

and hence the congruence

$$x^2 \equiv q \pmod p$$

is solvable.

19. Suppose that p and q are odd rational primes, $p \equiv 1 \pmod 4$, and that the congruence

$$x^2 \equiv p \pmod q$$

is solvable. Show that if $\mathbf{Q}(\sqrt{p})$ has the unique factorization property, then there are rational integers a and b such that

$$a^2 - pb^2 = \pm 4q.$$

Use this to show that the congruence

$$x^2 \equiv q \pmod p$$

is solvable.

In problems 13–15 we have shown that if p is an odd prime, then the congruence

$$x^2 \equiv 2 \pmod p$$

is solvable if $p \equiv 1,7 \pmod 8$ and insolvable if $p \equiv 3,5 \pmod 8$, provided certain side conditions on unique factorization are met. These side conditions are actually unnecessary. In problems 16–19 we have shown that *if p and q are odd primes, then the congruences*

$$x^2 \equiv p \pmod q, \qquad y^2 \equiv q \pmod p$$

are either both solvable or both insolvable unless $p \equiv q \equiv 3 \pmod 4$, in which case one congruence is solvable and the other insolvable, provided certain side conditions on unique factorization are met. Again, the side conditions are unnecessary; the italicized statement is true for all odd primes. This was first proved by Gauss and is known as the **law of quadratic reciprocity**.

20. Let $d = p_1 p_2 \cdots p_n$, where $p_1, p_2, \ldots, p_n$ are distinct rational primes, $n \geq 2$, and $p_1 \equiv 1 \pmod 4$, p_2 is odd. Let q be a prime, $q \equiv 1 \pmod 8$, which is a nonsquare $\pmod{p_j}$, $j = 1, 2$, and is a square $\pmod{p_j}$, $j = 3, \ldots, n$. (It can be shown that a prime q with these properties exists, see problem 22.) Use the law of quadratic reciprocity to show that the congruence

$$x^2 \equiv d \pmod q$$

is solvable. [If some $p_j = 2$, then the result quoted above on the congruence $x^2 \equiv 2 \pmod q$ should be used.] Show that if $\mathbf{Q}(\sqrt{d})$ has the unique factorization property, then [whether or not $d \equiv 1 \pmod 4$] there are rational integers a and b such that

$$a^2 - db^2 = \pm 4q.$$

Use this to show that the congruence

$$x^2 \equiv q \pmod{p_1}$$

is solvable. But this contradicts the definition of q, and hence $\mathbf{Q}(\sqrt{d})$ does not have the unique factorization property.

21. Let $d = 2p$, where p is a rational prime $\equiv 1 \pmod 4$. Let q be a prime $\equiv 5 \pmod 8$ such that q is a nonsquare $\pmod p$. Use the law of quadratic reciprocity to show that the congruence

$$x^2 \equiv 2p \pmod q$$

is solvable. Use this to show that if $\mathbf{Q}(\sqrt{d})$ has the unique factorization property, then there are rational integers a and b such that

$$a^2 - 2pb^2 = \pm q.$$

Show that this is impossible $\pmod p$ and hence $\mathbf{Q}(\sqrt{d})$ does not have the unique factorization property. Problems 20 and 21 combined show that if $\mathbf{Q}(\sqrt{d})$ has the unique factorization property, $d > 0$, then either d is a rational prime $\equiv 1 \pmod 4$ or d has no rational prime divisors $\equiv 1 \pmod 4$. Since $\mathbf{Q}(\sqrt{6})$ and $\mathbf{Q}(\sqrt{21})$ have the unique factorization property, it cannot be shown that d must always be a prime, but it can be shown that d is either a prime or a product of two primes.

22. Suppose that $p_1, p_2, \ldots, p_n$ are distinct odd rational primes and $a_0, a_1, \ldots, a_n$ are rational integers such that

$$(a_0, 8) = (a_1, p_1) = \cdots = (a_n, p_n) = 1.$$

Show that there are infinitely many integers q such that

$$q \equiv a_0 (\bmod 8), \quad q \equiv a_1 (\bmod p_1), \ldots, q \equiv a_n (\bmod p_n)$$

and if q_0 is one such integer, then all solutions are given by

$$q = q_0 + t \cdot 8 p_1 p_2 \cdots p_n \qquad (t \text{ in } \mathbf{Z}).$$

Show that $(q_0, 8 p_1 p_2 \cdots p_n) = 1$. A theorem of Dirichlet (1805–1859) can now be used to show that infinitely many of the q are rational primes. Dirichlet showed in two memoirs dated 1837 and 1840 that if $(a, d) = 1$, $d \neq 0$, then for infinitely many n, $a + dn$ is a prime. Although certain special cases such as $a = d = 1$ are simple, the general result is extremely difficult. By picking $a_0 = 1$, a_j a nonsquare $(\bmod p_j)$, $j = 1, 2$, and a_j a square $(\bmod p_j)$, $j = 3, \ldots, n$, we have the primes q of problem 20. A similar application is possible to problem 21.

23. Show that if x, y, and z are rational integers such that

$$x^2 + 3y^2 = z^3, \qquad (x, y) = 1, \qquad 3 \nmid x,$$

then there are rational integers a and b such that

$$a^2 + 3b^2 = z.$$

The result of problem 6 should be useful. Euler used this in his proof of Fermat's last theorem for third powers, but it is now generally thought that Euler did not prove this fact.

24. Suppose that $d > 0$. For any given positive integer n, show that $\mathbf{Q}(\sqrt{d})$ has infinitely many units of the form $a + b\sqrt{d}$, where $n \mid b$.

25. Let d and e be rational integers other than 0 and 1 having no square factors other than one. Suppose there is an integer α in $\mathbf{Q}(\sqrt{e})$ such that

$$N(\alpha) = d.$$

Show that if $\mathbf{Q}(\sqrt{d})$ has the unique factorization property, then there is an integer β in $\mathbf{Q}(\sqrt{d})$ such that

$$N(\beta) = e.$$

Thus if $\mathbf{Q}(\sqrt{d})$ and $\mathbf{Q}(\sqrt{e})$ both have the unique factorization property, then either both of the equations

$$N(\alpha) = d, \qquad \alpha \text{ an integer in } \mathbf{Q}(\sqrt{e}),$$

$$N(\beta) = e, \qquad \beta \text{ an integer in } \mathbf{Q}(\sqrt{d}),$$

in the unknowns α and β have solutions or neither has a solution.

26. Note that $N(18 + 7\sqrt{5}) = 79$. Show that there is no integer β in $Q(\sqrt{79})$ such that $N(\beta) = 5$, and use problem 25 to show that $Q(\sqrt{79})$ does not have the unique factorization property. Note that if the fundamental unit of $Q(\sqrt{79})$ were larger, there would be more special instances to check out, and thus, in some sense, unique factorization is more likely to occur with large fundamental units (see bibliography).

27. Suppose that $d > 0$ and ε_0 is the fundamental unit of $Q(\sqrt{d})$ (see Problem 3). Use Problem 25 to show that if d has no prime divisors $\equiv 3 \pmod 4$ and $Q(\sqrt{d})$ has the unique factorization property, then $N(\varepsilon_0) = -1$. Find the fundamental unit of $Q(\sqrt{34})$ and show that it has norm 1 [therefore $Q(\sqrt{34})$ does not have the unique factorization property]. This shows that the unique factorization hypothesis cannot be completely eliminated; unfortunately, we have seen in Problems 20 and 21 that it requires d to be a prime.

28. We have seen in Problem 27 that if p is a positive rational prime, $p = 2$ or $p \equiv 1 \pmod 4$, and $Q(\sqrt{p})$ has the unique factorization property, then the fundamental unit of $Q(\sqrt{p})$ has norm -1. It follows from this problem that the same result is true even if $Q(\sqrt{p})$ does not have the unique factorization property. Suppose that x and y are positive rational integers, $x^2 - py^2 = 1$. Show that $(x + 1, y) = 2a$, where a is in Z. Set

$$m = \left(\frac{x + 1}{2a}\right)^2 - p\left(\frac{y}{2a}\right)^2,$$

so that m is in Z. Show that $m = (x + 1)/(2a^2)$. Show that $m|(x + 1)/(2a)$ and $m|p[y/(2a)]^2$ and hence either $m = 1$ or $m = p$. In the latter case, $p|(x + 1)/(2a)$ and

$$\left(\frac{y}{2a}\right)^2 - p\left(\frac{x + 1}{2ap}\right)^2 = -1.$$

Put all this together to show that the Fermat–Pell equation $x^2 - py^2 = -1$ has solutions.

Appendix A

TWO-DIMENSIONAL VECTORS

Definition. A **two-dimensional vector** V is an ordered pair of numbers, $V = (a,b)$. The number a is called the first (or x) coordinate of V and b is called the second (or y) coordinate of V.

In order to be useful, we wish to combine vectors; in this book, we will need only to know how to add vectors and multiply them by constants.

Definition. If $V_1 = (a_1,b_1)$ and $V_2 = (a_2,b_2)$ are vectors, then we define the vector $V_1 + V_2$ to be

$$V_1 + V_2 = (a_1 + a_2, b_1 + b_2),$$

and if k is a real number, we define the vector $kV_1 = V_1k$ to be

$$kV_1 = V_1k = (ka_1, kb_1).$$

With these definitions, many of the rules of arithmetic are also true for vectors.

Theorem A.1. If V_1, V_2, V_3 are vectors and k_1, k_2 are real numbers, then

$$V_1 + V_2 = V_2 + V_1,$$
$$(V_1 + V_2) + V_3 = V_1 + (V_2 + V_3),$$
$$k_1(V_1 + V_2) = k_1V_1 + k_1V_2,$$
$$k_1(k_2V_1) = (k_1k_2)V_1.$$

Proof. Let $V_i = (a_i, b_i)$ for $i = 1, 2, 3$. We know that the corresponding theorems are true for real numbers and thus

$$V_1 + V_2 = (a_1 + a_2, b_1 + b_2) = (a_2 + a_1, b_2 + b_1) = V_2 + V_1,$$

$$(V_1 + V_2) + V_3 = (a_1 + a_2, b_1 + b_2) + (a_3, b_3) = (a_1 + a_2 + a_3, b_1 + b_2 + b_3)$$
$$= (a_1, b_1) + (a_2 + a_3, b_2 + b_3) = V_1 + (V_2 + V_3),$$

$$k_1(V_1 + V_2) = k_1(a_1 + a_2, b_1 + b_2) = (k_1 a_1 + k_1 a_2, k_1 b_1 + k_1 b_2)$$
$$= (k_1 a_1, k_1 b_1) + (k_1 a_2, k_1 b_2) = k_1 V_1 + k_1 V_2,$$

$$k_1(k_2 V_1) = k_1(k_2 a_1, k_2 b_1) = (k_1 k_2 a_1, k_1 k_2 b_1) = (k_1 k_2) V_1. \qquad \blacktriangle$$

It is because of Theorem A.1 that we may write something like

$$V_1 + V_2 + V_3 \qquad \text{or} \qquad k_1 k_2 V$$

without ambiguity. For example, while one person may interpret $V_1 + V_2 + V_3$ as $(V_1 + V_2) + V_3$ (that is, he first adds V_1 and V_2 and then adds V_3 to the sum) and another may interpret $V_1 + V_2 + V_3$ as $(V_3 + V_2) + V_1$, they will both get the same result. In this example, we have

$$(V_1 + V_2) + V_3 = V_3 + (V_1 + V_2) = V_3 + (V_2 + V_1) = (V_3 + V_2) + V_1$$

by the first two parts of Theorem A.1. There are other conceivable ambiguities that fortunately are not ambiguities at all. For example, we may desire to simplify $V + V + V + V + V$ by writing it in the more compact form, $5V$. But $5V$ already has the meaning of multiplying each coordinate of V by 5. Fortunately, $V + V + V + V + V$ also is the vector each of whose coordinates is five times the corresponding coordinate of V. In general, multiplication of a vector by a positive integer corresponds to that vector being added to itself several times.

Definition. We define the **zero vector** to be $O = (0,0)$.

Clearly, $O + V = V + O$ and $0V = O$ for all vectors V.

Definition. If $V = (a, b)$, we define $-V$ to be $-V = (-a, -b)$.

Thus $V + (-V) = O$ and $-(-V) = V$.

Definition. If V_1 and V_2 are vectors, we define $V_1 - V_2$ to be

$$V_1 - V_2 = V_1 + (-V_2).$$

Thus if $V_i = (a_i, b_i)$, then $V_1 - V_2 = (a_1 - a_2, b_1 - b_2)$. Our definition of subtraction of vectors also enables us to transpose vectors from one side of an equation to another by merely changing sign:

$$\text{If } V_1 + V_2 = V_3, \quad \text{then } V_1 = V_3 - V_2.$$

There is another possible ambiguity in the notation $-V$. The vectors $-V$ and $(-1)V$ have different definitions, but fortunately these definitions are such that

$$-V = (-1)V.$$

It is possible to give a geometrical interpretation of addition of vectors and multiplication of vectors by constants. If the vector $V = (a, b)$, then we may represent V in the XY plane by the point (a, b) (Figure A.1). If $V \neq 0$, then there is a line L passing through V and O and for all k, the point representing kV is on L. In fact, if $k > 0$, then the point representing kV is on the same side of O as V and k times the distance from O as V. If $k = -1$, then $kV = -V$ is directly opposite V from O (and the same distance from O as V). If $k < 0$, then since $kV = |k|(-V)$, we see that kV is on the opposite side of O from V and $|k|$ times as far away from O as V. These statements are all easily proved by means of similar triangles. A few examples are illustrated in Figure A.1.

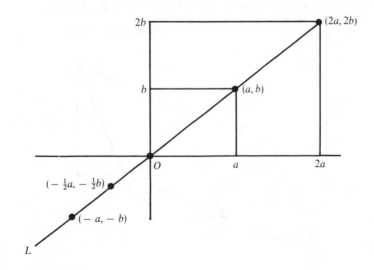

Figure A.1. Shown are $V = (a, b)$, $2V$, $-V$, and $-\frac{1}{2}V$

In Figure A.2, we see the geometrical interpretation of addition and sub-traction of two nonzero vectors. If the point A represents the vector (a,b) and B represents (c,d) and E represents $(a,b) + (c,d)$, then $OAEB$ is a parallelo-gram. This fact is known as the parallelogram rule for addition of vectors. It is most easily proved by introducing the auxiliary points $C = (c,0)$ and $D = (a + c, b)$. Then OC is parallel to AD (both are horizontal) and OC and AD are equal in length. Likewise, $BC \| ED$ and $BC = ED$ in length. Also

$$\angle OCB = \angle ADE \, (= 90°)$$

and thus $\triangle OCB \cong \triangle ADE$. Since the sides of these congruent right triangles are parallel, so are the hypotenuses, $OB \| AE$, and since the triangles are congruent, $OB = AE$. Thus two opposite sides of quadrilateral $OAEB$ are parallel and equal in length and hence $OAEB$ is a parallelogram. We may represent the difference of two vectors $V_1 - V_2$ geometrically by using the parallelogram rule for adding V_1 and $-V_2$. Thus in Figure A.2, if F represents $-(a,b)$ and $OBGF$ is a parallelogram, then G represents $(c,d) - (a,b)$.

EXERCISES

1. Let $V_1 = (2,3)$, $V_2 = (-1,2)$, $V_3 = (-3,-1)$, $V_4 = (7,-8)$. What are the coordinates of the following?

   (a) $V_1 + V_3$ (d) $[(-3V_3 + V_1) + 7V_4] - V_2$
   (b) $2V_1 - V_2$ (e) $-2(V_4 - V_1) + V_3 \cdot 3$
   (c) $(7V_4 - 3V_3) + (V_1 - V_2)$

2. Use Theorem A.1 to show that for all vectors V_1, V_2, V_3, V_4,

$$(7V_4 - 3V_3) + (V_1 - V_2) = [(-3V_3 + V_1) + 7V_4] - V_2.$$

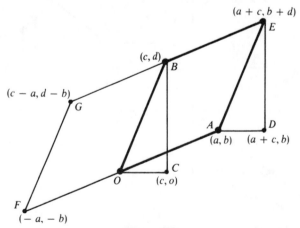

Figure A.2.

3. Show for all vectors V and constants k that $(-k)V = -(kV)$.
4. In Figure A.3, the XY plane has been partitioned into congruent parallelograms. Let A and B be the points representing the vectors A and B. Which points represent the following?

(a) $2A$

(b) $4B$

(c) $-2A$

(d) $A + B$

(e) $A + 2B$

(f) $B - A$

(g) $B - 2A$

(h) $3A + 2B$

(i) $4A - 3B$

(j) $-2A + 3B$

(k) $-A - B$

(l) $-2A - 2B$

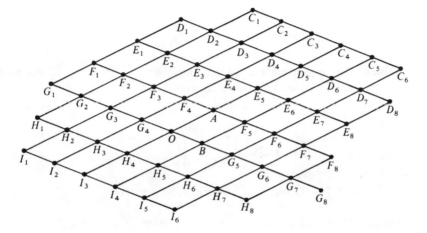

Figure A.3.

Appendix B

TWO BY TWO MATRICES

Definition. An $m \times n$ **matrix** (of real numbers) is an array of mn real numbers in a rectangle of m rows and n columns.

For example,

$$(\pi), \quad (1,3), \quad \begin{pmatrix} 7.1 \\ -8 \end{pmatrix}, \quad \begin{pmatrix} 3.7 & \sqrt{2} \\ -2 & 3 \end{pmatrix}$$

are 1×1, 1×2, 2×1, and 2×2 matrices, respectively.

Definition. If A and B are two $m \times n$ matrices, then we define $A + B$ to be the $m \times n$ matrix whose entries are the sums of the corresponding entries of A and B.

For example,

$$(1,4) + (-3,7) = (-2,11),$$

$$\begin{pmatrix} 1 & 5 \\ 7 & 2 \end{pmatrix} + \begin{pmatrix} 2 & -1 \\ -1 & 3 \end{pmatrix} = \begin{pmatrix} 3 & 4 \\ 6 & 5 \end{pmatrix}.$$

Definition. If A is an $m \times n$ matrix and k a real number, then we define $kA = Ak$ to be the $m \times n$ matrix whose entries are k times the corresponding entries of A.

For example,

$$3(1,4) = (1,4) \cdot 3 = (3,12),$$

$$7\begin{pmatrix} 2 & -1 \\ 1 & 8 \end{pmatrix} = \begin{pmatrix} 2 & -1 \\ 1 & 8 \end{pmatrix} \cdot 7 = \begin{pmatrix} 14 & -7 \\ 7 & 56 \end{pmatrix}.$$

Note that addition and multiplication by real numbers of 1×2 matrices is exactly the same as addition and multiplication by real numbers of two-dimensional vectors. Thus two-dimensional vectors and their arithmetic (as developed in Appendix A) are a special case of the arithmetic of matrices. As an analogue of Theorem A.1, we have the following theorem (its proof is exactly the same as that of Theorem A.1 and is left to the reader).

Theorem B.1. If A_1, A_2, A_3 are $m \times n$ matrices and k_1 and k_2 are real numbers, then

$$A_1 + A_2 = A_2 + A_1,$$

$$(A_1 + A_2) + A_3 = A_1 + (A_2 + A_3),$$

$$(k_1 + k_2)A_1 = k_1 A_1 + k_2 A_1,$$

$$k_1(k_2 A_1) = (k_1 k_2)A_1.$$

Because of Theorem B.1, there is no ambiguity to an expression such as

$$4A_1 + 3A_2 + A_3 + 5A_4;$$

no matter how these matrices are combined, the result will be the same in the end.

Definition. The **zero** $m \times n$ **matrix** is the $m \times n$ matrix all of whose entries are 0. If A is an $m \times n$ matrix, we define $-A$ to be the $m \times n$ matrix whose entries are minus the corresponding entries of A [and thus $-A = (-1)A$]. If B is also an $m \times n$ matrix, we define the matrix $A - B$ to be $A + (-B)$.

Theorem B.2. If A, B, C are $m \times n$ matrices and O is the zero $m \times n$ matrix, then

$$A + O = O + A = A,$$

$$A - A = O;$$

if $A + B = C$, then $A = C - B$.

Proof. The first two parts are immediate consequences of the definitions. The third part is also clear, but we give here a proof using the first two parts of the theorem and Theorem B.1. If

$$A + B = C,$$

then

$$A = A + O = A + (B - B) = (A + B) - B = C - B. \qquad \blacktriangle$$

The proof of the third part of Theorem B.2 shows what is really involved in transposing something from one side of an equation to the other. After Theorems B.1 and B.2, we see that addition and subtraction of matrices and multiplication of matrices by real numbers follow the same rules of arithmetic as the real numbers themselves. This is not true of multiplication of matrices, to which we now turn; however, multiplication of matrices will obey enough of the usual laws of arithmetic to be exceedingly useful.

Definition. We define the product of two 2×2 matrices to be a 2×2 matrix as follows:

$$\begin{pmatrix} a & b \\ c & d \end{pmatrix} \begin{pmatrix} e & f \\ g & h \end{pmatrix} = \begin{pmatrix} ae + bg & af + bh \\ ce + dg & cf + dh \end{pmatrix}.$$

We define the product of a 1×2 and a 2×2 matrix (in that order) to be a 1×2 matrix as follows:

$$(a,b) \begin{pmatrix} e & f \\ g & h \end{pmatrix} = (ae + bg, \quad af + bh).$$

The second definition is merely the top half of the first definition. In each definition, the element of the ith row and jth column of the product is found from the ith row of the first matrix and jth column of the second matrix; in fact, it is the product of the first entries in the ith row of the first matrix and jth column of the second matrix plus the product of the second entries in the ith row of the first matrix and jth column of the second matrix. For example, the entry in the first row and second column of the product is found from the first row of the first matrix, (a,b), and the second column of the second matrix, $\begin{pmatrix} f \\ h \end{pmatrix}$, by multiplying the first entries, af, and adding the product of the second entries, bh, thereby getting $af + bh$. The reader should remember to always move horizontally in the first matrix (from left to right) and vertically in the second matrix (from top to bottom).

It is possible to give a general definition along these lines of the product of an $m \times n$ matrix and an $n \times p$ matrix (the product being an $m \times p$ matrix), but such generality is unnecessary in this book and further, the computational proofs given below for 2×2 matrix products would be quite impossible for 100×100 matrix products. We first give two examples.

We see that

$$(2,3)\begin{pmatrix} 1 & -1 \\ 2 & 4 \end{pmatrix} = (8,10),$$

while

$$\begin{pmatrix} 1 & -1 \\ 2 & 4 \end{pmatrix}(2,3)$$

is not even defined. Further,

$$\begin{pmatrix} 2 & 1 \\ 3 & -1 \end{pmatrix}\begin{pmatrix} 1 & 2 \\ 3 & 4 \end{pmatrix} = \begin{pmatrix} 5 & 8 \\ 0 & 2 \end{pmatrix},$$

while

$$\begin{pmatrix} 1 & 2 \\ 3 & 4 \end{pmatrix}\begin{pmatrix} 2 & 1 \\ 3 & -1 \end{pmatrix} = \begin{pmatrix} 8 & -1 \\ 18 & -1 \end{pmatrix}.$$

Thus if A and B are matrices and AB is defined, BA may not be defined and, even if BA is defined, BA and AB are not necessarily the same. Thus one of the fundamental laws of multiplication fails to hold for matrices. However, the failure of this particular law is actually only a minor nuisance, it simply means that we must pay attention to the order in which things are written. The really important laws of multiplication remain valid for matrices.

Theorem B.3. If A and A_1 are either both 2×2 matrices or both 1×2 matrices, and if B and C are 2×2 matrices, then

$$(AB)C = A(BC),$$

$$A(B + C) = AB + AC,$$

$$(A + A_1)B = AB + A_1 B$$

(where all the above products are defined).

Proof. The first part of the theorem is the messiest of the three parts; we prove it for A being a 2×2 matrix, this being twice as messy as the case that A is a 1×2 matrix. The other case of the first part and the last two parts of the theorem are left to the reader as exercises. Let

$$A = \begin{pmatrix} a & b \\ c & d \end{pmatrix}, \qquad B = \begin{pmatrix} e & f \\ g & h \end{pmatrix}, \qquad C = \begin{pmatrix} i & j \\ k & l \end{pmatrix}.$$

Then

$$(AB)C = \left[\begin{pmatrix} a & b \\ c & d \end{pmatrix}\begin{pmatrix} e & f \\ g & h \end{pmatrix}\right]C = \begin{pmatrix} ae + bg & af + bh \\ ce + dg & cf + dh \end{pmatrix}\begin{pmatrix} i & j \\ k & l \end{pmatrix}$$

$$= \begin{pmatrix} aei + bgi + afk + bhk & aej + bgj + afl + bhl \\ cei + dgi + cfk + dhk & cej + dgj + cfl + dhl \end{pmatrix},$$

while

$$A(BC) = A\left[\begin{pmatrix} e & f \\ g & h \end{pmatrix}\begin{pmatrix} i & j \\ k & l \end{pmatrix}\right] = \begin{pmatrix} a & b \\ c & d \end{pmatrix}\begin{pmatrix} ei + fk & ej + fl \\ gi + hk & gj + hl \end{pmatrix}$$

$$= \begin{pmatrix} aei + afk + bgi + bhk & aej + afl + bgj + bhl \\ cei + cfk + dgi + dhk & cej + cfl + dgj + dhl \end{pmatrix}.$$

Thus by direct computation,

$$(AB)C = A(BC).$$

The other parts to the theorem may be handled similarly.   ▲

Because of the first part of Theorem B.3, there is no ambiguity in an expression such as $ABCD$, as long as we remember to multiply only adjacent matrices (and in the order given). Thus, for example, if one person interprets $ABCD$ as $[A(BC)]D$ and another interprets $ABCD$ as $A[B(CD)]$, then after the multiplying is over, they will both get the same result. This is because

$$[A(BC)]D = A[(BC)D] = A[B(CD)].$$

On the other hand, we cannot interpret $ABCD$ as $[(AC)B]D$, since this involves changing the order of B and C. However, we may put a real number into a product wherever we wish: If A is either a 1×2 or 2×2 matrix, B is a 2×2 matrix, k a real number, then

$$k(AB) = (kA)B = (Ak)B = A(kB).$$

Let us give an example of Theorem B.3 in action. Suppose we wish to solve the two equations

(1)
$$13x + 3y = 1,$$
$$5x + y = 2$$

for x and y. Let

$$A = \begin{pmatrix} 13 & 5 \\ 3 & 1 \end{pmatrix}, \qquad B = \begin{pmatrix} -\frac{1}{2} & \frac{5}{2} \\ \frac{3}{2} & -\frac{13}{2} \end{pmatrix}.$$

Since

$$(x,y)A = (x,y)\begin{pmatrix} 13 & 5 \\ 3 & 1 \end{pmatrix} = (13x + 3y, \quad 5x + y),$$

equation (1) may be put in the form

(2) $(x,y)A = (1,2).$

Thus the solutions to (1) have $(x,y)A$ and $(1,2)$ being the same matrix, and therefore

(3) $[(x,y)A]B = (1,2)B;$

in other words, we have just multiplied (2) by B, putting B on the right-hand side in each case since order matters. Now

(4) $(1,2)B = (1,2)\begin{pmatrix} -\frac{1}{2} & \frac{5}{2} \\ \frac{3}{2} & -\frac{13}{2} \end{pmatrix} = \left(\frac{5}{2}, \quad \frac{-21}{2}\right)$

and, by Theorem B.3,

$$[(x,y)A]B = (x,y)[AB] = (x,y)\left[\begin{pmatrix} 13 & 5 \\ 3 & 1 \end{pmatrix}\begin{pmatrix} -\frac{1}{2} & \frac{5}{2} \\ \frac{3}{2} & -\frac{13}{2} \end{pmatrix}\right]$$

$$= (x,y)\begin{pmatrix} 1 & 0 \\ 0 & 1 \end{pmatrix}$$

(5) $= (x,y).$

When we combine (3), (4), and (5), we see that

$$(x,y) = \left(\frac{5}{2}, \quad \frac{-21}{2}\right),$$

so that

$$x = \tfrac{5}{2}, \qquad y = -\tfrac{21}{2}$$

is the solution to (1). Without Theorem B.3, we could not have done this. However, we still are faced with the problem of finding B.

Definition. Let I denote the 2×2 **identity matrix,**

$$I = \begin{pmatrix} 1 & 0 \\ 0 & 1 \end{pmatrix}.$$

If

$$A = \begin{pmatrix} a & b \\ c & d \end{pmatrix} \qquad \text{with } ad - bc \neq 0,$$

then we define a matrix A^{-1}, called A **inverse**, by

$$A^{-1} = \begin{pmatrix} \dfrac{d}{ad - bc} & \dfrac{-b}{ad - bc} \\[2ex] \dfrac{-c}{ad - bc} & \dfrac{a}{ad - bc} \end{pmatrix} = \frac{1}{ad - bc} \begin{pmatrix} d & -b \\ -c & a \end{pmatrix}.$$

Note that A^{-1} is not defined for all 2×2 matrices.

Theorem B.4. If A is a 2×2 matrix, then

$$AI = IA = A.$$

If V is a 1×2 matrix,

$$VI = V.$$

If A is a 2×2 matrix and A^{-1} is defined, then

$$A^{-1}A = AA^{-1} = I.$$

The proof consists of performing the multiplications and is left to the reader. In the preceding example, to solve for x and y, we wish to somehow "divide" both sides of (2) by A. The method of dividing by A is to multiply by A^{-1} (which in the previous example is the matrix B). In order that A and A^{-1} cancel each other (that is, multiply out to I), they must be adjacent to each other, and this says that we must multiply equation (2) through by A^{-1} on the right-hand side. It is very important to note that A and A^{-1} will not in general cancel each other unless they are adjacent. For example, in general

$$ABA^{-1} \neq B.$$

(For certain choices of A and B, namely those A and B such that $AB = BA$, equality holds; the point is equality does not always hold.)

There are two other properties of 2×2 matrices that we use in the book.

Definition. Let

$$A = \begin{pmatrix} a & b \\ c & d \end{pmatrix}.$$

Then we define the **determinant** of A to be

$$\det A = ad - bc.$$

In terms of determinants, we have defined A^{-1} only for those A with $\det A \neq 0$.

Theorem B.5. If A and B are 2×2 matrices, then

$$\det(AB) = (\det A)(\det B).$$

The proof of Theorem B.5 is also straightforward and is left to the reader. Since $\det I = 1$, we see that

$$(\det A)(\det A^{-1}) = \det(AA^{-1}) = \det I = 1$$

and thus

$$\det A^{-1} = \frac{1}{\det A}.$$

We may also use Theorem B.5 to show the futility of trying to define A^{-1} when $\det A = 0$. Suppose $\det A = 0$ and

(6) $$AB = I.$$

Then

$$0 = 0 \cdot \det B = (\det A)(\det B) = \det(AB) = \det I = 1,$$

which is a contradiction. Thus there is no matrix B such that (6) holds.

Definition. If

$$A = \begin{pmatrix} a & b \\ c & d \end{pmatrix},$$

then we define a matrix A', called A **transpose**, as

$$A' = \begin{pmatrix} a & c \\ b & d \end{pmatrix}.$$

Theorem B.6. If A and B are 2×2 matrices, then

$$(AB)' = B'A'$$

(note the change in order).

The proof again consists of multiplying both sides out and is left to the exercises.

EXERCISES

1. Let

$$A = \begin{pmatrix} -1 & 2 \\ 3 & -7 \end{pmatrix}, \quad B = \begin{pmatrix} 2 & 4 \\ 1 & 5 \end{pmatrix}, \quad C = \begin{pmatrix} 3 & -2 \\ -7 & 1 \end{pmatrix}, \quad D = \begin{pmatrix} -5 & -4 \\ -3 & -1 \end{pmatrix}.$$

Find (a) AB, (b) BA, (c) AC, (d) $B + C$.
   (e) Use (a) and (c) to find $AB + AC$.
   (f) Use (d) to find $A(B + C)$.
   (g) Verify that $(AB)C = A(BC)$.
   (h) Verify that $[A(BC)]D = A[B(CD)]$.

2. Find A^{-1} and ABA^{-1}, where A and B are given in problem 1.

3. Find $\begin{pmatrix} 3 & 2 \\ 4 & 3 \end{pmatrix}^{-1}$ and use this to solve the equation

$$(x,y)\begin{pmatrix} 3 & 2 \\ 4 & 3 \end{pmatrix} = (2, -3).$$

4. Find $\begin{pmatrix} 2 & 3 \\ 8 & -2 \end{pmatrix}^{-1}$ and use this to solve the equation

$$(x,y)\begin{pmatrix} 2 & 3 \\ 8 & -2 \end{pmatrix} = (7,28).$$

5. Find det A, det B, det(AB), det(BA), where A and B are given in problem 1.

6. Verify Theorem B.6 for the matrices A and B of problem 1.

7. Prove the first part of Theorem B.3 when A is a 1×2 matrix.

8. Prove the second part of Theorem B.3.

9. Prove Theorem B.4.

10. Prove Theorem B.5.

11. Prove Theorem B.6.

Appendix C

FIELDS

Definition. Let F be a collection of objects which can be combined by two operations which we call plus $(+)$ and times $(\cdot)$. We say that F is a **field** if the following postulates (1–9) are all satisfied.

1. *Uniqueness.* If a and b are in F, then there is a unique object which is equal to $a + b$ and there is a unique object which is equal to $a \cdot b$.
2. *Closure.* If a and b are in F, then $a + b$ and $a \cdot b$ are in F.
3. *Commutative laws.* If a and b are in F, then

$$a + b = b + a, \qquad a \cdot b = b \cdot a.$$

4. *Associative laws.* If a, b, and c are in F, then

$$(a + b) + c = a + (b + c), \qquad (a \cdot b) \cdot c = a \cdot (b \cdot c).$$

5. *Distributive law.* If a, b, and c are in F, then

$$a \cdot (b + c) = (a \cdot b) + (a \cdot c).$$

6. *Zero element.* F contains an element 0 such that for all a in F,

$$a + 0 = a.$$

7. *Unity.* F contains an element $1 \neq 0$ such that for all a in F,

$$a \cdot 1 = a.$$

8. *Additive inverse.* If a is in F, then there is an element x of F such that

$$a + x = 0.$$

9. *Multiplicative inverse.* If a is in F, $a \neq 0$, then there is an element x of F such that

$$ax = 1.$$

From these postulates, we derive virtually all the laws of ordinary arithmetic. In particular, postulates 8 and 9 form the basis for subtraction and division in F.

Familiar examples of fields are the set of all rational numbers, $\mathbf{Q}$, the set of all real numbers, $\mathbf{R}$, and the set of all complex numbers, $\mathbf{C}$. Other examples of fields are given by arithmetic (mod p), where p is a prime, the quadratic fields, $\mathbf{Q}(\sqrt{d})$, of Chapter 8, and the set of four elements, $\{0,1,\alpha,\beta\}$, considered in problem 5, page 140. Examples of arithmetical systems that we have considered in this book that are not fields are given by arithmetic (mod n), where n is composite and the set of all 2×2 matrices. Arithmetic (mod n) satisfies all the postulates except for postulate 9. The arithmetic of 2×2 matrices satisfies all the postulates except the multiplication part of postulate 3 and postulate 9.

The great advantage of dealing with the abstract concept of a field lies with the fact that any theorem proved on the basis of postulates 1 through 9 is automatically satisfied by all fields. For example, one can prove that if $a_0 \neq 0, a_1, a_2, \ldots, a_n$ are in a field F, then the equation

$$a_0 x^n + a_1 x^{n-1} + \cdots + a_n = 0$$

has at most n different solutions in F. The reader has probably seen this statement for $\mathbf{Q}, \mathbf{R}$, and $\mathbf{C}$ in high school; we proved it for arithmetic (mod p) in Section 3.6. The proofs are virtually identical; in dealing with fields, we simply prove the theorem once and we are done with it.

As another example, one can show that if $a_0 \neq 0, a_1, a_2$ are in a field F and the equation

(1) $$a_0 x^2 + a_1 x + a_2 = 0$$

has no solutions in F, then there is a bigger field, $F(\alpha)$, such that every element of F is in $F(\alpha)$, α satisfies (1), and every element of $F(\alpha)$ may be written uniquely in the form

$$b + c\alpha,$$

where b and c are in F. When $F = \mathbf{Q}$, this is how we created $\mathbf{Q}(\sqrt{d})$ in Chapter 8. When $F = \mathbf{R}$, this is how the set of all complex numbers was created, equation (1) then being

$$x^2 + 1 = 0.$$

When F is the set of two elements $\{0,1\}$ with arithmetic (mod 2), and equation (1) is

$$x^2 + x + 1 = 0,$$

we get the field of four elements $\{0,1,\alpha,\beta\}$ on page 140. In each of these cases, $F(\alpha)$ is a field and thus all the usual laws of arithmetic will continue to hold. This one theorem on abstract fields thus justifies the creation of all three of the fields, $\mathbf{Q}(\sqrt{d})$, $\mathbf{C}$, and $\{0,1,\alpha,\beta\}$, and guarantees that the standard arithmetic laws continue to hold in these new creations. We can even create something like $\mathbf{Q}(\sqrt{2})(\sqrt{3})$ or $\{0,1,\alpha,\beta\}(\gamma)$, where

$$\gamma^2 + \gamma + \alpha = 0.$$

The concrete approach would force us to examine each of these new sets and verify their arithmetic properties; the abstract approach has already taken care of these details in the one theorem above.

BIBLIOGRAPHY

General References in Number Theory

Davenport, H., *The Higher Arithmetic*, London, Hutchinson, 1952.

Dickson, L. E., *History of the Theory of Numbers*, Washington, D.C., Carnegie Institution; vol. 1, 1919; vol. 2, 1920; vol. 3, 1923.

Hardy, G. H., and E. M. Wright, *An Introduction to the Theory of Numbers*, New York, Oxford University Press, 4th ed., 1960.

Landau, E., *Elementary Number Theory*, New York, Chelsea, 1958.

Lehmer, D. H., *Guide to Tables in the Theory of Numbers*, Washington D.C., National Research Council, Bulletin, 105, 1941.

LeVeque, W. J., *Elementary Theory of Numbers*, Reading, Mass., Addison-Wesley, 1962.

———, *Topics in Number Theory*, 2 vols., Reading, Mass., Addison-Wesley, 1956.

Mordell, L. J., *Three Lectures on Fermat's Last Theorem*, New York, Cambridge University Press, 1921.

Niven, I., and H. S. Zuckerman, *An Introduction to the Theory of Numbers*, New York, Wiley, 1960.

Olds, C. D., *Continued Fractions*, New York, Random House, 1963.

Sierpinski, W., *Elementary Theory of Numbers*, Warsaw, 1964.

———, *A Selection of Problems in the Theory of Numbers*, Oxford, Pergamon Press, 1964 (distributed in the United States by Macmillan, New York).

General Related References

Ball, W. W. Rouse, *Mathematical Recreations and Essays*, 12th ed., revised by H. S. M. Coxeter, New York, Macmillan, 1947.

Bell, E. T., *The Last Theorem*, New York, Simon and Schuster, 1961.

Courant, R., and H. Robbins, *What Is Mathematics*, New York, Oxford University Press, 1953.

Polya, G., *Mathematics and Plausible Reasoning*, Princeton, N.J., Princeton University Press, 1954.

Rademacher, H., and O. Toeplitz, *The Enjoyment of Mathematics*, Princeton, N.J., Princeton University Press, 1957.

Special References

(Some of these will be useful only for specialists in number theory.)

Chapter 1

page 2
D. N. Lehmer compiled two tables, List of Prime Numbers from 1 to 10 006 721, Washington, D.C., Carnegie Institution, No. 165, 1914, and Factor Tables for the First Ten Millions, Washington, D.C., Carnegie Institution, No. 105, 1909.

page 5
For Pipping's results on Goldbach's conjecture, see *Acta Acad. Aboensis* **11**, no. 4 (1938), and **12**, no. 11 (1940).

page 7
Haselgrove's disproof of the Polya conjecture is entitled, A Disproof of a Conjecture of Polya, and appeared in *Mathematika* **5** (1958), pp. 141–145. The counterexample at $n = 906\ 180\ 359$ was found by R. S. Lehman, On Liouville's Function, *Math. Computation* **14** (1960), pp. 311–320.

Chapter 2

page 43
The result that there are no odd perfect numbers less than 10^{20} is that of H. J. Kanold, Über Mehrfach Vollkommene Zahlen II, *J. Reine Angew. Math.* **197** (1957), pp. 82–96.

Chapter 3

page 81
The first example of a pseudoprime was given by Sarrus, *Ann. Math.* **9** (1818–1919), p. 320, his example being 341. Since then many other examples have been given. A paper by D. H. Lehmer, On the Converse of Fermat's Theorem II, *Amer. Math. Monthly* **56** (1949), pp. 300–309, lists all those pseudoprimes between 10^8 and $2 \cdot 10^8$ whose smallest prime factor is greater than 313. Lehmer's paper gives references to other tables. Lehmer's earlier paper with the same title, *Amer. Math. Monthly* **43** (1936), pp. 347–354, was written before the advent of the electronic computer and illustrates the methods of performing difficult calculations by hand.

The proof that there are infinitely many even pseudoprimes was given by N. G. W. H. Beeger, *Amer. Math. Monthly* **58** (1951), pp. 553–555.

pp. 108–109   problem 8. Banachiewicz's conjectured explanation of why Fermat stated his conjecture on the primality of all F_n is given in a paper by W. Sierpinski, L'Induction incomplète dans la théorie des nombres, *Scripta Math.* **28** (1967), pp. 5–13.

Chapter 4

Chapter 7 of Ball's book cited above contains various methods of constructing magic squares. D. N. Lehmer was the first to analyze the uniform step method from the point of view of congruence equations; see his article, On the Congruences Connected with Certain Magic Squares, *Trans. Amer. Math. Soc.* **31** (1929), pp. 529–551.

page 138   McClintock's paper, On the Most Perfect Forms of Magic Squares, with Methods for Their Production, *Amer. J. Math.* **19** (1897), pp. 99–120, should be quite readable to the student lucky enough to have access to it. McClintock attributes the name "diabolic square" to Lucas.

page 139   R. C. Bose, S. S. Shrikhande, and E. T. Parker, Further Results on the Construction of Mutually Orthogonal Latin Squares and the Falsity of Euler's Conjecture, *Can. J. Math.* **12** (1960), pp. 189–203.

Chapter 5

page 146   The result for the first case of Fermat's last theorem is that of D. H. and Emma Lehmer, On the First Case of Fermat's Last Theorem, *Bull. Amer. Math. Soc.* **47** (1941), pp. 139–142. The result for the second case of Fermat's last theorem is that of J. L. Selfridge, C. A. Nicol, and H. S. Vandiver, Proof of Fermat's Last Theorem for All Prime Exponents Less Than 4002, *Proc. Natl. Acad. Sci. (U.S.)* **41** (1955), pp. 970–973. The result of Lander and Parkin was announced in the *Bull. Amer. Math. Soc.* **72** (1966), p. 1079.

Chapter 6

pp. 176–177   The result of K. F. Roth is given in his paper, Rational Approximations to Algebraic Numbers, *Mathematika* **2**

(1955), pp. 1–20 and 168. The first effective improvement of Liouville's theorem is due to A. Baker, Contributions to the Theory of Diophantine Equations, *Phil. Trans. Roy. Soc. London* **A263** (1968), pp. 173–208. Baker shows how to calculate a value of δ such that

$$\left| \alpha - \frac{p}{q} \right| > \frac{\delta e^{(\log q)^{1/\kappa}}}{q^n}$$

Here κ is a fixed number $> n + 1$ and δ depends on κ as well as α. This result still does not enable us to calculate a value of δ when $\theta < n$, but it is sufficient to deal with certain types of Diophantine equations effectively.

page 180   Baker's result on $\sqrt[3]{2}$ does not seem to extend to the general algebraic number. It may be found in his paper, Rational Approximations to $\sqrt[3]{2}$ and Other Algebraic Numbers, *Quart. J. Math.* **15** (1964), pp. 375–383.

Chapter 7

Geometrical interpretations of continued fractions and even some of the theorems have occurred in the past. The closest approach to this chapter that I have seen may be found in the early sections of the book by F. Klein and A. Sommerfeld, *Ausgewählte Kapitel der Zahlentheorie I* Göttingen, 1896 (see especially pp. 46–47).

page 244   There are now two very extensive tables of continued fraction expansions of $\sqrt{d}$. The first is that of Wilhelm Patz, *Tafel der Regelmässigen Kettenbrüche und Ihrer Vollständigen Quotienten für die Quadratwurzeln aus den Natürlichen Zahlen von 1–10 000*, Berlin, Akademie-Verlag, 1955. The second is that of R. Kortum and G. McNiel, *A Table of Periodic Continued Fractions*, Sunnyvale, Calif., Lockheed Missiles and Space Division, 1961. The Kortum and McNiel tables also include those d in the range 1 to 10 000. In addition, the Kortum and McNiel tables include the smallest solution to the Fermat–Pell equation; this information is not given in the Patz tables.

Chapter 8

page 294   Davenport's result that there are no Euclidean quadratic fields with $d > 2^{14}$ appeared in *Proc. London Math. Soc.* (2)

53 (1951), pp. 65–82, and Chatland and Davenport's paper appeared in the *Can. J. Math.* **2** (1950), pp. 289–296.

page 295 The Heilbronn and Linfoot paper appeared in the *Quart. J. Math. Oxford* **5** (1934), pp. 293–301. For the Stark and Baker results, see H. M. Stark, A Complete Determination of the Complex Quadratic Fields of Class-Number One, *Mich. Math. J.* **14** (1967), pp. 1–27, and A. Baker, Linear Forms in the Logarithms of Algebraic Numbers, *Mathematica* **13** (1966), pp. 204–216.

page 316 Problem 26: The vague feeling that unique factorization is more likely with large fundamental units has been substantiated by C. L. Siegel, who showed that the fundamental units of fields with unique factorization are larger than those of fields without unique factorization. The precise statement of his result is that if $\varepsilon_0(d)$ is the fundamental unit of $\mathbf{Q}(\sqrt{d})$ and $h(d)$ the class number of $\mathbf{Q}(\sqrt{d})$ [$h(d)$ is a positive integer and equals 1 if and only if $\mathbf{Q}(\sqrt{d})$ has the unique factorization property], then

$$\lim_{d \to \infty} \frac{\ln[h(d)\ln \varepsilon_0(d)]}{\ln \sqrt{d}} = 1.$$

Siegel's paper appeared in *Acta Arithmetica* **1** (1935), pp. 83–86.

Table 1. Greek Alphabet

A	α	Alpha		N	ν	Nu
B	β	Beta		Ξ	ξ	Xi
Γ	γ	Gamma		O	o	Omicron
Δ	δ	Delta		Π	π	Pi
E	ε	Epsilon		P	ρ	Rho
Z	ζ	Zeta		Σ	σ	Sigma
H	η	Eta		T	τ	Tau
Θ	θ	Theta		Υ	υ	Upsilon
I	ι	Iota		Φ	ϕ, φ	Phi
K	κ	Kappa		X	χ	Chi
Λ	λ	Lambda		Ψ	ψ	Psi
M	μ	Mu		Ω	ω	Omega

Table 2. Primes Less Than 500 and Their Smallest Positive Primitive Roots

p^a	g^a	p	g	p	g	p	g	p	g
2	1	71	7	167	5	271	6	389	2
3	2	73	5	173	2	277	5	397	5
5	2	79	3	179	2	281	3	401	3
7	3	83	2	181	2	283	3	409	21
11	2	89	3	191	19	293	2	419	2
13	2	97	5	193	5	307	5	421	2
17	3	101	2	197	2	311	17	431	7
19	2	103	5	199	3	313	10	433	5
23	5	107	2	211	2	317	2	439	15
29	2	109	6	223	3	331	3	443	2
31	3	113	3	227	2	337	10	449	3
37	2	127	3	229	6	347	2	457	13
41	6	131	2	233	3	349	2	461	2
43	3	137	3	239	7	353	3	463	3
47	5	139	2	241	7	359	7	467	2
53	2	149	2	251	6	367	6	479	13
59	2	151	6	257	3	373	2	487	3
61	2	157	5	263	5	379	2	491	2
67	2	163	2	269	2	383	5	499	7

^{a} p stands for prime and g stands for the smallest positive primitive root of p.

Table 3.   Continued Fraction Expansions of $\sqrt{d}$ for $d < 100$ (and Not a Square)

$\sqrt{2} = \langle 1,\overline{2} \rangle$	$\sqrt{53} = \langle 7,\overline{3,1,1,3,14} \rangle$
$\sqrt{3} = \langle 1,\overline{1,2} \rangle$	$\sqrt{54} = \langle 7,\overline{2,1,6,1,2,14} \rangle$
$\sqrt{5} = \langle 2,\overline{4} \rangle$	$\sqrt{55} = \langle 7,\overline{2,2,2,14} \rangle$
$\sqrt{6} = \langle 2,\overline{2,4} \rangle$	$\sqrt{56} = \langle 7,\overline{2,14} \rangle$
$\sqrt{7} = \langle 2,\overline{1,1,1,4} \rangle$	$\sqrt{57} = \langle 7,\overline{1,1,4,1,1,14} \rangle$
$\sqrt{8} = \langle 2,\overline{1,4} \rangle$	$\sqrt{58} = \langle 7,\overline{1,1,1,1,1,1,14} \rangle$
$\sqrt{10} = \langle 3,\overline{6} \rangle$	$\sqrt{59} = \langle 7,\overline{1,2,7,2,1,14} \rangle$
$\sqrt{11} = \langle 3,\overline{3,6} \rangle$	$\sqrt{60} = \langle 7,\overline{1,2,1,14} \rangle$
$\sqrt{12} = \langle 3,\overline{2,6} \rangle$	$\sqrt{61} = \langle 7,\overline{1,4,3,1,2,2,1,3,4,1,14} \rangle$
$\sqrt{13} = \langle 3,\overline{1,1,1,1,6} \rangle$	$\sqrt{62} = \langle 7,\overline{1,6,1,14} \rangle$
$\sqrt{14} = \langle 3,\overline{1,2,1,6} \rangle$	$\sqrt{63} = \langle 7,\overline{1,14} \rangle$
$\sqrt{15} = \langle 3,\overline{1,6} \rangle$	$\sqrt{65} = \langle 8,\overline{16} \rangle$
$\sqrt{17} = \langle 4,\overline{8} \rangle$	$\sqrt{66} = \langle 8,\overline{8,16} \rangle$
$\sqrt{18} = \langle 4,\overline{4,8} \rangle$	$\sqrt{67} = \langle 8,\overline{5,2,1,1,7,1,1,2,5,16} \rangle$
$\sqrt{19} = \langle 4,\overline{2,1,3,1,2,8} \rangle$	$\sqrt{68} = \langle 8,\overline{4,16} \rangle$
$\sqrt{20} = \langle 4,\overline{2,8} \rangle$	$\sqrt{69} = \langle 8,\overline{3,3,1,4,1,3,3,16} \rangle$
$\sqrt{21} = \langle 4,\overline{1,1,2,1,1,8} \rangle$	$\sqrt{70} = \langle 8,\overline{2,1,2,1,2,16} \rangle$
$\sqrt{22} = \langle 4,\overline{1,2,4,2,1,8} \rangle$	$\sqrt{71} = \langle 8,\overline{2,2,1,7,1,2,2,16} \rangle$
$\sqrt{23} = \langle 4,\overline{1,3,1,8} \rangle$	$\sqrt{72} = \langle 8,\overline{2,16} \rangle$
$\sqrt{24} = \langle 4,\overline{1,8} \rangle$	$\sqrt{73} = \langle 8,\overline{1,1,5,5,1,1,16} \rangle$
$\sqrt{26} = \langle 5,\overline{10} \rangle$	$\sqrt{74} = \langle 8,\overline{1,1,1,1,16} \rangle$
$\sqrt{27} = \langle 5,\overline{5,10} \rangle$	$\sqrt{75} = \langle 8,\overline{1,1,1,16} \rangle$
$\sqrt{28} = \langle 5,\overline{3,2,3,10} \rangle$	$\sqrt{76} = \langle 8,\overline{1,2,1,1,5,4,5,1,1,2,1,16} \rangle$
$\sqrt{29} = \langle 5,\overline{2,1,1,2,10} \rangle$	$\sqrt{77} = \langle 8,\overline{1,3,2,3,1,16} \rangle$
$\sqrt{30} = \langle 5,\overline{2,10} \rangle$	$\sqrt{78} = \langle 8,\overline{1,4,1,16} \rangle$
$\sqrt{31} = \langle 5,\overline{1,1,3,5,3,1,1,10} \rangle$	$\sqrt{79} = \langle 8,\overline{1,7,1,16} \rangle$
$\sqrt{32} = \langle 5,\overline{1,1,1,10} \rangle$	$\sqrt{80} = \langle 8,\overline{1,16} \rangle$
$\sqrt{33} = \langle 5,\overline{1,2,1,10} \rangle$	$\sqrt{82} = \langle 9,\overline{18} \rangle$
$\sqrt{34} = \langle 5,\overline{1,4,1,10} \rangle$	$\sqrt{83} = \langle 9,\overline{9,18} \rangle$
$\sqrt{35} = \langle 5,\overline{1,10} \rangle$	$\sqrt{84} = \langle 9,\overline{6,18} \rangle$
$\sqrt{37} = \langle 6,\overline{12} \rangle$	$\sqrt{85} = \langle 9,\overline{4,1,1,4,18} \rangle$
$\sqrt{38} = \langle 6,\overline{6,12} \rangle$	$\sqrt{86} = \langle 9,\overline{3,1,1,1,8,1,1,1,3,18} \rangle$
$\sqrt{39} = \langle 6,\overline{4,12} \rangle$	$\sqrt{87} = \langle 9,\overline{3,18} \rangle$
$\sqrt{40} = \langle 6,\overline{3,12} \rangle$	$\sqrt{88} = \langle 9,\overline{2,1,1,1,2,18} \rangle$
$\sqrt{41} = \langle 6,\overline{2,2,12} \rangle$	$\sqrt{89} = \langle 9,\overline{2,3,3,2,18} \rangle$
$\sqrt{42} = \langle 6,\overline{2,12} \rangle$	$\sqrt{90} = \langle 9,\overline{2,18} \rangle$
$\sqrt{43} = \langle 6,\overline{1,1,3,1,5,1,3,1,1,12} \rangle$	$\sqrt{91} = \langle 9,\overline{1,1,5,1,5,1,1,18} \rangle$
$\sqrt{44} = \langle 6,\overline{1,1,1,2,1,1,1,12} \rangle$	$\sqrt{92} = \langle 9,\overline{1,1,2,4,2,1,1,18} \rangle$
$\sqrt{45} = \langle 6,\overline{1,2,2,2,1,12} \rangle$	$\sqrt{93} = \langle 9,\overline{1,1,1,4,6,4,1,1,1,18} \rangle$
$\sqrt{46} = \langle 6,\overline{1,3,1,1,2,6,2,1,1,3,1,12} \rangle$	$\sqrt{94} = \langle 9,\overline{1,2,3,1,1,5,1,8,1,5,1,1,3,2,1,18} \rangle$
$\sqrt{47} = \langle 6,\overline{1,5,1,12} \rangle$	$\sqrt{95} = \langle 9,\overline{1,2,1,18} \rangle$
$\sqrt{48} = \langle 6,\overline{1,12} \rangle$	$\sqrt{96} = \langle 9,\overline{1,3,1,18} \rangle$
$\sqrt{50} = \langle 7,\overline{14} \rangle$	$\sqrt{97} = \langle 9,\overline{1,5,1,1,1,1,1,1,5,1,18} \rangle$
$\sqrt{51} = \langle 7,\overline{7,14} \rangle$	$\sqrt{98} = \langle 9,\overline{1,8,1,18} \rangle$
$\sqrt{52} = \langle 7,\overline{4,1,2,1,4,14} \rangle$	$\sqrt{99} = \langle 9,\overline{1,18} \rangle$

ANSWERS TO SELECTED EXERCISES

Section 1.2. 3: One is the smallest positive integer so that there are no integers between 0 and 1. 5: ± 1, ± 2, ± 3, ± 4, ± 6, ± 12. 6: ± 1, ± 2, ± 3, ± 4, ± 6, ± 12.

Section 2.1. 2: (a) 7; (b) 26; (c) 1. There are other correct solutions to problems 3–7 but the Euclidean algorithm leads to those given. 3: $6 \cdot 37 - 13 \cdot 17 = 1$. 4: $4 \cdot 703 - 7 \cdot 399 = 19$. 5: $r = 171$, $s = -148$. 6: $r = -101$, $s = 67$. 7: $r = 3892$, $s = -1659$. 8: No. 11: No.

Section 2.2. 1: 11. 2: $3 \cdot 11$. 3: $2 \cdot 7 \cdot 7$. 4: $2 \cdot 3 \cdot 5 \cdot 11$. 5: 1009. 6: 1. 7: None of these numbers can be a prime. See also miscellaneous exercise 7. 14: 24. 15: $2 \cdot 3 \cdot 5$, $2 \cdot 3 \cdot 7$, $2 \cdot 5 \cdot 7$, $3 \cdot 5 \cdot 7$. There are naturally other examples.

Section 2.3. 5: No, $d_1 = 1$ and $d_2 = 11$ work fine.

Section 2.4. 3: 1210 and 1184, respectively. 4: 14 288, 15 472, 14 536, 14 264, and 12 496, respectively. 6: 60. 7: 180, 234, 362, or 369.

Section 2.5. 1: $x = 1 + 2t$, $y = -1 - 3t$. 2: $x = 1 + 2t$, $y = 1 + 3t$. 3: $x = 20 + 14t$, $y = -24 - 17t$. 4: $x = -3 + 4t$, $y = -9 + 11t$. 5: None. 6: $x = 2860 + 503t$, $y = -2280 - 401t$. 7: $x = -3 + 7t$, $y = -10 + 23t$, $t \ge 1$. 8: None. 9: $(x,y) =$ (4,85), (51,46), (98,7). 10: $x = 4$, $y = 5$. 11: $4H + 5B$ will do. 12: 40 gallon and 78 half-gallon containers. 13: $d = \$3550$. The students' answer is unique.

Section 3.1. 1: True, true, false. 3: The first and second are true. 4: $a = 6$. 5: $a = 70$. 10: 17.

Section 3.2. 3: 4. 13: $x^2 + y^2 \equiv 0$, 1 or 2 (mod 4) (but never 3). 15: *Hint:* $5^2 \equiv 2^3 \pmod{17}$. 32: 2.

Section 3.3. 1: $x \equiv 3 \pmod 7$. 2: $x \equiv 10 \pmod{45}$. 3: $x \equiv 46 \pmod{87}$. 4: None. 5: $x \equiv 4 \pmod 7$. 6: $x \equiv 1 \pmod 4$. 7: $x \equiv -1 \pmod{12}$. 8: $x \equiv 151 \pmod{414}$. 9: $x \equiv -1 \pmod{35}$. 10: $x \equiv 13 \pmod{55}$. 11: $x \equiv 1 \pmod 7$, $y \equiv 2 \pmod 7$. 12: $x \equiv 4 \pmod{13}$, $y \equiv 0 \pmod{13}$. 13: $x \equiv 2 \pmod{11}$, $y \equiv 3 \pmod{11}$, $z \equiv 5 \pmod{11}$. 14: $(x,y) \equiv$ (5,0), (6,1), (0,2), (1,3), (2,4), (3,5), (4,6) (mod 7). 15: $(x,y) \equiv$ (0,6), (1,1), (2,4), (3,7), (4,2), (5,5), (6,0), (7,3) (mod 8). 17: $(x,y) \equiv$ (2,1), (2,3), (2,5), (5,1), (5,3), (5,5) (mod 6). 18: none. 19: $(x,y) \equiv$ (1,2), (5,10), (9,6) (mod 12). 20: 1, 211, 421, 631, and 841. 21: 990 away. 22: 9. Note that casting out nines does not distinguish between 0 and 9.

Section 3.4. 1: $\{1,3,7,9,11,13,17,19\}$, $\phi(20) = 8$. 2: $\{1,7,11,13,17,19,23,29\}$, $\phi(30) = 8$. 3: $n = 6$ and $n = 8$ illustrate both possibilities. 7: $n = 6$ and $n = 561$ work.

Section 3.5. 1: 8, 16, 36, 300, and 320. 7: $x = 6$. 8: $x = 12$. 9: 143.

341

Section 3.6. 1: $(x-1)(x-2)$. 2: $(x-2)(2x+1)$. 3: $(x-5)(x+5)$. 4: $(x-4)(x+4)$.
5: $(x-1)(x-2)(x-3)$. 6: $(x-1)^3$. 7: no. 11: $\sum_{j=1}^{p-1} j \equiv 0 \pmod{p}$ for $p > 2$.
12: $\sum_{i=2}^{p-1} \sum_{j=1}^{i-1} ij \equiv 0 \pmod{p}$ for $p > 3$. 13: $x \equiv 1, 3 \pmod{5}$. 14: none. 15: $x \equiv$
$\pm 2, \pm 7 \pmod{15}$. 16: $x \equiv 3, 4 \pmod{8}$. 17: $x \equiv 3 \pmod{11}$.

Section 3.7. 1: No. 2: One. 9: $x \equiv \pm 8, \pm 18 \pmod{65}$. 12: $x \equiv \pm 1 \pmod{14}$.

Section 4.1. 1: First row: 7, 0, 5; second row: 2, 4, 6; third row: 3, 8, 1; filled and magic.
2: Magic but not filled. 3: Filled and magic. 4: Filled but not magic. 5: Neither.
6: Filled and magic. 7: $a = 5, b = 3, c = -3, d = -2, e = 1, f = 2$. 9: No. 10:
$[n(n^2 - 1)/2]$. 11: No.

Section 4.2. 1: 9, 1, -81, -2, 472. 2: 4. 5: Filled and magic. 6: Filled and magic. 7:
Column magic. 8: Magic. 9: Filled and magic. 12: $a = 4, b = 5, c = 1, d = 1$,
$e = 4, f = 5$.

Section 4.3. 1: Magic, magic in the negative diagonals, symmetric. 2: Filled, column
magic. 3: Filled, magic, magic in the negative diagonals, symmetric. 4: Nothing
whatsoever. 5: Filled, magic in the negative diagonals, symmetric. 6: $x_0 = (n+1)/2$,
$y_0 = n$. 10: $n = 3, n = 19, n = 57$. 12: $r + t$.

Section 5.1. 1: $x = y = 2$. 2: No solutions. 3: $x = 3, y = 2$. 4: $x = \pm 1, y = 0$.

Section 5.2. 3: There are no solutions. This can be decided (mod 3) and (mod 4). 4: No
solutions. This can be decided (mod 3) and (mod 8). 5: $x = y = z = 0$. This can be
decided (mod 3). 6: $x = y = z = 0$. This can be decided (mod 5), (mod 8), (mod 13).
7: No solutions. 8: $x = y = z = 0$. 9: No solutions. 10: No solutions.

Section 5.3. 1: $x = |d(u^2 - 2v^2)|$, $y = 2duv$, $w = d(u^2 + 2v^2)$, $u > 0$, $v > 0$, $d > 0$.
2: $x = |d[(u^2 - 3v^2)/2]|$, $y = duv$, $z = d[(u^2 + 3v^2)/2]$, $u > 0$, $v > 0$, $d > 0$, d even
if u and v are not both odd. 3: $x = |d[(u^2 - pv^2)/2]|$, $y = duv$, $z = d[(u^2 + pv^2)/2]$,
$u > 0$, $v > 0$, $d > 0$, d even if u and v are not both odd. 4: $x = 2(u^2 - v^2)$, $y = 4uv$,
$z = u^2 + v^2$ (or the same thing with x and y interchanged), $u > v > 0$, $(u,v) = 1$,
one of u and v is even and the other odd. 7: $x = d|u^3 - dv^3|/2|$, $y = duv$, $z =$
$d[(u^3 + dv^3)/2]$, where either $d = 2, u > 0, v > 0, (u,v) = 1, u$ odd or $d = 1, u > v >$
0, $(u,v) = 1$, u odd, v odd.

Section 6.1. 3: 1/9. 4: 1/37. 5: 1343/4950. 7: 1.

Section 6.2. 1: Irrational and rational, respectively.

Section 7.1. 3: (1,1), (1,2), (2,3).

Section 7.2. 5: $V_0 = (1,1)$, $V_1 = (1,2)$, $V_2 = (2,3)$, $V_3 = (3,5)$, $V_4 = (5,8)$, $V_5 = (8,13)$,
$a_0 = a_1 = a_2 = a_3 = a_4 = a_5 = 1$. 6: $V_0 = (1,1)$, $V_1 = (2,3)$, $V_2 = (5,7)$, $V_3 =$
$(12,17)$, $a_0 = 1$, $a_1 = a_2 = a_3 = 2$. 7: $V_0 = (1,1)$, $V_1 = (7,8)$, $a_0 = 1$, $a_1 = 7$.
8: $V_0 = (1,0)$, $V_1 = (1,1)$, $V_2 = (4,3)$, $V_3 = (17,13)$, $a_0 = 0$, $a_1 = 1$, $a_2 = 3$, $a_3 = 4$.

Section 7.3. 1: $\langle 2,2,2,1,5 \rangle$, $V_0 = (1, 2)$, $V_1 = \underline{(2,5)}$, $V_2 = (5,12)$, $V_3 = (7,17)$, $V_4 = (40,97)$.
2: $\langle \bar{1} \rangle = (1,1,1,1,\ldots)$. 3: $\langle 1, \bar{2} \rangle$. 4: $\langle 2, \overline{1,1,1,4} \rangle$. 5: 193/71. 8: 196/185. 10(b): 0/1,
1/3, 3/10, 28/93, 59/196. 11: $p_2/q_2 = 3/4$, $p_5/q_5 = 228/293$.

Section 7.4. 4: (127,16) already has a vertical distance to the line less than .001. 7: (1,15),
(5,76), (151,2295), (760,11 551). 8: (1,22), (2,45), (5,112), (112,2509).

Section 7.6. 2: $2 + \sqrt{7}$. 3: $\sqrt{115}$. 4: $(17 + \sqrt{3})/13$. 5: $2 - \sqrt{2}$. 6: $(13 - \sqrt{2})/7$. 7:
$(325 + 2\sqrt{39})/247$. 8: $(2 - \sqrt{2})/4$. 9: $(3 - \sqrt{3})/24$. 10: $(21 - \sqrt{2})/8$. Note also
that $\alpha = \langle 2,2,4,\overline{3,22,3,5} \rangle$. 11: $\langle \overline{1,3,1} \rangle$. 12: $\langle 2,4,\overline{22,5,1,1,5} \rangle$. 13: $\langle 88,1,\bar{2} \rangle$. 14:
$\langle 7,11,\overline{2,1,2,1,2,10} \rangle$. 15: $\langle 0,1,\overline{1,8,1,18} \rangle$. 16: $\langle n,\overline{2n/k,2n} \rangle$, and $k|2n$ is sufficient.

Section 7.7. 1: $x = 5, y = 2$; $x = 49, y = 20$. 2: $x = 10, y = 3$; $x = 199, y = 60$.
3: $x = 24, y = 5$; $x = 1151, y = 240$. 4: $x = 15, y = 4$; $x = 449, y = 120$. 5: $x = 9$,
$y = 4$; $x = 161, y = 72$. 6: $x = 70, y = 13$. 7: $x = 99, y = 13$. 8: $x = 1068, y = 125$.

9: $x = 4005$, $y = 389$. 10: $x = 5604$, $y = 569$. 11: $x = 3$, $y = 1$; $x^2 - 10y^2 = -1$.
12: $x = 182$, $y = 25$; $x^2 - 53y^2 = -1$. 13: $x = 24\,335$, $y = 3588$; $x^2 - 46y^2 = 1$.
14: $x = 57$, $y = 5$; $x^2 - 130y^2 = -1$. 15: $x = 682$, $y = 61$; $x^2 - 125y^2 = -1$.

Section 8.1. 1: $x = u^2 - 2v^2$, $y = 2uv$, $z = u^2 + 2v^2$.

Section 8.2. 1: $9 + \sqrt{7}$, $1 - 2\sqrt{-7}$, 10, $(9 + 15\sqrt{-7})/2$, $(9 + 15\sqrt{-7})/6$ $[= (3 + 5\sqrt{-7})/2]$, $(10 + \sqrt{-108})/4$, $(3 + 2\sqrt{6})/(1 - \sqrt{6})$. 2: $3 - 2\sqrt{-2}$, $1 - \sqrt{-2}$. 3: 7,24,25; 117,44,125; 33,56,65; 16,63,65.

Section 8.3. 3: $(7 + \sqrt{53})/2$, $(39 + 5\sqrt{61})/2$. 7: $(2 + \sqrt{-7})(2 - \sqrt{-7})[(-1 - \sqrt{-7})/2][(1 - \sqrt{-7})/2]^3$. 9: $(4 + \sqrt{14})(-7 + 2\sqrt{14})$. Other factorizations can be converted to this one with the aid of appropriate units.

Section 8.4. 5: $3 \cdot 7 = (1 + 2\sqrt{-5})(1 - 2\sqrt{-5})$. 7: $7 + 13i$.

Section 8.5. 1: One such family is given by $x = u^3 - 3uv^2$, $y = 3u^2v - v^3$, $z = u^2 + v^2$. 2: $x = \pm 5$, $y = 3$. 3: $x = \pm 2$, $y = 2$; $x = \pm 11$, $y = 5$. 4: $x = \pm 4$, $y = 3$; $x = \pm 58$, $y = 15$. 6: $x = \pm d(u^2 - 2uv - v^2)$, $y = \pm d(u^2 + 2uv - v^2)$, $z = \pm d(u^2 + v^2)$ (or the same thing with x and y interchanged).

Appendix A. 1: (a) $(-1,2)$; (b) $(5,4)$; (c) $(61,-52)$; (d) $(61,-52)$; (e) $(-19,19)$. 4: (a) E_5; (b) G_7; (c) I_3; (d) F_5; (e) F_6; (f) H_5; (g) I_4; (h) D_7; (i) C_1; (j) I_6; (k) H_3; (l) I_1.

Appendix B. 1: (a) $\begin{pmatrix} 0 & 6 \\ -1 & -23 \end{pmatrix}$; (b) $\begin{pmatrix} 10 & -24 \\ 14 & -33 \end{pmatrix}$; (c) $\begin{pmatrix} -17 & 4 \\ 58 & -13 \end{pmatrix}$; (d) $\begin{pmatrix} 5 & 2 \\ -6 & 6 \end{pmatrix}$;

(e) $\begin{pmatrix} -17 & 10 \\ 57 & -36 \end{pmatrix}$; (f) $\begin{pmatrix} -17 & 10 \\ 57 & -36 \end{pmatrix}$; (g) both sides $= \begin{pmatrix} -42 & 6 \\ 158 & -21 \end{pmatrix}$; (h) both

sides $= \begin{pmatrix} 192 & 162 \\ -727 & -611 \end{pmatrix}$. 2: $A^{-1} = \begin{pmatrix} -7 & -2 \\ -3 & -1 \end{pmatrix}$, $ABA^{-1} = \begin{pmatrix} -18 & -6 \\ 76 & 25 \end{pmatrix}$. 3: $\begin{pmatrix} 3 & -2 \\ -4 & 3 \end{pmatrix}$, $(x,y) = (18,-13)$. 4: $-\frac{1}{28}\begin{pmatrix} -2 & -3 \\ -8 & 2 \end{pmatrix}$, $(x,y) = (\frac{17}{2},-\frac{5}{4})$. 5: 1, 6, 6, 6.

6: Both sides $= \begin{pmatrix} 0 & -1 \\ 6 & -23 \end{pmatrix}$.

LIST OF SYMBOLS

$a\|b,\ a \nmid b$	1
(a,b)	16
$d(n),\ \sigma(n)$	36
$\displaystyle\sum_{d\|n}$	38
$a \equiv b(\bmod\ n),\ a \not\equiv b(\bmod\ n)$	51
$\phi(n)$	78
$f(x) \equiv g(x)(\text{poly}\bmod\ n)$	86
$f(x) \equiv g(x)(\bmod\ n)$	87
$\text{ord}_n(a)$	98
$\mu(n)$	112
$[\alpha]$	120
$1.1\overline{216}$	164
$\alpha,\ L,\ a_n,\ V_n,\ q_n,\ p_n$	188
d_n	196
α_n	198

$\langle a_0, a_1, \ldots, a_n \rangle,\ \langle a_0, a_1, \ldots \rangle$	203
$\langle 1,\overline{1,2} \rangle$	206
$A_n,\ M_n,\ \gamma_n$	226
$\|A\|$	233
d (for Section 7.7)	239
$\langle x_0, x_1, \ldots \rangle$	249
$\mathbf{Q},\ \mathbf{Z},\ \mathbf{Q}(\sqrt{d})$	258
d (for Chapter 8)	260
$\overline{\alpha}$	261
$\mathbf{N}(\alpha)$	263
$\alpha\|\beta$	270
UFD	281
$\mathbf{Z}[\sqrt{d}]$	296
$\mathbf{Q}(\rho)$	305
$\det A,\ A'$	329

INDEX

Above a line, 181
Additive questions, 3
Algebraic number, 172
Algebraic number theory, 258, 305
Amicable numbers, 43
Answers to selected exercises, 341
Approximation by rationals, 8, 187, 209
Arithmetic (mod n), 59
Ascent, proof by, 158, 160
Associates, 272, 278, 297

Baker, A., 180, 295, 296, 337, 338
Banachiewicz, 109, 336
Base n, 125, 129, 177
Beeger, N. G. W. H., 336
Below a line, 181
Bibliography, 334
Bose, R. C., 139, 336

Casting out elevens, 63
  nines, 62
  ninety nines, 65
  one thousand and ones, 65
Cell, 118
Chatland, 294, 338
Chinese remainder theorem, 72
Class number, 306
Column magic, 130
Complete residue system, 58
Completely multiplicative function, 37
Complex quadratic field, 259
Composite, 2, 277

Congruence, 51
  equations, 66, 74
Congruent, 51
  as polynomials, 86
Conjugate, 261
  complex, 261
Continued fraction, 181, 202
  algorithm, 185, 187, 202
  periodic, 206, 226
Convergent, 195

Davenport, H., 294, 337, 338
Dedekind, J. W. R., 308
Defining equation, 262
Degree (mod n) of a polynomial, 93
Descent, proof by, 155
Determinant of a matrix, 329
Diabolic square, 133
  sum, 133
Diophantine equations, 4, 145, 177, 257, 299
  linear, 44
Diophantus, 4, 145
Dirichlet, 315
Divides, 1, 270, 296
Division algorithm, 11

Eratosthenes, sieve of, 14
Euclid, 13, 20, 42, 43, 290
Euclidean algorithm, 16, 20, 206
  field, 290
Euler, L., 2, 6, 8, 50, 78, 80, 109, 110, 138, 139, 146, 305, 315

345

Euler's theorem, 80
  ϕ function, 77, 78, 82

Factorization into primes, 12, 279
Fermat, P., 2, 5, 6, 109, 111, 145–148, 155,
  156, 163, 336
Fermat numbers, F_n, 2, 108, 336
Fermat–Pell equation, 147, 149, 239, 253, 275,
  316, 337
Fermat's last theorem, 5, 145, 146, 163, 305–
  307, 315, 336
Fermat's theorem, 80
Field, 140, 258, 305, 331
  quadratic, 259
Filled square, 122
First quadrant, 181
Frenicle, 147
Fundamental theorem of algebra, 172
  theorem of arithmetic, 26, 28
  unit, 309

Gauss, C. F., 3, 6, 51, 64, 78, 105, 106, 172,
  295, 305, 313
Gaussian integer, 305
Gelfond, 179
Goldbach conjecture, 5, 335
Greatest common divisor, 16, 24
  integer functions, 120
Greco-Latin square, 139

Hadamard, J., 3
Haselgrove, 7, 335
Heilbronn, H., 295, 338

Ideal numbers, 306
Identity matrix, 327
Imaginary quadratic field, 259
Infinite continued fraction, 203
Integer, 1, 265, 306
  Gaussian, 305
  quadratic, 265
  rational, 265
Inverse of a matrix, 328
Irrational number, 34, 164, 170
Irregular prime, 307

Kanold, H. J., 335
Klein, F., 337
Kortum, R., 337
Kummer, E., 305–308

Lagrange, J. L., 233
Lander, L. J., 146, 336
Lattice, 210
Law of quadratic reciprocity, 313
Least common multiple, 48
Legendre, 2
Lehman, R. S., 7, 335
Lehmer, D. H., 81, 335, 336
Lehmer, D. N., 2, 118, 335, 336
Lehmer, Emma, 336
Length of the period, of a continued fraction,
  240
  of a decimal expansion, 164
Linfoot, E. H., 295, 296, 338
Liouville, J., 172, 176, 335
Liouville's theorem, 172, 337
Loubère, De la, 10, 118, 119, 137
Loubère method, 10, 123
Lucas, 336

Magic in the columns, 130
  negative diagonals, 133
  positive diagonals, 133
  rows, 130
Magic square, 118, 130
  sum, 118, 130
Matrix, 322
McClintock, 138, 336
McNiel, G., 337
Mersenne, 111
  numbers, M_m, 111
Möbius function, $\mu(n)$, 112
  inversion formula, 113
Modular arithmetic, 59
Multiplicative functions, 36, 37
  properties of integers, 1

Negative diagonal, 133
Newton, I., 146
Nicol, C. A., 336
Nonsquare (mod p), 311
Norm, 263
  of a matrix, 233

Order (mod n), 98
Over a line, 181

Pairwise relatively prime, 25
Parker, E. T., 139, 336
Parkin, T. R., 146, 336
Partial quotients, a_n, 188, 198
Patz, W., 337
Pell's equation, 149
Perfect number, 42, 335
Period of a decimal expansion, 164
Periodic continued fractions, 206, 226
    decimal expansions, 165
Pipping, 5, 335
Points closest to a line, 185, 214
Polya conjecture, 7, 335
Positive diagonal, 132
Poussin, de la Vallée, 3
Prime, 2, 42, 276, 297
    rational, 277
    relatively, 21
Prime number theorem, 3
Primitive root, 97, 98, 108, 168
    of unity, 98, 105, 305
Primitive triangle (triplet), 151, 258
Pseudoprime, 81, 109, 111, 335
Purely periodic decimal expansions, 167
Pythagoras, 43
Pythagorean triple, 4, 151, 299

Quadratic field, 257, 259
    complex or imaginary, 259
    real, 259
Quadratic integer, 265
Quadratic reciprocity law, 313

Rabinovitch, 310
Rational integer, 265
    number, 49, 164
Real quadratic field, 259
Reduced residue system, 77
Regular prime, 307
    square, 138
Relatively prime, 21
    pairwise, 25
Residues, complete system of, 58
    reduced system of, 77
Roth, K. F., 176, 177, 180, 336
Row magic, 130

Sarrus, 335
Schneider, 179
Selfridge, J. L., 336
Set of points closest to a line, 185, 214
Shrikhande, S. S., 139, 336
Siegel, C. L., 176, 177, 338
Sierpinski, W., 111, 336
Sieve of Eratosthenes, 14
Simple field, 281
Slope, 181
Sommerfeld, A., 337
Square (mod p), 311
Stark, H. M., 295, 296, 338
Storer, T., 55
Symmetric square, 135
    sum, 135

Tables, 339
Tarry, G., 138
Thue, A., 176, 177, 179, 180
Totient function of Euler, 78
Transcendental number, 172, 179
Transpose of a matrix, 329
Two-dimensional vector, 317

Under a line, 181
Uniform step method, 118, 121
Unique (mod n), 66
Unique factorization domain (UFD), 281,
    297
    property, 281, 306
    theorem, 28, 272
Unit, 272, 273, 296
    fundamental, 309

Vandiver, H. S., 336
Vector, 186, 317
Vertical distance from a line, 214

Wallis, J., 146–148
Well-ordering principle, 11
Wilson's theorem, 96

Zero matrix, 323
    vector, 318